Advances in Physical Geochemistry

Volume 4

Advances in Physical Geochemistry

Series Editor: Surendra K. Saxena

Metamorphic Reactions
Kinetics, Textures, and Deformation

Edited by
Alan Bruce Thompson and David C. Rubie

With Contributions by
B. Bayly   K. H. Brodie   M. A. Carpenter
R. J. Knipe   E. L. McLellan   S. A. F. Murrell
A. Putnis   J. Ridley   D. C. Rubie
E. H. Rutter   B. K. Smith   A. B. Thompson
R. J. Tracy   R. P. Wintsch

With 81 Illustrations

Springer-Verlag
New York Berlin Heidelberg Tokyo

Alan Bruce Thompson
Departement für Erdwissenschaften
ETH-Zentrum
CH-8092 Zürich
Switzerland

David C. Rubie
Department of Geology
The University of Manchester
Manchester, M13 9PL
Great Britain

Series Editor
Surendra K. Saxena
Department of Geology
Brooklyn College
City University of New York
Brooklyn, New York 11210
U.S.A.

Library of Congress Cataloging in Publication Data
Main entry under title:
Metamorphic reactions.
   (Advances in physical geochemistry; v. 4)
   Includes bibliographies and index.
   1. Metamorphism (Geology)   I. Thompson,
Alan Bruce, 1947–   . II. Rubie, David C.
III. Series.
QE475.A2M4745 1985   552'.4   84-20307

Typeset by Bi-Comp, Incorporated, York, Pennsylvania.

9 8 7 6 5 4 3 2 1

ISBN-13: 978-1-4612-9548-8   e-ISBN-13: 978-1-4612-5066-1
DOI: 10.1007/978-1-4612-5066-1

Preface

The fourth volume in this series consists of eleven chapters. The first five deal with more theoretical aspects of the kinetics and mechanisms of metamorphic reactions, and the next six consider the interdependence of deformation and metamorphism. All papers deal with natural processes that interact on various time scales and with different degrees of mass and heat transfer. Consequently, many fundamental axioms of metamorphic petrology and structural geology are questioned both for their accuracy and their usefulness. In raising such questions, most contributors have pointed to ways in which the answers could be forthcoming from appropriate experimental studies or observations on natural materials.

In their discussion of how order/disorder can influence mineral assemblages, Carpenter and Putnis emphasize that metastable crystal growth is common in metamorphic systems and state "there may be some reluctance (among many earth scientists) to accept that significant departures from equilibrium could occur." On the basis of presented evidence, they question whether reactions ever occur close to an equilibrium boundary. The necessity for pressure or temperature overstepping is also required by nucleation-rate theory. In any case, the degree of order is severely influenced by these kinetic effects in igneous, sedimentary, and metamorphic environments.

From an evaluation of available mineral kinetic data, Rubie and Thompson present some guidelines for quantitative approaches to experimental kinetic studies. Proposals are made as to how the common variables that can affect mineral kinetics may be controlled in experiments so that such data may be more realistically applied to natural situations.

In examining the effect of reaction enthalpy on the progress of a metamorphic reaction, Ridley has considered the relationships among kinetics, heat flow, and microstructures of metamorphic rocks. He provides insight into how the inverse problem of observation of grain size and porphyroblast development could be used to deduce the rate-controlling step during meta-morphic reaction.

In the fourth chapter, Smith illustrates how information on dislocation types and their densities, obtained mainly by transmission electron micros-copy, can be used to learn much about the mechanisms of mineral reactions. The progressive effects of deformation are especially well shown by the successive motion of dislocations in the chosen orthosilicate examples for garnet to chlorite and olivine to spinel transformations.

Tracy and McLellan present kinetic interpretations of textural and com-positional phenomena in contact metamorphosed emeries. They discuss cri-teria by which diffusion-controlled and interface-controlled processes may be distinguished. Because the silica diffusion rate must have been very slow during their contact metamorphism, diffusion of SiO_2 through crystals or along dry grain boundaries is inferred to have occurred.

Brodie and Rutter review the minerology and microstructures of meta-morphosed basaltic rocks and examine theoretically some of the interrela-tionships between metamorphism and deformation. They emphasize the value of shear zones to such studies because deformed and undeformed rocks are easily compared and conclude that in shear zones the deformabil-ity may be reaction enhanced.

From their examination of polyphase mylonites, Knipe and Wintsch pro-pose an evolutionary model involving the cyclic redistribution of deforma-tion between different mineralogical domains. The flow pattern changes from a more continuous and distributed ductile deformation to a more loca-lized deformation at lower temperatures characterized by the development of mica from the syntectonic breakdown of feldspar.

Murrell has used experimental observations of a sharp change in rock strength when the sample dehydrates to evaluate the geological significance of elevated pore pressure especially as regards changing styles of deforma-tion.

Rutter and Brodie discuss several aspects of the permeability of medium to high-grade metamorphic rocks, especially in shear zones. They deduce apparent permeabilities of 10^{-26} to 10^{-28} m^2 for metamorphic conditions, which are several orders of magnitude below any feasible measurement limit in the laboratory. Volatile permeation along shear zones is 10 to 100 times faster than across them.

Wintsch has examined some possible effects of deformation in displacing mineral equilibrium boundaries. His calculations show that a significant P–T displacement of an equilibrium boundary could occur in mylonites, repeat-edly deformed rocks, and fault zones where very high dislocation densities are produced ($>10^{11}$ cm^{-2}).

In the final chapter, Bayly returns to the somewhat controversial issue of the nature of chemical potential in a material under nonhydrostatic stress. The paper is presented as a series of postulates that may be capable of geological investigation.

In addition to specific details of how kinetic factors may be investigated, many contributors have pointed towards larger scale or more general geological problems and their possible solution. The observed association of high strain deformation with regional metamorphism underlines the necessity to quantify the energies of deformation and metamorphism as well as their relative timing and duration. Likewise, it has become clear that we possess very few criteria that may be used to determine if equilibrium is ever achieved on the scales assumed by many petrological studies which never consider kinetic factors and assume that fluids are ubiquitous during medium to high-grade metamorphism. Several consequences of fluid-absent metamorphism are discussed and some suggestions given of the different reaction mechanisms compared to when fluid is present. We clearly must consider further the related problem of exactly how metamorphic reactions and deformation influence rock permeability.

Unfortunately, most experimental studies lag behind field observations and theoretical studies. There is very little information on how the flow laws of polyphase metamorphic rocks are changed when metamorphic reactions occur. New types of experiments are indicated in several papers and would clearly be easier to apply to natural situations if the microstructures of naturally and experimentally deformed materials were understood in terms of reaction mechanisms.

We realize that the subject coverage is not as broad as originally planned and items such as the influence of fluid species on defect chemistry and diffusion, or the regional scale relations between momentum transfer and heat or mass transfer, are only briefly treated. The authors have provided extensive bibliographies that hopefully will enable the reader to pursue the directions not covered in detail by the papers themselves. Thanks are due to M. Casey, N. Mancktelow, E. H. Perkins, S. Saxena, S. Schmid, J. Tullis, B. Yardley, and R. A. Yund, in addition to the contributors themselves, for their reviews of several chapters. We are grateful to K. Malmstroem for editorial assistance and to the staff at Springer-Verlag for their cooperation and patience during the preparation of the volume.

ALAN BRUCE THOMPSON
DAVID C. RUBIE

Contents

Contributors

BAYLY, B.   Department of Geology, Rensselaer Polytechnic In-
                          stitute, Troy, New York 12181, USA

BRODIE, K. H.   Geology Department, Imperial College, London,
                          SW7 2BP, Great Britain

CARPENTER, M. A.   Department of Earth Sciences, University of Cam-
                          bridge, Downing Street, Cambridge, CB2 3EQ,
                          Great Britain

KNIPE, R. J.   Department of Earth Sciences, The University of
                          Leeds, Leeds, LS2 9JT, Great Britain

McLELLAN, E. L.   Department of Geology and Geophysics, Yale Uni-
                          versity, New Haven, Connecticut 06520, USA

MURRELL, S. A. F.   Department of Geology, University College Lon-
                          don, Gower Street, London, WC1E 6BT, Great
                          Britain

PUTNIS, A.   Department of Earth Sciences, University of Cam-
                          bridge, Downing Street, Cambridge, CB2 3EQ,
                          Great Britain

RIDLEY, J.   Departement für Erdwissenschaften, ETH-Zen-
                          trum, CH-8092 Zürich, Switzerland

RUBIE, D. C.   Department of Geology, The University of Manchester, Manchester, M13 9P1, Great Britain, and Mineralogisches Institut, Universität Bern, CH-3012 Bern, Switzerland

RUTTER, E. H.   Geology Department, Imperial College, London, SW7 2BP, Great Britain

SMITH, B. K.   Department of Geology, Arizona State University, Tempe, Arizona 85287, USA

THOMPSON, A. B.   Departement für Erdwissenschaften, ETH-Zentrum, CH-8092 Zürich, Switzerland

TRACY, R. J.   Department of Geology and Geophysics, Yale University, New Haven, Connecticut 06520, USA

WINTSCH, R. P.   Department of Geology, Indiana University, Bloomington, Indiana 47405, USA

Chapter 1
Cation Order and Disorder during Crystal Growth: Some Implications for Natural Mineral Assemblages

M. A. Carpenter and A. Putnis

Introduction

It is very common to find that in mineral synthesis experiments the crystals that form first have disordered cations, even when the synthesis conditions are well within the stability field of the ordered state. Some examples are the crystallization of albite from glass starting material (MacKenzie, 1957) or a flux (Woensdregt, 1983), cordierite from glass (Schreyer and Schairer, 1961; Putnis, 1980a), and plagioclase from glass (Eberhard, 1967; Kroll and Müller, 1980). The uncertainties that might arise in the interpretations of natural mineral assemblages (in terms of prevailing PT conditions), if the same metastable disordered states develop also during crystallization processes in nature, could be quite serious (see, for example, Putnis, 1980b). The problem is one of kinetics. It appears, in general, that at conditions of sufficient supersaturation the disordered phase is kinetically favoured over its ordered equivalent, irrespective of their relative thermodynamic stabilities. It so happens that this phenomenon has been studied in other, nonmineralogical situations. Thus we may quote Chernov and Lewis (1967) in this context: "The composition and structure of a crystal formed in a multicomponent system are determined by the equilibrium diagram only if the conditions of crystal growth are close to those of equilibrium. If the departure from equilibrium is considerable, both the composition and actual structure of the crystal will depend on the crystallization kinetics. . . . There is, consequently, a broad class of actual crystallization conditions in which a purely thermodynamic approach may lead to inaccurate or simply incorrect results." The questions we wish to address in this paper are whether and how this class of crystallization conditions could extend to the natural environment in relation to the nucleation and growth of ordered minerals.

We have used, as our starting point, the qualitative results of modelling experiments on crystal growth in binary systems (Chernov and Lewis, 1967; Cherepanova *et al.*, 1978; Baikov *et al.*, 1980; Pfeiffer, 1981; and see Haubenreisser and Pfeiffer, 1983). These models use rather specific growth mechanisms, treating, in particular, attachment of atoms at ledge sites on the crystal surfaces. While they highlight the most relevant aspects of order/disorder during crystal growth for the present discussion, the resulting formulations are not directly applicable to silicates, which have possibly more complex, and certainly less well-understood, growth surfaces and growth mechanisms (Kirkpatrick, 1975, 1983). We therefore have chosen to use the classical transition state theory of nucleation and growth to identify, in the most general terms, the principal constraints on the crystallization behavior. Under conditions of supersaturation, disordered phases are kinetically favored over ordered phases because of the entropy of activation for nucleation and growth. This term depends on the number of possible alternative configurations for the activated step, which is larger for the case of a disordered nucleus than for an ordered nucleus. Unfortunately, transition state theory gives rather inaccurate predictions of nucleation rates and uses parameters that are not easy to quantify (Kirkpatrick, 1981, 1983). It does, however, allow some qualitative prediction of the effects of pressure and temperature and, in the absence of correct atomistic models, we can at least make order of magnitude estimates of the kinetic effects, based on guesses of the activation energies.

In the second part of the paper we outline some of the criteria that may be used to recognize the operation of metastable, disordered crystallization in minerals and review some known examples. These examples are taken from authigenic, hydrothermal, metamorphic, and igneous environments and include feldspars, omphacite, dolomite, cordierite, aluminous sillimanite, and a Pb, Be-rich amphibole, joesmithite. In most cases, annealing subsequent to crystal growth has caused the degree of order to change and no textural evidence, on a thin section scale, of the original metastable crystallization behavior need be preserved. Only where we have microstructural evidence of an ordering transformation can we really appraise this aspect of the crystallization history. In the final section, we consider petrogenetic implications of the relationship between cation order and crystal growth. We suggest that without kinetic analysis the interpretation of equilibrium assemblages, when ordered and disordered phases are involved, can be precarious. This applies particularly to interpretations of miscibility gaps in plagioclase feldspars, which are highly sensitive to ordering, to the crystallization of dolomite and omphacite, and to the placing of metamorphic reaction boundaries.

Models of Nucleation and Growth for Ordered and Disordered Crystals

One of the simplest possible models for crystal growth in a binary system involves A and B atoms being added to a single chain. Chernov and Lewis (1967) determined the growth velocity and state of order of such a chain in terms of the interaction energies of $A-A$, $B-B$, and $A-B$ pairs, the probabilities of A or B atoms being attached to it, and variations in the degree of supersaturation. Their two-dimensional model then treated the single chain as a continuous coil, with each new atom attaching both to the end of the chain and to the previous layer in the coil. Equilibrium between the two-dimensional crystal and the gas phase from which it was grown, i.e., the condition of zero growth rate, occurs at a higher vapor pressure for a disordered seed than for an ordered one. This is exactly as should be expected, since the equilibrium vapor pressure over an ordered crystal is less than over a disordered one at temperatures below the equilibrium order/disorder temperature (T_c) of the crystal. At a small supersaturation the crystal grows in an ordered state with only isolated mistakes in the ordering sequence. At higher supersaturations antiphase boundaries appear, which line up normal to the growth face. At the highest supersaturations, even using an ordered seed, the density of domain boundaries is large and almost no long-range order remains. For a three-dimensional crystal, the same features are preserved. Thus for the growth of a crystal below T_c "there must exist a critical supersaturation at which the structure of the crystal changes from order to disorder" (Chernov and Lewis, 1967).

The physical origin of this change in order, which Chernov and Lewis termed a "kinetic phase transformation" is in the cooperative interactions between the atoms during crystallization. Each atom has a residence time at the growth point with a finite probability, depending on its attachment energy, of being detached and then replaced. The attachment energy depends on interactions of the arriving atom with the preexisting local structure. If the supersaturation is high there is an increased probability that the "wrong" atom would be trapped in place by succeeding atoms, because the growth rate is high (relative to lower supersaturations). In other words, the higher the supersaturation the faster the growth rate and the greater the probability of mistakes being incorporated even though the interaction energies would favor ordered arrays of atoms.

The effect of increasing supersaturation at a constant temperature is to increase the growth rate and therefore reduce the degree of long-range order in the growing crystal. The effect of reducing temperature in most cases will also be to increase the supersaturation, but, in addition, it will reduce the rate of exchange of atoms between growth surface and growth medium and therefore increase the residence time of atoms at the growth point. In this case the absolute growth rate may show a reduction, but "wrong" atoms would still have a high probability of being retained.

More sophisticated models have been described (Cherepanova *et al.*, 1978; Baikov *et al.*, 1980; Saito and Müller-Krumbhaar, 1981; Pfeiffer, 1981), but they show, qualitatively, the same essential features summarized in Fig. 1 (after Baikov *et al.*, 1980). If crystallization occurs at $T > T_c$ the crystals grow in an equilibrium disordered state and there is no problem. If they grow at $T < T_c$ but under equilibrium growth conditions (at T_e), they will incorporate an equilibrium degree of order. If they grow at $T < T_c$ but under conditions of supersaturation ($T < T_e$), they will grow with less than the equilibrium degree of long-range order. Below a certain temperature (T_k, where

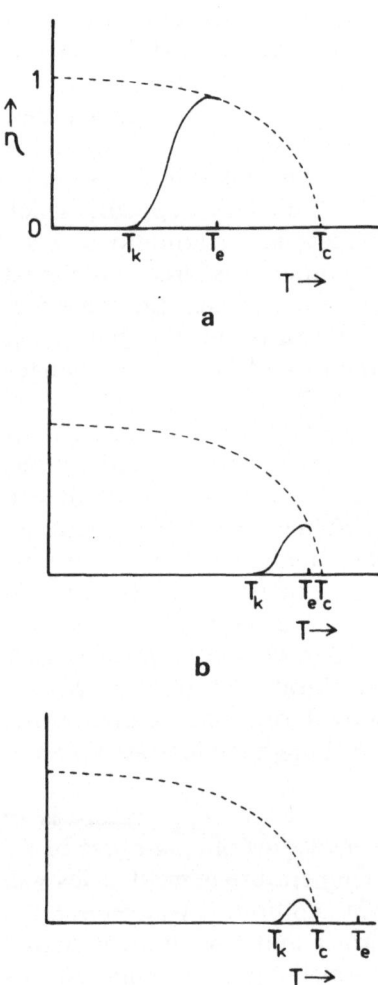

a

b

c

Fig. 1. Schematic relationship between degree of long-range order (η) of AB crystals growing at different temperatures (T) in relation to their equilibrium crystallization temperature (T_e) and their equilibrium order/disorder temperature (T_c) (slightly modified after Baikov *et al.*, 1980, to show the most generalized result of their model). Dashed line represents equilibrium, thermodynamic ordering; solid line represents kinetic ordering, i.e., the degree of order actually attained during crystallization at temperatures below T_e. T_k is the "kinetic phase transition" temperature (Chernov and Lewis, 1967; Baikov *et al.*, 1980). At $T = T_k$ and $T < T_k$ no long-range ordering of A and B atoms would be incorporated into the growing AB crystals. (a) $T_k < T_e < T_c$. (b) $T_k < T_e \lesssim T_c$. (c) $T_k < T_c \lesssim T_e$.

$T_k < T_e$), only disordered crystals will grow. Figure 1 shows the qualitative variation of long-range order for different growth temperatures (T), when $T_k < T_e < T_c$ (Fig. 1(a)); $T_k < T_e \lesssim T_c$ (Fig. 1(b)); $T_k < T_c \lesssim T_e$ (Fig. 1(c)).

Thus we have physical models that reproduce and justify observed growth behavior of metastable, disordered structures at high supersaturations. More phenomenological treatments should display the same gross features.

Activated State Model

In the activated state model (Glasstone *et al.*, 1941; and reviewed recently by Kirkpatrick, 1981) the rate of nucleation (I) of crystals from some arbitrary parent phase (sea water, hydrothermal solution, intergranular fluid, melt, etc.) is given by:

$$I \propto e^{-\Delta G^*/RT} \cdot e^{-\Delta G^a/RT} \qquad (1)$$

where ΔG^* is effectively the free energy barrier for the formation of a critical nucleus, ΔG^a is effectively the free energy of activation for diffusion across the crystal/fluid interface, R is the gas constant, and T is absolute temperature.

For the simplest treatments, ΔG^* consists of three energy terms, the chemical free energy difference between the parent and the nucleating phase (ΔG_v), plus a surface and a strain energy term. ΔG_v represents the driving force for nucleation and depends on the degree of supersaturation.

It may easily be shown (Putnis and McConnell, 1980; Kirkpatrick, 1981) that $\Delta G^* \propto 1/(\Delta G_v)^2$. Within the stability field of the ordered phase (i.e., at $T < T_c$), ΔG_v for the nucleation of ordered crystals will obviously be greater than for the nucleation of disordered crystals. At, or only just below, the equilibrium crystallization temperature $(T \lesssim T_e$, i.e., for very small supersaturations) only ordered nuclei can form, and it is only at some lower temperature (higher degree of supersaturation) that disordered nuclei can become stable relative to the starting material. Thus there is a temperature interval close to T_e (at $T < T_c$) over which only ordered crystals could possibly nucleate and grow. At the other extreme, for $T \ll T_e$, ΔG_v would be large for both phases and the free energy barrier will become extremely small, though still marginally favoring the ordered polymorph. At large supersaturation, the kinetics of nucleation and growth are dominated by the second exponential term $(e^{-\Delta G^a/RT})$ in the rate equation because ΔG^* becomes extremely small. The concentration of atoms with sufficient energy to diffuse across the crystal/fluid interface decreases with falling temperature. As a result of this, the overall nucleation rate also decreases and we obtain the familiar C-shaped curve for nucleation shown on *TTT* diagrams (Fig. 2, and see Putnis and McConnell, 1980). If we consider only high supersaturations and assume that ΔG^* is negligibly small for both ordered and disordered crystals the

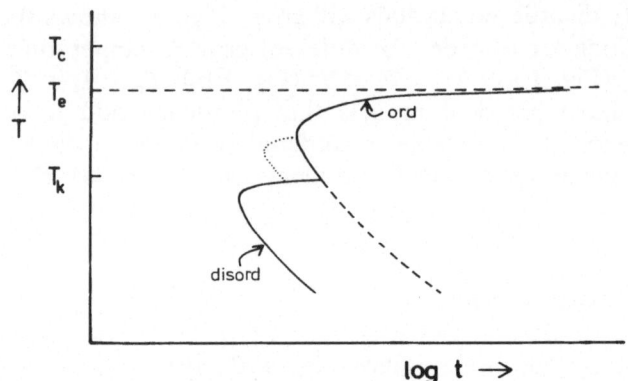

Fig. 2. Schematic *TTT* diagram showing onset of nucleation and growth of ordered (ord) and disordered (disord) crystals (solid lines). T = temperature; t = time; T_c, T_e, T_k have the same meaning as in Fig. 1. The dotted line shows the onset of nucleation and growth for some intermediate state of long-range order; in principle, curves should be shown for a complete range of intermediate states but only one is shown here.

origin of variations in crystallization behavior must be incorporated in the ΔG^a term.

We may write $\Delta G^a = \Delta H^a - T\Delta S^a + P\Delta V^a$, and hence (for $T < T_e$):

$$I \propto e^{-\Delta H^a/RT} \cdot e^{\Delta S^a/R} \cdot e^{-P\Delta V^a/RT} \tag{2}$$

where ΔH^a, ΔS^a, and ΔV^a are, respectively, the enthalpy, entropy, and volume of activation for the activated step during attachment of atoms to the surface of the nucleus. The activated states leading to the growth of ordered and disordered nuclei will be slightly different and hence values of the kinetic parameters for each case will also be slightly different. In particular, for crystallization below T_c, ΔH^a is a positive term that will be smaller (less positive) for an ordered activated state than for a disordered one, by an amount ΔH^a_{ord}. This is because of favorable interactions between the atoms in the case of an ordered nucleus. ΔS^a is probably negative and will be smaller (less negative) for the disordered nuclei than for the ordered nuclei, by an amount ΔS^a_{ord}, because a smaller increase in atomic order is required. ΔV^a will be negative and smaller (less negative) by ΔV^a_{ord} for disordered nuclei than for ordered nuclei, since disordered crystals usually have larger volumes than the equivalent ordered crystals. Thus for the disordered nuclei, $\Delta H^a_{disordered\ nucleus} = \Delta H^a_{disordered\ nucleus} + \Delta H^a_{ord}$, etc., and we may rewrite the kinetic Eq. (2) in terms of the relative rates of nucleation and growth of ordered and disordered crystals:

$$\frac{I_{disordered}}{I_{ordered}} \propto e^{-\Delta H^a_{ord}/RT} \cdot e^{\Delta S^a_{ord}/R} \cdot e^{-P\Delta V^a_{ord}/RT} \tag{3}$$

The signs of ΔH_{ord}^a, ΔS_{ord}^a, and ΔV_{ord}^a are such that the enthalpy and volume exponential terms favor ordered nuclei and the entropy term favors disordered nuclei.

We can now reproduce observed crystallization behavior and the effects predicted by the atomistic models discussed above if ΔS^a dominates the kinetics at high supersaturations. This result is easy to rationalize because ΔS^a can be understood, physically, in terms of the number of possible configurations of the activated state that will lead to a particular product. To produce an ordered nucleus a very specific configuration of atoms is required, but for a disordered nucleus the number of possible alternatives will clearly be substantially greater.

As pointed out by Chernov and Lewis (1967), we are dealing merely with a special case of the Ostwald step rule for kinetic control of reactions proceeding under conditions deviating substantially from equilibrium. It is also a special case of what Goldsmith (1953) referred to as a "symplexity principle," that crystals with disorder, structural simplicity, or high entropy will be easier to nucleate and grow than crystals with more complex structures, order, or low entropy. We could have scrutinized ΔG^* in the same way and suggest that ΔS^* would contribute to the same effect. The arguments extend also to growth, after the initial nucleation event, because the activated step will be very similar to the step controlled by ΔG^a. Figure 2 shows the predicted behavior on a TTT plot for the extreme cases of completely ordered and disordered nucleation and growth. For convenience, only one case of an intermediate state of ordering is shown but a full range of partly ordered phases would be anticipated (as in Fig. 1(a)). Subsequent changes in order may or may not occur in the crystal behind the growth face, depending only on the kinetics of exchange between neighboring atomic sites in the solid state.

As has already been stressed, a severe limitation of the transition state theory is that absolute values for the activation energy parameters are extremely difficult to predict. We are now in a position, however, to assess qualitatively the effects of P and T and identify the crystallization conditions under which we might expect to find metastable, disordered crystal growth in nature. At crystallization conditions close to equilibrium ($T \simeq T_e$, $P \simeq P_e$) ordered crystals will grow, but only slowly (actually at T_e, P_e the growth rate is zero). Ordered crystals will also be favored when ΔH_{ord}^a is large, i.e., when the interaction energies of $A-B$ type pairs are substantially greater (more favorable to ordering) than $A-A$ and $B-B$ type pairs, and if ΔV_{ord}^a is significant, at high pressure through the $P\Delta V_{\text{ord}}^a$ term. Disordered crystals will form at large supersaturations, when growth will be rapid (relative to growth of ordered crystals at the same P and T). They will be favored by low P and small ΔH_{ord}^a values. The importance of temperature is related to the circumstances in which high degrees of supersaturation will develop. If, during changing conditions, the equilibrium phase boundary for a new phase is crossed at a low temperature, the absolute rate of nucleation and growth will be slow. There will be a much greater possibility of progressing into a field of substantial supersaturation than if the boundary was crossed at high temperatures, where a much faster rate of nucleation and growth would prevail. Nonequilibrium crystallization conditions are, therefore, more likely to occur at low temperatures during metamorphism, in marine diagenesis, or during fast cooling of igneous systems.

It is worth guessing values of ΔH^a_{ord}, ΔS^a_{ord}, and ΔV^a_{ord} if only to predict order of magnitude kinetic effects. If we assume that the activated state during crystallization is rather similar to the final state, ΔH^a_{ord} will be somewhat less than the enthalpy of ordering in the bulk crystal, given that long-range interactions cannot occur in all directions. A value of $\Delta H^a_{ord} = 1$ kcal/mole gives $e^{-\Delta H^a_{ord}/RT} \approx 1/1.6$ at 1000 K. Similarly, ΔV^a_{ord} may be slightly less than the volume change associated with ordering in the bulk crystal. If we put $\Delta V^a_{ord} = 0.1$ cm^3/mole (= 0.1/41.84 cal/bar/mole), $e^{-P\Delta V^a_{ord}/RT} \approx 1/1.000001$ at 1 bar, 1000 K, and $\approx 1/1.01$ at 5 kb, 1000 K. If we put ΔS^a_{ord} approximately equivalent to the entropy difference between ordered and disordered crystals, a value of ~ 5 cal/mole (as for Al/Si ordering in feldspars) gives $e^{\Delta S^a_{ord}/RT} \approx 12$. Under this set of arbitrary conditions (5 kbar, 1000 K) disordered crystals would nucleate and grow faster than ordered crystals by a factor of $1/1.6 \times 12 \times 1/1.01 \approx 7.5$, and the contributions of ΔH^a_{ord} and ΔV^a_{ord} are clearly less important than the contribution of ΔS^a_{ord}. The $P\Delta V^a_{ord}/RT$ term is insignificant.

Evidence for Growth of Metastable Disordered Minerals

The recognition that an ordered phase crystallized initially with disordered cations is made easier if the process of ordering, subsequently, involves a symmetry change. If the ordered structure belongs to a lower symmetry crystal class, twin domains may result from the transformation. These may be observed either optically or by transmission electron microscopy (TEM). If the ordered structure has lost translational symmetry, antiphase domains (APDs) can be generated and may be observed by TEM. Thus a study of the relict microstructures of minerals may provide important information regarding transformations that have taken place. If ordering involves the continuous exchange of cations between sites, but with no symmetry change, recognition of the original structural state may not be possible. At low temperatures (i.e., less than a few hundred °C) ordering in silicates is usually extremely slow, however, even relative to a geological time scale, and the first formed crystals may be preserved.

Authigenic Environments

There is no doubt that in the authigenic environment minerals can grow with metastable disordered cation distributions. Kastner and Siever (1979) have reviewed the occurrence, composition, and structural state of authigenic feldspars. The crystals can have almost any state of Al/Si order, from fully ordered, triclinic albites and microclines to highly disordered monoclinic sanidines. Albites tend to be ordered but there is an apparent correlation between rock type and structural state for the K-feldspars, which Kastner and Siever (1979) suggest is due to kinetic factors. In carbonate rocks, the K-feldspars are only present in very small amounts, are fully ordered, and do not have cross-hatch twinning. Slow, ordered growth close to equilibrium is implied. In ash beds the K-feldspar fraction is much greater and the crystals tend to be highly disordered. It is easy to see that in the latter case supersaturation of ions in solution could readily be achieved because of the availability of highly soluble ash material. Abundant, relatively rapid growth of crystals would result. In sandstones and shales a wide variation of degree of order is found, presumably reflecting variations in supersaturation of the pore solutions from which the crystals grow.

Dolomite is an even more important authigenic mineral in which the problem of cation ordering plays a key role in controlling crystallization behavior. The literature on "the dolomite problem," i.e., the failure of dolomite to precipitate from sea water under normal marine conditions, is vast. In the present context it is only relevant to note that the dolomite structure has an ordered distribution of Ca^{2+} and Mg^{2+} ions and that the need to achieve this order during crystallization could seriously inhibit its formation (Goldsmith, 1953, 1969; Fyfe and Bischoff, 1965; Bathurst, 1975; Lippman, 1973; Folk and Land, 1975; Land, 1982; and others). Sea water is grossly oversaturated with respect to dolomite formation and therefore favors the precipitation of disordered crystals. Dolomite with severe cation disorder, however, is probably much less stable than other carbonates that may also be kinetically favored and is therefore unlikely to form. Folk and Land (1975) identify solutions appropriate for primary dolomite precipitation as being dilute, allowing crystallization under conditions closer to equilibrium and thus favoring slow, ordered nucleation and growth. Experimental synthesis conditions for the preparation of ordered dolomite crystals at low temperatures would similarly need to be close to equilibrium and would therefore result in extremely slow nucleation and growth rates. In practice, primary ordered dolomite can be readily synthesized only down to ~200°C (Graf and Goldsmith, 1956). Below this temperature the synthetic phases with dolomite compositions tend to have disordered or partially disordered Ca/Mg distributions (Graf and Goldsmith, 1956; Glover and Sippel, 1967; Schneider, 1976).

The effects of order/disorder properties in authigenesis and diagenesis are not solely restricted to the distribution of cations. Exactly analogous issues and arguments arise in the case of minerals that can have stacking disorder. The diagenesis of silica in marine environments involves the dissolution of amorphous biogenic silica (Opal A) to produce highly supersaturated pore fluids. Under these conditions the silica phase precipitated is porcelanite (Opal-CT), which is generally accepted as having a cristobalite structure with considerable stacking disorder in the [111] direction. Subsequent diagenesis results in progressive changes in X-ray diffraction patterns and has been interpreted as solid-state ordering within the stacking sequence (Murata and Nakata, 1974; Murata and Larson, 1975). Although the mechanism of this transformation is not well understood, the precipitation of a disordered phase that subsequently tends to order is the expected behavior under such conditions. Similar observations have also been made in hydrothermal experiments (Mizutani, 1977).

The growth of sheet silicates with stacking disorder is governed by the same general principle (Baronnet, 1980).

Hydrothermal Environments: Adularia

Adularias are K-feldspars distinguished principally by their habit and paragenesis. They are typically found in low-temperature hydrothermal veins and are remarkable for their extreme structural diversity, from high sanidine to maximum microcline (Smith, 1974; Černý and Chapman, in press). In the electron microscope some samples show a fine cross-hatched modulated texture that represents the intermediate stage of a transformation from the monoclinic, Al/Si disordered, state to triclinic, Al/Si ordered microcline (McConnell, 1965, 1971). The presence of this texture clearly indicates that the crystals grew initially with monoclinic symmetry. The equilibrium order/disorder temperature is thought to be ~450°C (Bambauer and Bernotat, 1982), which is probably above the crystallization temperature of most adularias. Metastable growth with more or less disordered cations is implicated (Laves, 1952; Goldsmith and Laves, 1954; Bambauer and Laves, 1960; Smith, 1974). One occurrence is particularly illuminating. Blasi et al. (1983) found overgrowths on amazonite, in the pocket zone of pegmatites from the Pike's Peak batholith, U.S.A., which were untwinned, maximum microcline. The overgrowths had evidently grown slowly near equilibrium with full Al/Si order. In rare instances, however, a precipitate of fine-grained high sanidine was found to coat the pocket minerals and was attributed to the effects of a rapid loss of pressure from the pocket (Martin and Foord, 1983). Decompression, possibly resulting from the development of cracks, would cause an increase in the supersaturation of the pocket solution and lead to relatively rapid growth of crystals with highly disordered Al/Si distributions. The wide variation in structural state of adularias presumably also reflects differences in supersaturation and growth conditions of pegmatites and hydrothermal veins (Černý and Chapman, in press).

Some postcrystallization structural changes occur in these hydrothermal environments, but the adularia microstructures and microcline twinning preserve good evidence of this. On the other hand, ordering in albite does not cause the development of obvious microstructures. There is, nevertheless, indirect evidence from phase equilibrium studies that albite too can grow with Al/Si disorder in Alpine fissures (Poty et al., 1974).

Metamorphic Environments

The stability field of cation ordering in omphacite may extend to temperatures as high as ~865°C (Carpenter, 1981a). The limits will vary with composition, but ~800°C is well above the crystallization temperature of most blueschist pyroxenes. Every blueschist omphacite so far examined, however, contains antiphase domains, implying that an ordering transformation has occurred (Carpenter, 1981b). On the basis of this observation, Champness (1973) and Carpenter (1978, 1980, 1981b) have argued that the crystals grow metastably, either with disordered cations or with only short-range ordering. The same applies to omphacites that crystallized at ~700°C (Carpenter and Smith, 1981). The rate of ordering appears to be sufficiently fast for the Na^+/Ca^{2+}, $Mg^{2+}/Al^{3+}/Fe^{2+}/Fe^{3+}$ cations to order between different M1 and M2 sites after crystallization. APDs develop during the resulting symmetry change, $C2/c \rightarrow P2/n$. At temperatures as low as ~350°C, the final long-range order does not exceed a scale of ~50 Å, however (Carpenter, 1981b).

This contrasts with the mineral joesmithite, a Be,Pb-rich amphibole that has ordered Be^{2+} and Si^{4+} on tetrahedral sites (Moore, 1969). In the electron microscope the crystals do contain APDs, but they are extremely irregular and are often closely associated with dislocations (Carpenter, 1982). There is no real evidence of an ordering transformation (which would have given a symmetry change from $C2/m$ to $P2/a$), so we may conclude that the crystals grew in an ordered state. The antiphase boundaries could be due to infrequent growth mistakes, to the passage of partial dislocations, or both.

Many metamorphic plagioclases must grow well within the stability field of Al/Si order. This applies particularly to the more calcium-rich crystals, because the boundary of $I\bar{1}$ ordering runs approximately through the points An_{59}, 1000°C and An_{77}, 1440°C (Carpenter and McConnell, 1984). The microstructural evidence for postcrystallization phase transformations is abundant (see, for example, Smith, 1974; McLaren and Marshall, 1974; Grove, 1977; Wenk and Nakajima, 1980), but there are serious difficulties in reconstructing the exact sequence of changes. Pure anorthite, itself, in slowly cooled rocks does not appear to have type b APDs (Müller et al., 1972) and so there is no evidence of an initial metastable, $C\bar{1}$ state with disordered

Al/Si. Sodium-rich crystals tend to contain exsolution lamellae or have the *e* ordered structure. The equilibrium phase relations are not well-enough defined to show whether a stable field of *e* ordering could exist or extend to temperatures above the crystallization temperatures but, in any case, it is unlikely that the highly regular type *e* domains could form during crystal growth. Al/Si ordering in albite does not cause the development of obvious microstructures because there is no symmetry change associated purely with the ordering. In spite of these uncertainties, some evidence of nonequilibrium variations in Al/Si order is provided by the range of compositions shown by metamorphic plagioclases, as discussed below.

Cordierite is another phase that has ordered Al/Si distributions and develops characteristic microstructures if it crystallizes in an initially disordered state (Schreyer and Schairer, 1961; Putnis, 1980a). T_c for the hexagonal (disordered) to orthorhombic (ordered) transformation is ~1450°C (Schreyer and Schairer, 1961; Smart and Glasser, 1977) and is therefore well above normal geological temperatures. Occasionally disordered cordierite (termed indialite) is preserved in rocks (Miyashiro *et al.*, 1955; Kitamura and Hiroi, 1982), but most metamorphic cordierites appear to have well-ordered Al/Si. However, a postcrystallization ordering transformation from hexagonal to orthorhombic symmetry would produce twinning in which individual twin domains were related by 60° rotations about [001] producing sector trilling (Zeck, 1972). Such trilling is a conspicuous feature of many natural cordierites and has also been observed to form in experimental cordierites undergoing Al/Si ordering (Schreyer and Yoder, 1964; Armbruster and Bloss, 1981). The implication, therefore, is that sector trilling indicates that the cordierite grew to essentially its full size with hexagonal symmetry before transforming to the ordered orthorhombic state.

Wenk (1983) has described ordering and exsolution microstructures in an aluminous sillimanite. The suggestion, once again, is that the crystals grew in a metastable state, with Al and Si disordered between the tetrahedral sites, and then ordered (Wenk, 1983). The sample came from a pelitic xenolith in tonalite. Similar features have not yet been described in sillimanites from regionally metamorphosed rocks.

Igneous Environments

Igneous anorthites show some signs that they can grow in a metastable $C\bar{1}$ state. Nord *et al.* (1973) and Nord (1983) have described type *b* APDs in lunar anorthites that crystallized well within the $I\bar{1}$ stability field, at temperatures of between 1200 and 1400°C. The most calcic parts of the crystals (i.e., An > 96% at 1400°C, and An > 90% at 1200–1300°C) are free of APDs and clearly grew fully ordered. The less calcic zones show regular type *b* domains that Nord (1983) suggested are the product of a $C\bar{1} \rightarrow I\bar{1}$ transformation. The changeover, from $I\bar{1}$ to $C\bar{1}$ growth, could correspond to Chernov and Lewis's (1967) "kinetic phase transition." With decreasing An content of the plagioclase during crystallization, either the melt was becoming over-saturated with respect to feldspar crystallization, or the contribution of ΔH°_{ord} to the kinetic factors (Eq. (3)) was being reduced (the enthalpy of ordering for the complete $C\bar{1} \rightleftharpoons I\bar{1}$ transformation would decrease as the composition changes towards albite). Even the elongate domain textures present between the two zones (Nord et al., 1973) are predicted by Chernov and Lewis for small supersaturation in their two-dimensional computer model.

Cordierites nucleated from melts often show complex transformation twins (e.g., Venkatesh, 1954; Zeck, 1972) indicating initially disordered growth at temperatures where well-ordered cordierite is the stable phase.

We would like to stress that in this short review we have only included the examples of minerals that give real evidence of nonequilibrium ordering behavior during nucleation and growth. Exactly the same order/disorder problems could arise in other minerals that have variable states of cation order but that do not undergo discrete phase transformations to produce relict microstructures. There is no a priori reason to suppose that metastable states of disorder could not be produced during the crystallization of epidote (Fe^{3+}/Al), orthopyroxene and anthophyllite (Fe^{2+}/Mg), etc.

Petrological Implications

The most disconcerting feature of the kinetic control of ordering during crystallization is that petrographic textures need not show any evidence of the nonequilibrium behavior. Variations in order could be accompanied by significant variations in free energy and could therefore lead to some misin-terpretation of what appear to be equilibrium assemblages. For example, in the estimation of palaeo-pressures and temperatures it is clearly necessary to ensure that the state of order used for model thermodynamic properties corresponds exactly to that of the minerals when their compositions became fixed. Textural relations and a metastable equilibrium partitioning of ele-ments between one or more metastably disordered phases may have been established at the time of crystallization. If bulk diffusion rates are very slow and intragrain equilibration relatively fast, the state of order would be changed but the "wrong" distribution of elements preserved (see, for exam-ple, Christie, 1962; Rutland, 1962; Brown and Parsons, 1981). Similarly, of course, the state of order in minerals produced in the laboratory should be rather well characterized if the resulting experimental data are to be applied with confidence to rocks.

We can illustrate a number of other areas where the role of order/disorder may play an important role in determining mineral assemblages.

Order-Dependent Miscibility Gaps

Miscibility gaps in the plagioclase feldspars appear to be dependent directly on the Al/Si ordering transformations, as there is little evidence for nonideal mixing in the disordered ($C\bar{1}$) solid solution (Orville, 1972; Windom and Boettcher, 1976; Seil and Blencoe, 1979; Kotelnikov *et al.*, 1981; Carpenter and McConnell, 1984). It follows that one effect of metastable variations in Al/Si order developed during crystal growth would be to cause variations in the apparent limits of solid solubility. For example, in greenschist/amphibolite facies rocks two feldspars, albite and oligoclase, often crystallize together across the peristerite gap (Evans, 1964; Crawford, 1966; Maruyama *et al.*, 1982; Grapes and Otsuki, 1983; and reviewed in Goldsmith, 1982a). However, the precise limits of the miscibility gaps, given by the compositions of the coexisting feldspars, are not always perfectly consistent (Crawford, 1966; Cooper, 1972) and metastable growth well within the supposed gap can occur (Nord *et al.*, 1978). Figure 3 (after Carpenter, 1981c) demonstrates how different Al/Si order in the albite crystals would give different (metastable) equilibrium limits to the miscibility gap. If the albites grow

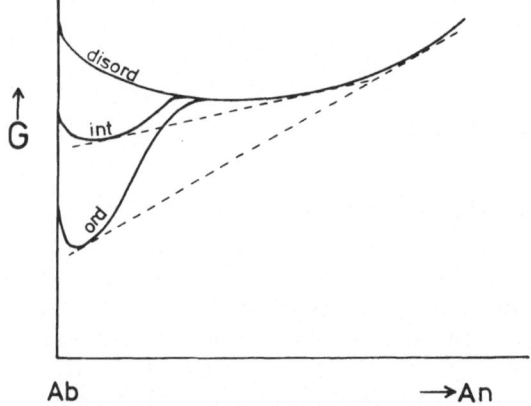

Ab   →An

Fig. 3. Schematic free energy (G)–composition (Ab = albite, An = anorthite) curves defining the peristerite gap in plagioclase feldspars at some low temperature (after Carpenter, 1981c). The curve for disordered feldspars is indicated by "disord"; albite solid solution with an equilibrium degree of Al/Si order is represented by the curve marked "ord"; the "int" curve represents the albite solid solution with some intermediate degree of order constrained by kinetics. Dashed lines are the common tangents that give the compositions of coexisting, equilibrium ordered albite and disordered oligoclase and of intermediate, metastable ordered albite and disordered oligoclase. In the absence of ordering, as a result of rapid nucleation and growth, there is no effective miscibility gap. Under equilibrium nucleation and growth conditions, which allow maximum ordering, the gap is widest. At intermediate degrees of ordering, constrained by the rate of crystal growth, the peristerite gap will appear narrow if judged purely on the basis of the composition of coexisting crystals.

partially ordered, the coexisting albite and oligoclase would be Ca- and Na-richer, respectively, than if maximum order is achieved. The experiments of Goldsmith (1982b) are consistent with this logic since they show no sign of the peristerite immiscibility region and probably involve very little Al/Si ordering during crystallization of the albite.

Based on natural assemblages, the peristerite gap also appears to be markedly pressure dependent (Brown, 1962; Crawford, 1970; Orville, 1974; Maruyama et al., 1982). It extends to peak temperatures of around 550–600°C at high pressures ($\sim$7–10 kbar) but only to $\sim$420°C under low-pressure regional or contact metamorphic conditions (Maruyama et al., 1982). Brown (1962) has made a rough estimate of the thermodynamic contribution of changing pressure and suggested a shift in the solvus peak of $\sim$10–40°C per 10 kbar. It may be necessary to postulate an additional kinetic contribution (e.g., Orville, 1974). During thermal metamorphism, temperature changes are relatively rapid and conditions of oversaturation, favoring crystallization of more disordered albite, are more likely to develop. The ΔV^a_{ord} term in Eq. (3) also favors less ordered growth at low pressure since the result of increasing pressure is to increasingly suppress disordered nuclei, but the magnitude of this effect may be rather small (as in the rough calculations at the end of the first section, above).

Exactly the same issues arise in the interpretation of two-plagioclase assemblages delineating other possible miscibility gaps in the solid solution. The contribution of both thermodynamic and kinetic parameters must be considered and, because of the sensitivity of the miscibility limits to ordering, the kinetic effects may predominate. Almost any combination of coexisting plagioclases can be found (Crawford, 1972; Wenk and Wenk, 1977; Garrison, 1978; H-R. Wenk, 1979; Grove et al., 1983; and others reviewed in Goldsmith, 1982a), and they certainly cannot all represent equilibrium crystallization conditions. Some additional criteria for equilibrium are needed, other than just textural relations. For reasons already outlined, an apparently regular partitioning of elements between the feldspar and its coexisting phases (Spear, 1980) or a regular dependence of plagioclase composition on whole rock composition (E. Wenk, 1979) are necessary but need not be sufficient for true, stable equilibrium.

Altogether in the plagioclase solid solution there are three ordered structures—low albite, ordered anorthite, and e plagioclase—to deal with. Of these we know that albite and anorthite can grow metastably disordered, at least in the laboratory. We might reasonably expect plagioclases with the e structure to have a large entropy of activation for nucleation and growth, if it is in fact possible for them to grow as primary phases at all. Clearly without a proper evaluation of the kinetic problems associated with nucleation and growth in this system, arguments about equilibrium behavior based on observed assemblages would seem to be rather precarious.

An analogous system to the plagioclases is $Al_2SiO_5-Al_2O_3$. Sillimanite is a stoichiometric ordered phase at geological temperatures (like anorthite or low albite), mullite has a modulated, ordered structure (like *e* plagioclase), and there is a miscibility gap between them (Cameron, 1977; Wenk, 1983) (like the Huttenlocher or peristerite gap?). Again, the limits of solid solution are probably extremely dependent on the ordering.

Order-Independent Miscibility Gaps

Plagioclase feldspars represent an extreme case of the sensitivity of mineral assemblages and compositions to order/disorder problems associated with nucleation and growth. Some other systems, notably jadeite–augite (Jd–Aug) and calcite–magnesite (Ct–Mag), have mixing properties that depend on ordering but would probably have miscibility gaps at low temperatures in any case. The slightly different circumstances give rise to different but, nevertheless, important effects.

Free energy–composition relations for both systems are shown for some arbitrary low temperature in Fig. 4 (after Carpenter, 1980, and Reeder, 1981). In the disordered solid solution (Jd–Aug or Ct–Mag), a positive deviation from ideality is shown that is sufficient to give a broad miscibility gap. The intermediate compositions (omphacite or dolomite) are only stable because of the free energy of ordering. In each system, if there is any kinetic difficulty associated with the nucleation and growth of the ordered phase, the alternative would not be simply a disordered phase of similar composition (as for K-feldspar, say) but an assemblage of Jd + Aug or Ct + Mag. Crystallization conditions that favor metastable disorder would therefore lead to the failure of omphacite or dolomite to appear at all. In the case of dolomite these extreme conditions do appear to prevail for sea water at low

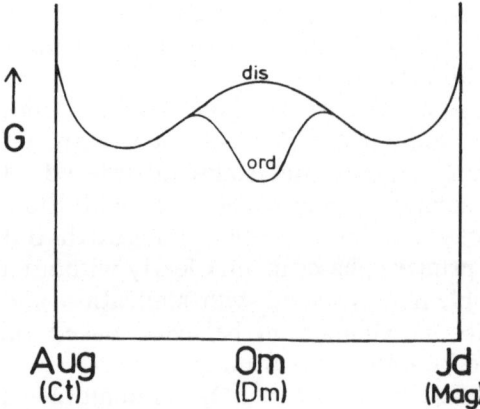

Fig. 4. Schematic free energy (G)–composition relations, at some low temperature, for jadeite (Jd)–augite (Aug) and magnesite (Mag)–calcite (Ct) (dis) with the ordered (ord) intermediate phases omphacite (Om) and dolomite (Dm) (after Carpenter, 1980, and Reeder, 1981). In the absence of cation ordering crystals of Om or Dm composition would not be stable relative to a two-phase mixture of disordered crystals.

temperatures. The available thermodynamic data are consistent with disordered dolomite being less stable than Ct_{ss} + Mag_{ss} (Helgeson *et al.*, 1978; A. B. Carpenter, 1980). For Jd–Aug, the fact that omphacite occurs in low-temperature blueschist rocks at all implies that the crystals do not grow completely disordered but have sufficient short-range order to stabilize them relative to an alternative pyroxene assemblage.

The final case to consider is that of the alkali feldspars. Both albite and K-feldspar can be ordered at low temperatures, but the miscibility gap between them is not very sensitive to variations in the degree of order actually attained (Yund and Tullis, 1983). The role of metastable disorder in controlling the compositions of coexisting crystals could therefore only be slight.

Displacement of Metamorphic Reaction Boundaries

Let us consider a reaction in a metamorphic rock of the type $A + B \rightarrow C$ where A, B, and C are three minerals and C can have ordered (C) or disordered (C') cations. At temperatures below the equilibrium order/disorder temperature for $C \rightleftharpoons C'$, there will be one set of physical conditions for the equilibrium $A + B \rightleftharpoons C$ and a second set for the metastable equilibrium $A + B \rightleftharpoons C'$, as illustrated in Figs. 5(a) and 5(b). The latter requires an overstep of the equilibrium boundary by a temperature ΔT and pressure ΔP, and this could occur either during prograde or retrograde metamorphism, i.e., for $A + B \rightarrow C$ with increasing P, T (Fig. 5(c)) or $A + B \rightarrow C$ with decreasing P, T (Fig. 5(d)). For an extreme case, where the new phase can only be fully ordered or fully disordered and the rate of nucleation of C is small relative to the rate of nucleation of C', the full overstep $(\Delta T, \Delta P)$ might be achieved before new crystals (of C') nucleate and grow. In practice, a whole range of (metastable) states of disorder may be possible and the actual degree of order of the new phases, and the overstep at which they first appear, will depend on a balance between the kinetic factors and the rate of crossing the equilibrium reaction boundary. Relatively rapid changes in temperature will obviously promote nonequilibrium, since the rate of nucleation of new phases could become insufficient to keep up with the changing conditions (Ridley, this volume). The important kinetic factors are, first, the absolute rate of nucleation and growth of C and, second, the values of ΔH_{ord}^{a}, ΔV_{ord}^{a}, ΔS_{ord}^{a}, which promote the nucleation and growth of C' in its place. Similarly, continuous equilibrium with changing P and T is less likely under retrograde conditions than under prograde conditions because nucleation and growth rates generally decrease with falling temperature, and equilibration therefore becomes progressively more difficult as cooling proceeds. The catalytic effects of a hydrous fluid in promoting metamorphic reactions are well known

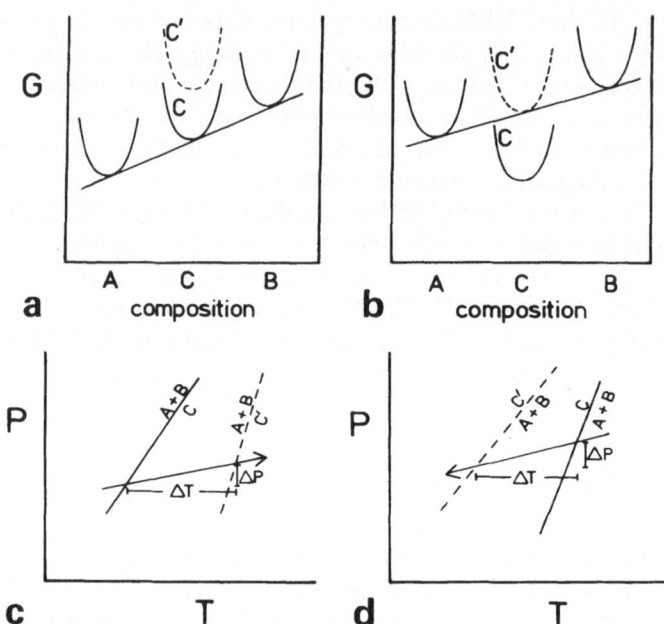

Fig. 5. Schematic illustration of the overstepping of a metamorphic reaction bound-ary ($A + B \rightarrow C$) in association with a change in the degree of cation order of the phase being produced (C = ordered, C' = disordered). Each diagram refers to *PT* conditions where the ordered phase C is stable relative to the disordered phase C'. (a) Free energy–composition diagram showing the equilibrium conditions for $A + B \rightleftharpoons C$. A will not react with B to produce C' since this involves an increase in free energy. (b) At some overstep (ΔT, ΔP), whether for the reaction occurring as a retrograde or prograde reaction, $A + B \rightarrow C'$ becomes possible if, for kinetic rea-sons, the ordered phase C fails to nucleate. The diagram shows the metastable equilibrium conditions for $A + B \rightleftharpoons C'$. (c) $A + B \rightarrow C$ as a prograde reaction. The equilibrium reaction is $A + B \rightarrow C$ (solid line). A certain overstep (ΔT, ΔP) is required to reach the $A + B \rightarrow C'$ metastable equilibrium boundary (dashed line). (d) For a quite different retrograde reaction ($A + B \rightarrow C$ with cooling and decompres-sion) the overstep beyond the equilibrium boundary involving the disordered phase C' is shown by ΔT, ΔP. These reactions are shown only in the most general way and refer to the extremes of complete cation order and complete cation disorder. If intermediate states of order between C and C' are possible, the overstep required to produce them will be less than the oversteps shown in (c) and (d). The actual over-step that occurs in a rock will depend on the rate of change of physical conditions relative to the rates of nucleation and growth of new crystals and the kinetic parame-ters ΔH^a_{ord}, ΔS^a_{ord}, and ΔV^a_{ord}.

and nonequilibrium crystallization is more likely to take place under low fluid pressures than at high P_{fluid}.

This metastable crystallization might be difficult to prove if, during postcrystallization annealing, C' undergoes an ordering transformation to C, unless some microstructures resulting from ordering are preserved. The tendency is to assume that, because geological time scales are so long, no such overstepping can occur, but the microstructural evidence, as discussed above, suggests that conditions of supersaturation can arise even in regional metamorphism. The possibility of this type of behavior also adds an additional source of uncertainty to the application of experimentally derived P, T data for the modelling of mineral reactions in nature. For example, measurements of the heats of solution of Mg–cordierites with different degrees of order have shown (Carpenter et al., 1983) that the enthalpy effect associated with Al/Si ordering can be substantial ($\gtrsim 10$ kcal/mole). Using this value and the fact that Mg–cordierite has a first-order transformation at $\sim 1450°C$ we may estimate (using $\Delta H_{ord} = T \cdot \Delta S_{ord}$) that the configurational entropy contribution caused by Al/Si disorder is at least ~ 6 cal K^{-1} mole. In many metamorphic reactions involving cordierite, the overall enthalpy and entropy changes are very small (Newton et al., 1974) so that large errors could accrue in calculated equilibration conditions if either the experimental phases involved some Al/Si disorder, or such disorder occurred during growth of the natural materials (Putnis, 1980b). Cordierite is by no means an isolated example of such behavior, although the effect of disorder on enthalpy and entropy is larger than would be expected for many other minerals.

Liquidus Relations and Crystallization

Kirkpatrick (1983) has recently reviewed the equilibrium and metastable crystallization sequences of silicates in igneous systems during continuous cooling. By considering the activation energy term in the classical theory he was able to explain why the nucleation of some structures can be suppressed more than that of others. Our contribution to this discussion is relevant only to special cases where crystals growing directly from the melt can have ordered cations (e.g., cordierite and anorthite). The need to have cation ordering during equilibrium crystal growth would lead to a further kinetic constraint on the crystallization properties, but the effect would probably be small relative to the grosser effects of structure as discussed by Kirkpatrick (1983).

Conclusion

In this paper our intentions have been: first, to show that it is easy to rationalize the metastable growth behavior of crystals and show why disordered growth should frequently be favored over equilibrium, ordered growth; second, to show that there is plenty of evidence for such metastable behavior in nature; and, finally, to outline the circumstances in which this behavior could have a real control on observed mineral assemblages. Most of the ideas and examples that we have presented are not new and can be found scattered around the literature. We hope, however, that by bringing them together in this form it will be apparent how widespread the effects could be and how little data we have for interpreting the kinetic as opposed to thermodynamic properties of minerals.

That metastable crystallization behavior is the norm in synthesis experiments involving highly reactive starting materials does not appear to be at issue. Nor is the fact that authigenic crystallization of some minerals in low-temperature environments may result in a wide range of structural states, although the factors affecting the state of order have never really been rigorously examined. Similarly, in higher temperature hydrothermal environments there is sufficient evidence for disordered growth to be confident that nonequilibrium conditions can prevail. In metamorphic systems, however, where the petrological implications of metastable crystal growth are most serious, there may be some reluctance to accept that significant departures from equilibrium could occur. For example, Wood and Walther (1983) have recently argued that, if the rate-controlling step in a metamorphic reaction such as prograde dehydration is the rate of surface reaction at activated sites, it is unlikely that any significant overstepping would occur in nature. They require the prior existence of all phases involved in the reaction and the presence of an excess of fluid, however. We have argued that the microstructural record of postcrystallization ordering transformations provides abundant evidence of nonequilibrium crystal growth conditions, though, without the appropriate kinetic data (nucleation rates and the activation energy parameters, at least), we are unable to predict whether this implies overstepping of equilibrium reaction boundaries by 0.5°C, 5°C, or 50°C. The detailed characterisation of mineral microstructures, coupled with studies of ordering mechanisms, therefore leads us to question the assumption of equilibrium for mineral nucleation and growth processes, even in regional metamorphism. Whether such questions will ultimately turn out to be significant, of course, remains to be seen, and further research in this area should be instructive.

Acknowledgments

Some of the work outlined in this review was funded by the Natural Environment Research Council of Great Britain (grant nos. GR3/3728 to A.P. and GR3/4404 to Dr. J. D. C. McConnell), and this support is gratefully acknowledged. We wish also to thank Dr. D. C. Rubie, Prof. A. B. Thompson, Dr. G. D. Price, Prof. R. J. Kirkpatrick, and C. Capobianco for their critical comments on successive versions of the manuscript. This is Cambridge Earth Sciences contribution no. 466.

References

Armbruster, T., and Bloss, F. D. (1981) Mg-cordierite: Si/Al ordering, optical properties and distortion. *Contrib. Mineral. Petrol.* **77**, 332–336.

Baikov, Y. A., Zelenev, Y. A., Haubenreisser, W., and Pfeiffer, H. (1980) On the analytical theory of kinetic order–disorder phase transition in binary crystals. *Phys. Stat. Sol.* **61**, 435–445.

Bambauer, H. U., and Bernotat, W. H. (1982) The microcline/sanidine transformation isograd in metamorphic regions. I. Composition and structural state of alkali feldspars from granitoid rocks of two N-S traverses across the Aar Massif and Gotthard ⟨⟨Masif⟩⟩, Swiss Alps. *Schweiz. Min. Pet. Mitt.* **62**, 185–230.

Bambauer, H. U., and Laves, F. (1960) Zum Adular problem. *Schweiz. Min. Pet. Mitt.* **40**, 177–205.

Baronnet, A. (1980) Polytypism in micas: A survey with emphasis on the crystal growth aspect, in *Current Topics in Materials Science,* Vol. 5, edited by E. Kaldis, pp. 447–548. North Holland, Amsterdam/New York/Oxford.

Bathurst, R. G. C. (1975) Carbonate sediments and their diagenesis, in *Developments in Sedimentology,* Vol. 12, 2nd Ed. Elsevier, Amsterdam/London/New York.

Blasi, A., Brajkovic, A., De Pol Blasi, C., Foord, E. E., Martin, R. F., and Zarazzi, P. F. (1983) Structure refinement and genetic aspects of a microcline overgrowth on amazonite from Pikes Peak batholith, Colorado, U.S.A. (abstract) Third NATO Advanced Study Institute on Feldspars, Feldspathoids and Their Parageneses, p. 96.

Brown, W. L. (1962) Peristerite unmixing in the plagioclases and metamorphic facies series. *Norsk. Geol. Tidsk.* **42**(2), 354–382.

Brown, W. L., and Parsons, I. (1981) Towards a more practical two-feldspar geothermometer. *Contrib. Mineral. Petrol.* **76**, 369–377.

Cameron, W. E. (1977) Mullite: A substituted alumina. *Amer. Mineral.* **62**, 747–755.

Carpenter, A. B. (1980) The chemistry of dolomite formation I: The stability of dolomite, in *Concepts and Models of Dolomitization,* edited by D. H. Zenger, J. B. Dunham, and R. L. Ethington. Soc. Econ. Pal. Min. Spec. Pub. No. 28, 111–121.

Carpenter, M. A. (1978) Kinetic control of ordering and exsolution in omphacite. *Contrib. Mineral. Petrol.* **67**, 17–24.

Carpenter, M. A. (1980) Mechanisms of exsolution in sodic pyroxenes. *Contrib. Mineral. Petrol.* **71**, 289–300.

Carpenter, M. A. (1981a) Time–temperature–transformation (TTT) analysis of cation disordering in omphacite. *Contrib. Mineral. Petrol.* **78**, 433–440.

Carpenter, M. A. (1981b) Omphacite microstructures as time-temperature indicators of blueschist and eclogite-facies metamorphism. *Contrib. Mineral. Petrol.* **78**, 441–451.

Carpenter, M. A. (1981c) A "conditional spinodal" within the peristerite miscibility gap of plagioclase feldspars. *Amer. Mineral.* **66**, 553–560.

Carpenter, M. A. (1982) Amphibole microstructures: Some analogies with phase transformations in pyroxenes. *Mineral. Mag.* **46**, 395–397.

Carpenter, M. A., and McConnell, J. D. C. (1984). Experimental delineation of the

$C\bar{1} \rightleftharpoons I\bar{1}$ transformation in intermediate plagioclase feldspars. *Amer. Mineral.* **69**, 112–121.

Carpenter, M. A., Putnis, A., Navrotsky, A., and McConnell, J. D. C. (1983) Enthalpy effects associated with Al/Si ordering in anhydrous Mg-cordierite. *Geochim. Cosmochim. Acta* **47**, 899–906.

Carpenter, M. A., and Smith, D. C. (1981) Solid solution and cation ordering limits in high temperature sodic pyroxenes from the Nybö eclogite pod. Norway. *Mineral. Mag.* **44**, 37–44.

Černý, P., and Chapman, R. (in press) Paragenesis, chemistry and structural state of adularia from granitic pegmatites. Proceedings, Third NATO Advanced Study Institute on Feldspars, Feldspathoids and their Parageneses.

Champness, P. E. (1973) Speculation on an order–disorder transformation in omphacite. *Amer. Mineral.* **58**, 540–542.

Cherepanova, T. A., Van der Eerden, J. P., and Bennema, P. (1978) Fast growth of ordered AB crystals: A Monte Carlo simulation for ionic crystal growth. *J. Crystal Growth* **44**, 537–544.

Chernov, A. A., and Lewis, J. (1967) Computer model of crystallization of binary systems: kinetic phase transitions. *J. Phys. Chem. Solids* **28**, 2185–2198.

Christie, O. J. (1962) Feldspar structure and the equilibrium between plagioclase and epidote. Discussion. *Amer. J. Sci.* **260**, 149–157.

Cooper, A. F. (1972) Progressive metamorphism of metabasic rocks from the Haast schist group of Southern New Zealand. *J. Petrol.* **13**, 457–492.

Crawford, M. L. (1966) Composition of plagioclase and associated minerals in some schists from Vermont, U.S.A., and South Westland, New Zealand, with references about the peristerite solvus. *Contrib. Mineral. Petrol.* **13**, 269–294.

Crawford, M. L. (1970) Phase relations of feldspars in contact metamorphic rocks (abstract). *Geol. Soc. Am. Abstr. Progs.* **2**, 528–529.

Crawford, M. L. (1972) Plagioclase and other mineral equilibria in a contact metamorphic aureole. *Contrib. Mineral. Petrol.* **36**, 293–314.

Eberhard, E. (1967) Zur Synthese der Plagioklase. *Schweiz. Min. Pet. Mitt.* **47**, 385–398.

Evans, B. W. (1964) Coexisting albite and oligoclase in some schists from New Zealand. *Amer. Mineral.* **49**, 173–179.

Folk, R. L., and Land, L. S. (1975) Mg/Ca ratio and salinity: Two controls over crystallization of dolomite. *Bull. Amer. Assoc. Petrol. Geol.* **59**, 60–68.

Fyfe, W. S., and Bischoff, J. L. (1965) The calcite-aragonite problem, in *Dolomitization and Limestone Diagenesis. A symposium,* edited by L. C. Pray and R. C. Murray. Soc. Econ. Pal. Min. Spec. Pub. No. 13, 3–13.

Garrison, Jr., J. R. (1978) Plagioclase compositions from metabasalt, Southeastern Llano uplift: Plagioclase unmixing during amphibolite grade metamorphism. *Amer. Mineral.* **63**, 143–147.

Glasstone, S., Laidler, K. J., and Eyring, H. (1941) *The Theory of Rate Processes.* McGraw-Hill, New York/London.

Glover, E. D., and Sippel, R. F. (1967) Synthesis of magnesium calcites. *Geochim. Cosmochim. Acta* **31**, 603–613.

Goldsmith, J. R. (1953) A "symplexity principle" and its relation to "ease" of crystallization. *J. Geol.* **61**, 439–451.

Goldsmith, J. R. (1969) Some aspects of the geochemistry of carbonates, in *Researches in Geochemistry*, edited by P. H. Abelson, pp. 336–358. Wiley, New York.

Goldsmith, J. R. (1982a) Review of the behavior of plagioclase under metamorphic conditions. *Amer. Mineral.* **67**, 643–652.

Goldsmith, J. R. (1982b) Plagioclase stability at elevated temperatures and water pressures. *Amer. Mineral.* **67**, 653–675.

Goldsmith, J. R., and Laves, F. (1954) Potassium feldspars structurally intermediate between microcline and sanidine. *Geochim. Cosmochim. Acta* **6**, 100–118.

Graf, D. L., and Goldsmith, J. R. (1956) Some hydrothermal syntheses of dolomite and protodolomite. *J. Geol.* **64**, 173–186.

Grapes, R., and Otsuki, M. (1983) Peristerite compositions in quartzofeldspathic schists, Franz Josef–Fox Glacier Area, New Zealand. *J. Metamorphic Geol.* **1**, 47–61.

Grove, T. L. (1977) Structural characterisation of labradorite–bytownite plagioclase from volcanic, plutonic and metamorphic environments. *Contrib. Mineral. Petrol.* **64**, 273–302.

Grove, T. L., Ferry, J. M., and Spear, F. S. (1983) Phase transitions and decomposition relations in calcic plagioclase. *Amer. Mineral.* **68**, 41–59.

Haubenreisser, W., and Pfeiffer, H. (1983) Microscopic theory of the growth of two-component crystals, in *Crystals, Growth, Properties and Applications, 9, Modern Theory of Crystal Growth I*, edited by H. Müller-Krumbhaar, and A. A. Chernov, pp. 43–73. Springer Verlag, Berlin/Heidelberg/New York.

Helgeson, H. C., Delaney, J. M., Nesbitt, H. W., and Bird, D. K. (1978) Summary and critique of the thermodynamic properties of rock forming minerals. *Amer. J. Sci.* **278A**, 229 p.

Kastner, M., and Siever, R. (1979) Low temperature feldspars in sedimentary rocks. *Amer. J. Sci.* **279**, 435–479.

Kirkpatrick, R. J. (1975) Crystal growth from the melt: A review. *Amer. Mineral.* **60**, 798–814.

Kirkpatrick, R. J. (1981) Kinetics of crystallization of igneous rocks, in *Kinetics of Geochemical Processes, Reviews in Mineralogy* **8**, edited by A. C. Lasaga, and R. J. Kirkpatrick. Mineral. Soc. America, 321–398.

Kirkpatrick, R. J. (1983) Theory of nucleation in silicate melts. *Amer. Mineral.* **68**, 66–77.

Kitamura, M., and Hiroi, Y. (1982) Indialite from Unazuki pelitic schist, Japan, and its transition texture to cordierite. *Contrib. Mineral. Petrol.* **80**, 110–116.

Kotelnikov, A. R., Bychkov, A. M., and Chenavina, N. I. (1981) The distribution of calcium between plagioclase and a water–salt fluid at 700°C and Pfl = 1000 kg/cm^2. *Geochem. Intern.* **18**, 61–75.

Kroll, H., and Müller, W. F. (1980) X-ray and electron optical investigation of synthetic high-temperature plagioclases. *Phys. Chem. Minerals* **5**, 255–277.

Land, L. S. (1982) Dolomitization. *Education Course Note Series* **24**, Amer. Assoc. Petrol. Geol., 1–20.

Laves, F. (1952) Phase relations of the alkali feldspars I. Introductory remarks. *J. Geol.* **60**, 436–450.

Lippman, F. (1973) *Sedimentary Carbonate Minerals.* Springer Verlag, New York/ Heidelberg/Berlin.

MacKenzie, W. S. (1957) The crystalline modifications of $NaAlSi_3O_8$. *Amer. J. Sci.* **255**, 481–516.

Martin, R. F., and Foord, E. E. (1983) Contrasting feldspar assemblages in granitic pegmatites of orogenic and nonorogenic affiliations (abstract). Third NATO Advanced Study Institute on Feldspars, Feldspathoids and Their Parageneses, p. 97.

Maruyama, S., Liou, J. G., and Suzuki, K. (1982) The peristerite gap in low grade metamorphic rocks. *Contrib. Mineral. Petrol.* **81**, 268–276.

McConnell, J. D. C. (1965) Electron optical study of effects associated with inversion in a silicate phase. *Phil. Mag.* **11**, 1289–1301.

McConnell, J. D. C. (1971) Electron optical study of phase transformations. *Mineral. Mag.* **38**, 1–20.

McLaren, A. C., and Marshall, D. B. (1974) Transmission electron microscope study of the domain structure associated with b-, c-, d-, e- and f-reflections in plagioclase feldspars. *Contrib. Mineral. Petrol.* **44**, 237–249.

Miyashiro, A., Iiyama, T., Yamasaki, M., and Miyashiro, T. (1955) The polymorphism of cordierite and indialite. *Amer. J. Sci.* **253**, 185–208.

Mizutani, S. (1977) Progressive ordering of cristobalitic silica in the early stage of diagenesis. *Contrib. Mineral. Petrol.* **61**, 129–140.

Moore, P. B. (1969) Joesmithite: A novel amphibole crystal chemistry. *Mineral. Soc. Amer. Spec. Pap.* **2**, 111–115.

Müller, W. F., Wenk, H-R., and Thomas, G. (1972) Structural variations in anorthites. *Contrib. Mineral. Petrol.* **34**, 304–314.

Murata, K. J., and Larsen, R. R. (1975) Diagenesis of Miocene siliceous shales, Temblor Range, California. *J. Res. U.S. Geol. Surv.* **3**, 553–566.

Murata, K. J., and Nakata, J. K. (1974) Cristobalitic stage in the diagenesis of diatomaceous shale. *Science* **184**, 567–568.

Newton, R. C., Charlu, T. V., and Kleppa, O. J. (1974) A calorimetric investigation of the stability of anhydrous Mg-cordierite with application to granulite facies metamorphism. *Contrib. Mineral. Petrol.* **44**, 295–311.

Nord, Jr., G. L. (1983) The $C\bar{1} \rightarrow I\bar{1}$ transition in lunar and synthetic anorthite (abstract). Third NATO Advanced Study Institute on Feldspars, Feldspathoids and Their Parageneses, p. 28.

Nord, Jr., G. L., Hammarstrom, J., and Zen, E-An (1978) Zoned plagioclase and peristerite formation in phyllites from southwestern Massachusetts. *Amer. Mineral.* **63**, 947–955.

Nord, Jr., G. L., Lally, J. S., Heuer, A. H., Christie, J. M., Radcliffe, S. V., Griggs, D. T., and Fisher, R. M. (1973) Petrologic study of igneous and metaigneous rocks from Apollo 15 and 16 using high-voltage transmission electron microscopy. *Proc. 4th Lunar Sci. Conf.* **1**, 953–970.

Orville, P. M. (1972) Plagioclase cation exchange equilibria with aqueous chloride solution: results at 700°C and 2000 bars in the presence of quartz. *Amer. J. Sci.* **272**, 234–272.

Orville, P. M. (1974) The "peristerite gap" as an equilibrium between ordered albite and disordered plagioclase solid solution. *Bull. Soc. Mineral. Cristallogr.* **97**, 386–392.

Pfeiffer, H. (1981) A simple approach to reactive processes on the interface during crystal growth. *Phys. Stat. Sol.* **65**, 637–642.

Poty, B. P., Stalder, H. A., and Weisbrod, A. M. (1974) Fluid inclusions studies in quartz from fissures of Western and Central Alps. *Schweiz. Min. Pet. Mitt.* **54**, 717–752.

Putnis, A. (1980a) The distortion index in anhydrous Mg-cordierite. *Contrib. Mineral. Petrol.* **74**, 135–141.

Putnis, A. (1980b) Order-modulated structures and the thermodynamics of cordierite reactions. *Nature* **287**, 128–131.

Putnis, A., and McConnell, J. D. C. (1980) *Principles of Mineral Behaviour.* Blackwell, Oxford/London/Edinburgh/Boston/Melbourne.

Reeder, R. J. (1981) Electron optical investigation of sedimentary dolomites. *Contrib. Mineral. Petrol.* **76**, 148–157.

Rutland, D. W. R. (1962) Feldspar structure and the equilibrium between plagioclase and epidote: A reply. *Amer. J. Sci.* **260**, 153–157.

Saito, Y., and Müller-Krumbhaar, H. (1981) Antiferromagnetic spin—1 Ising model. II. Interface structure and kinetic phase transition. *J. Chem. Phys.* **74**, 721–727.

Schneider, H. (1976) The progressive crystallization and ordering of low-temperature dolomites. *Mineral. Mag.* **40**, 579–487.

Schreyer, W., and Schairer, J. F. (1961) Compositions and structural states of anhydrous Mg-cordierites: A re-investigation of the central part of the system MgO-Al_2O_3-SiO_2. *J. Petrol.* **2**, 324–406.

Schreyer, W., and Yoder, Jr., H. S. (1964) The system Mg-cordierite-H_2O and related rocks. *N. Jb. Miner. Abh.* **101**, 271–342.

Seil, M. K., and Blencoe, J. G. (1979) Activity-composition relations of $NaAlSi_3O_8$-$CaAl_2Si_2O_8$ feldspars at 2kb, 600–800°C. *Geol. Soc. Amer. Abstr. Progs.* **11**, 513.

Smart, R. M., and Glasser, F. P. (1977) Stable cordierite solid solutions in the MgO-Al_2O_3-SiO_2 system: Composition, polymorphism and thermal expansion. *Science of Ceramics* **9**, 256–263.

Smith, J. V. (1974) *Feldspar Minerals*, Vol. 1, *Crystal Structure and Physical Properties.* Springer-Verlag, Berlin/Heidelberg/New York.

Spear, F. S. (1980) NaSi $\rightleftharpoons$ CaAl exchange equilibrium between plagioclase and amphibole. *Contrib. Mineral. Petrol.* **72**, 33–41.

Venkatesh, V. (1954) Twinning in cordierite. *Amer. Mineral.* **39**, 636–646.

Wenk, E. (1979) Bevorzugte Zusammensetzung und Variabilität der Plagioklase von Gesteinsserein der Verzasca. *N. Jb. Miner. Mh.* **1979**, 525–541.

Wenk, E., and Wenk, H-R. (1977) An-variation and intergrowths of plagioclases in banded metamorphic rocks from Val Carecchio (Central Alps). *Schweiz. Min. Pet. Mitt.* **57**, 41–57.

Wenk, H-R. (1979) Superstructure variations in metamorphic intermediate plagioclase. *Amer. Mineral.* **64**, 71–76.

Wenk, H-R (1980) Structure, formation and decomposition of APB's in calcic plagioclase. *Phys. Chem. Minerals* **6**, 169–186.

Wenk, H-R. (1983) Mullite-sillimanite intergrowth from pelitic inclusions in tonalite. *N. Jb. Miner. Abh.* **146,** 1–14.

Wenk, H-R., and Nakajima, Y. (1980) Structure, formation, and decomposition of APB's in calcic plagioclase. *Phys. Chem. Minerals* **6,** 169–196.

Windom, K. E., and Boettcher, A. L. (1976) The effects of reduced activity of anorthite on the reaction grossular +quartz = anorthite + wollastonite: A model for plagioclase in the earth's lower crust and upper mantle. *Amer. Mineral.* **61,** 889–896.

Woensdregt, C. F. (1983) Crystal growth and morphology of alkali feldspars grown from tungstate flux (abstract). Third NATO Advanced Study Institute on Feldspars, Feldspathoids and Their Parageneses, pp. 41–42.

Wood, B. J., and Walther, J. V. (1983) Rates of hydrothermal reactions. *Science* **222,** 413–415.

Yund, R. A., and Tullis, J. (1983) Subsolidus phase relations in alkali feldspars with emphasis on coherent phases, in *Feldspar Mineralogy, Reviews in Mineralogy,* Vol. 2, 2nd Ed., edited by P. H. Ribbe. Mineral. Soc. America, 141–176.

Zeck, H. P. (1972) Transformation trillings in cordierite. *J. Petrol.* **13,** 367–380.

Chapter 2
Kinetics of Metamorphic Reactions at Elevated Temperatures and Pressures: An Appraisal of Available Experimental Data

D. C. Rubie and A. B. Thompson

Introduction

Experimental kinetic studies and related theory for solid-state transformations relevant to metamorphic processes unfortunately lag far behind kinetic investigations of mineral–fluid reactions, related to natural waters and their precipitates, and mineral–melt reactions, related to crystallization of igneous rocks (see volume edited by Lasaga and Kirkpatrick, 1981). While metamorphic rocks have experienced successive temperature (T) and pressure (P) changes during their evolution, only rarely can we constrain the magnitude and direction of these P–T changes as functions of time (t). It is at present difficult to say whether characteristic disequilibrium features, observed petrographically, mean that a certain temperature was never achieved, or was only achieved for a short time. Such features might include the coexistence of two polymorphs or any complete reactant plus product assemblage as well as thermodynamically incompatible phases, irregular grain boundaries, and oriented fabrics. A great variety of environmental factors (presence of fluid, deformation, local chemical catalysts, etc.) can influence the kinetics of most mineral transformations. However, it is not often possible to see to what degree such factors have influenced an observed mineral assemblage from petrographic data alone.

Perhaps the best understood solid-state mineral transformations, relevant to the uplift part of a metamorphic PTt path, are exsolution in some feldspars and pyroxenes and cation ordering in these minerals and some amphiboles (see Putnis and McConnell, 1980; Ganguly, 1982; Yund, 1983). Many mineral decomposition reactions of interest to the mineral industry have been the subject of extensive kinetic studies at elevated temperatures but usually only at atmospheric pressure (see Kingery *et al.*, 1976; Schmalzried, 1974). It is to be expected that reaction mechanisms, as well as overall transformation kinetics, would not be the same at elevated pressure as at one atmosphere (Hanneman, 1969).

In this chapter we have attempted to review available experimental kinetic data on mineral reactions, partly to evaluate the relative effects of the many experimental variables and partly to understand how such data may be applied to interpreting microstructures in rocks. Our coverage is far from complete, and we have deliberately used the many data on the well-studied CaCO$_3$ (aragonite–calcite) transformations to discuss the unfortunate ambiguity of many kinetic investigations.

Symbols Used

C	Concentration
C_g	Number of grain corners per unit volume
d	Grain diameter
$\bar{d}$	Average grain diameter
D	Mass diffusivity
ΔG^*	Activation energy for nucleation
ΔG_r	Gibb's molar free energy change of reaction
ΔG_v	Gibb's free energy change of reaction per unit volume
h	Planck's constant
H^*	Activation enthalpy
k	Boltzmann's constant
K	Constant in a rate equation
L_g	Length of grain edge per unit volume
n	Constant in a rate equation
n_v	Number of nucleation sites per unit volume
N_v	Nucleation rate per unit volume
P	Pressure
ΔP	Pressure overstepping of equilibrium conditions
Q	Activation energy for growth
r	Radius
$\dot{r}$	Rate of change of radius
R	Gas constant
S_g	Area of grain boundary per unit volume
t	Time
T	Temperature
T_m	Temperature of melting
ΔT	Temperature overstepping of equilibrium conditions
ΔV	Molar volume change of reaction
ΔV^*	Activation volume

x Distance, or thickness of a layer
$\dot{x}$ Growth rate
z Ratio of molar volume of products to molar volume of reactants
α Thermal diffusivity
δ Interplanar spacing
ε Strain energy
$\dot{\varepsilon}$ Strain rate
ξ Extent of reaction (volume fraction transformed)
$\dot{\xi}$ Rate of overall transformation
σ Interfacial free energy per unit area
τ Time at which a nucleus forms

Mineral Reaction Mechanisms and Rate-Controlling Steps

In Table 1 we have compiled available experimental data on mineral transformation kinetics. These are grouped according to whether the reaction is polymorphic, a multiphase solid–solid reaction, a devolatilization reaction, or a volatilization reaction, without any attempt to specify reaction mechanisms. There are several classifications of phase transformations in the literature of which the summary by Christian (1975, p. 4) serves as an illustration.

Types of Heterogeneous Solid-State Mineral Transformations

Most solid-state transformations involve nucleation and growth (interphase boundary migration). The rate of such *reconstructive* transformations is a function of temperature, and at given temperature (isothermal) the volume fraction transformed (ξ) increases with time. The rate of *displacive* transformations (diffusionless, shear, or martensitic transformations) is almost independent of temperature (discrete regions transform rapidly), and the amount of transformation (ξ) is characteristic of a certain temperature range but does not necessarily increase with time. The type of boundary migration for displacive transformations is *glissile,* because the interphase boundaries (e.g., martensite, mechanical twin or low-angle grain boundaries) can migrate easily with a suitable driving stress even at very low temperatures and apparently do not require thermal activation. For reconstructive transformations, the migration of grain boundaries is *nonglissile* and their mobility is T-dependent. Nonglissile grain boundary migration, where no compositional change is involved (polymorphism, order–disorder, and single-phase recrystallization) describes *interface-controlled reactions.* Nonglissile grain

Table 1. A summary of experimental investigations of kinetics of heterogeneous reactions of geological interest.

	T range	P range	t_{max}	Other comments	Rate Eq. (Table 2)
A. Polymorphic reactions					
A.1. Aragonite → Calcite					
1. Chaudron (1954)	380–420°C	1 bar	9 hr	(P)* dry, rate measured by dilatometer	n.d.
2. Brown et al. (1962)	200–440°C	1–5000 bar	7 days	(P) dry and wet	Eq. (14)
3. Davis and Adams (1965)	340–500°C	1–10,000 bar	18 hr	(SC) dry, varied experimental apparatus	Eq. (13) ($n = 1.2$–10.0)
4. Bischoff and Fyfe (1968)	108°C	1 bar	8 days	(P) wet, varied fluid chemistry	Eq. (12) ($n = 2$ assumed)
5. Bischoff (1969)	50–120°C	1 bar	5 days	(P) wet ($CaCl_2$ aqueous solution)	Eq. (12) ($n = 2$ assumed)
6. Kunzler and Goodall (1970)	320–455°C	1 bar	17 hr	(P) dry, varied grain size, trace element chemistry, crystal morphology	Eq. (7.2) ($n = 0.6$–4.0)
7. Rao (1973)	400–440°C	1 bar	10 hr	(P) dry	Eq. (7.2) ($n = 1.3$), $1 - (1 - \xi)^{1/3} = Kt$
8. Tolokonnikova et al. (1974)	n.g.	1 bar	n.g.	n.g.	Eq. (7.2) ($n = 1$)
9. Carlson and Rosenfeld (1981)	375–455°C	1 bar		(SC) dry, determined growth rate $\dot{x}$	Eq. (2)
A.2. Calcite → Aragonite					
1. Davis and Adams (1965)	375–405°C	15 kbar	100 min	(P) dry	Eq. (13) ($n = 3.0$)
2. Brar and Schloessin (1979, 1980)	300–600°C	17–20 kbar	16 hr	(P and SC) dry	Eq. (7.2) ($n = 0.57$–1.03)

3. Wenk et al. (1973)	25–800°C	5–22 kbar	45 days	(R) dry, deformation experiments	n.d.
A.3. Quartz → Cristobalite					
1. Grimshaw et al. (1956)	1270–1450°C	1 bar	264 hr	(P) dry, various catalysts	Eq. (7.2) ($n = 1$ assumed)
2. Chaklader and Roberts (1961)	1500–1650°C	1 bar	24 hr	(P) dry, intermediate phase deduced	Eq. (13) ($n = 1$ assumed)
A.4. Sillimanite ⇌ Andalusite					
1. Holdaway (1971)	651–829°C	1.8 kbar	55 days	(P/SC) wet, no nucleation barrier (seeded runs)	n.d.; see Wood and Walther (1983)
A.5. Kyanite ⇌ Andalusite					
1. Holdaway (1971)	433–585°C	3.6–4.8 kbar	60 days	(P/SC) wet, no nucleation barrier (seeded runs)	n.d., see Wood and Walther (1983)
A.6. Olivine → Spinel ($(Mg,Fe)_2SiO_4$)					
1. Sung (1979)	400–800°C	110–220 kbar	208 days	(P) dry, diamond anvil apparatus–high differential stress	Eq. (7.2) ($n = 1$ assumed)
A.7. Olivine → Spinel (Ni_2SiO_4)					
1. Kasahara and Tsukahara (1971)	800–1000°C	35–40 kbar	2 hr	(P) dry	Eq. (13) ($n = 1$ assumed
2. Hamaya and Akimoto (1982)	800–1100°C	34–40 kbar	3 hr	(SC) dry	Eq. (7.2) ($n = 1.8$)
A.8. Quartz → Rutile (GeO_2)					
1. Zeto and Roy (1969)	380–600°C	10–30 kbar	45 hr	(P) dry, determined effect of high P on kinetics of a powder reaction	Eq. (7.2) ($n = 0.2$–0.5) $n = f(\xi)$

Table 1. (Continued)

	T range	P range	t_{max}	Other comments	Rate Eq. (Table 2)
B. Multiphase solid–solid reactions					
B.1. Jadeite + Quartz → Albite					
1. Matthews (1980)	454°C	1 kbar	9 days	(P) wet, varied fluid composition, metastable analcime	Eq. (12) ($n = 0.51$–0.61)
2. Rubie (unpublished)	410–720°C	1–5 kbar	50 days	(P) wet, metastable nepheline at 1 kbar	n.d.
B.2. Albite → Jadeite + Quartz					
1. Rubie (in preparation)	610°C	20 kbar	40 days	(A) wet	n.d.
	1000–1130°C	35 kbar	10 days	(P) dry, little reaction	n.d.
B.3. Quartz + Periclase → Forsterite					
1. Shaw (Fyfe *et al.*, 1958, p. 84)	425–1200°C	n.g.	n.g.	(P) wet and dry	n.g.
2. Brady (1983)	650–700°C	1 kbar	56 days	(P/SC) wet	Eq. (3), estimated grain boundary diffusivity
C. Dehydration and decarbonation reactions					
C.1. Talc → Anthophyllite + Quartz + H_2O					
Anthophyllite → Enstatite + Quartz + H_2O					
Talc → Enstatite + Quartz + H_2O					

	Temperature	Pressure	Time	Conditions	Rate law
1. Greenwood (1963)	805–830°C	1 kbar	128 days	(P) wet, metastable anthophyllite	Eq. (13) ($n = 1$ assumed)
C.2. Serpentine + Brucite → Forsterite + H_2O					
1. Wegner and Ernst (1983)	380–480°C	1–3 kbar	20 days	(P) wet	Eq. (7.2) ($n = 2$)
C.3. Calcite + Quartz → Wollastonite + CO_2					
1. Jander (1927)	830°C	1 bar	2 hr	(P) $X_{CO_2} = 1$	Eq. (9)
2. Gordon (1971)	650°C	2 kbar	35 days	(P) wet, $X_{H_2O} = 0.82$	n.d.
3. Kridelbaugh (1973)	800–950°C	1–1800 bar	11 days	(P) $X_{CO_2} = 1$, traces of H_2O changed reaction mechanism	Eq. (12) ($n = 0.17$–0.29)
4. Tanner et al. (1983)	500–850°C	1–3 kbar	n.g.	(P) $X_{CO_2} = 0.95$	n.g.
C.4. Brucite → Periclase + H_2O					
1. Fyfe et al. (1958)	310–450°C	1 bar	6 hr	(P)	Eq. (13) ($n = 1$ assumed), Eq. (14)
C.5. Sericite → Kaolinite					
1. Tsuzuki and Mizutani (1971)	130–270°C	V.P	158 days	(P) wet (acid solutions of varied composition)	Eq. (13) ($n = 1$ assumed)
C.6. Dehydration of Kaolinite					
1. Fyfe et al. (1958)	470–530°C	1 bar	4 hr	(P)	Eq. (13) ($n = 1$ assumed
2. Tsuzuki and Mizutani (1971)	255–270°C	V.P	158 days	(P) wet (acid solutions of varied composition)	Eq. (13) ($n = 1$ assumed

Table 1. (Continued)

	T range	P range	t_{max}	Other comments	Rate Eq. (Table 2)
C.7. Kaolinite + Quartz → Pyrophyllite + H_2O					
1. Thompson (1970)	250–400°C	1–4 kbar	3 months	(P/SC) wet, no nucleation barrier (seeded runs)	n.d., (see Fyfe *et al.* 1978, p. 124; Wood and Walther, 1983)
C.8. Analcime + Quartz → Albite + H_2O					
1. Thompson (1971)	130–250°C	2–6 kbar	14 days	(P/SC) wet, no nucleation barrier (seeded runs)	n.d., (see Fyfe *et al.*, 1978, p. 124; Wood and Walther, 1983)
2. Matthews (1980)	345–474°C	1 kbar	10 days	(P) wet, varied fluid composition	Eq. (12) ($n = 1.4$–1.7)
C.9. Muscovite + Quartz → K-Feldspar + Al_2SiO_5 + H_2O					
1. Evans (1965)	465–710°C	1–4 kbar	7 days	(P/SC) wet, no nucleation barrier (seeded runs)	n.d., see Wood and Walther (1983)
2. Kerrick (1972)	540–660°C	2 kbar	58 days	(P/SC) wet, no nucleation barrier (seeded runs), varied X_{CO_2}, X_{H_2O}	n.d., see Wood and Walther (1983)
3. Schramke *et al.* (1983)	n.g.	n.g.	n.g.	wet	n.g.
C.10. Dehydration of Chlorite					
1. Fyfe *et al.* (1958)	550–850°C	1 bar	200 min	(P)	n.d.

Reaction / Reference	Temperature	Pressure	Time	Conditions	Notes
D. Hydration and carbonation reactions					
D.1. Forsterite + H_2O → Brucite + Serpentine					
Enstatite + H_2O → Talc + Forsterite					
Enstatite + H_2O → Talc + Forsterite + Serpentine					
Enstatite + Forsterite + H_2O → Serpentine					
1. Martin and Fyfe (1970)	80–560°C	0.7–2.8 kbar	22 days	(P) wet	Eq. (9)
2. Wegner and Ernst (1983)	188–388°C	1–3 kbar	62 days	(P) wet, varied grain size, and % H_2O	Eq. (7.2) ($n = 1$)
D.2. Andalusite + Sanidine + H_2O → Muscovite + Quartz					
1. Schramke et al. (1983)	n.g.	n.g.	n.g.	wet	n.g.
D.3. Wollastonite + CO_2 → Calcite + Quartz					
1. Tanner et al. (1983)	500–850°C	1–3 kbar	n.g.	(P) $X_{CO_2} = 1$	n.g.

* Abbreviations: P–powdered sample; SC–single crystal; P/SC–mixed powder and single crystal; R–rock sample; A–annealed aggregate; V.P.–equilibrium vapour pressure; n.d.–not determined; n.g.–not given.

boundary migration where a compositional change is involved (eutectoid decompositions, $A \rightarrow B + C$; precipitation from a supersaturated crystalline solution, $\alpha \rightarrow \alpha_1 + \alpha_2$) requires a long-range transport of appropriate species to promote interface migration and is primarily *diffusion-controlled* but may also be interface or heat flow-controlled (Fisher, 1978).

Reaction Mechanisms, Pathways, and Progress

The progress of *homogeneous* reactions involving one elementary step is often described by the variation in concentration of one or more reactants and products. Homogeneous reactions are very rare in solids but may include migration of diffuse domains or structural modulations (Boehm, 1983). As discussed below, the *order of reaction* equation commonly used to describe homogeneous kinetics cannot realistically be applied to heterogeneous mineral transformations. In the case of the latter, an overall transformation is characterized either by *sequential* reaction mechanisms, where the slowest step is rate controlling, or by *parallel* reaction mechanisms, where the fastest step is rate controlling (Lasaga, 1981, p. 12).

During polymorphic reactions, the new stable phase may grow within the old unstable phase (on grain or subgrain boundaries or on other defects) or grow in the matrix at a site removed from the old phase, depending upon rates of nucleation at different sites. For any type of transformation, the reaction pathway may be quite different when the transformation, for example, proceeds in the presence of H_2O (e.g., Brown *et al.*, 1962; Fisher, 1970) or with a significant shear stress. In particular, water can allow solid–fluid pathways to dominate over slower solid–solid pathways (see also Yund *et al.*, 1972). In any case, depending on the reaction pathway, intermediate metastable phases may form during reaction, and these may or may not be later consumed (e.g., Table 1, B.1.1, B.1.2, C.1.1; Greenwood, 1963; Green, 1972; Loomis, 1979; Koons and Rubie, 1983).

For multiphase solid–solid reactions, the reaction mechanisms will be different for the opposing reactions. For example, the reaction forsterite to periclase plus quartz is a "forward" eutectoid decomposition. The reverse reaction ("backward" eutectoid), where the double oxide spatially separates the reactants, can only proceed if the reactant components are available to diffuse through the product phase (Schmalzried, 1978, p. 280). In rocks, such reactions may involve intermediate steps, some of which could be de- or revolatilization reactions.

For multiphase devolatilization reactions, reaction mechanisms will in general be more complicated than for solid–solid reactions. At the time of dehydration and decarbonation at low pressures, the volume of solid products plus fluid is greater than the volume of reactants (within the earth's crust for most reactions). This will induce local stress anisotropies and eventually lead to microfracturing. If the devolatilization fluid is able to collect and leave the generation site, the remaining product minerals occupy a smaller volume than the original minerals. Such reactions have greater enthalpies than solid–solid reactions and their rates may be heat flow-controlled (Fisher, 1978; Ridley, this volume). The relative importance of diffusion-control versus heat flow-control may be seen in Fig. 1, where distance x is plotted against $\sqrt{\alpha t}$ and $\sqrt{Dt}$, where α is thermal diffusivity and D is mass diffusivity. The rate of heat flow becomes the controlling factor over large distances for a given time.

The actual reactions occurring during metamorphism are usually fewer in

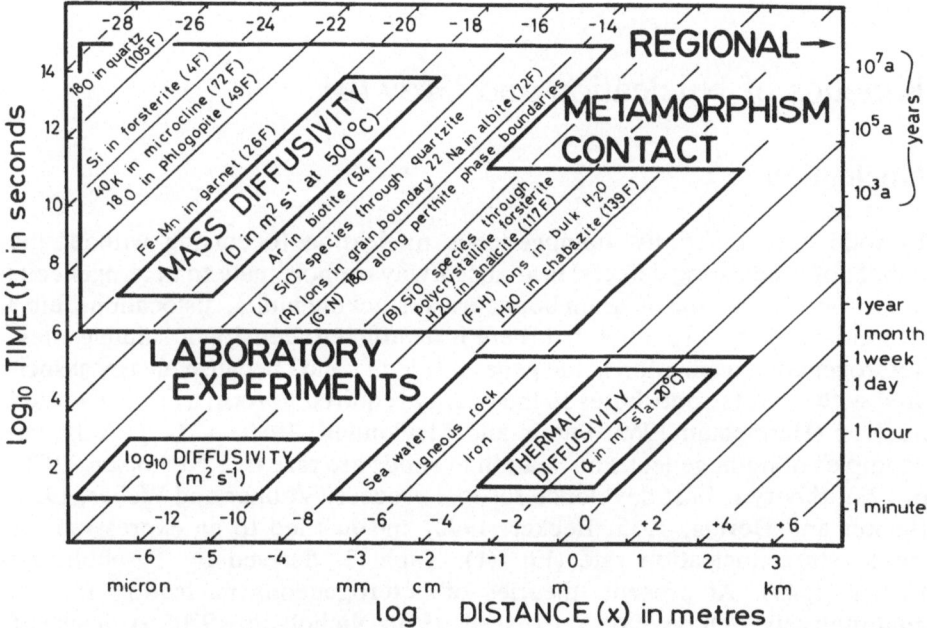

Fig. 1. Thermal and mass diffusivity data for some rocks, minerals, and water plotted as the relationships x (distance) $= \sqrt{Dt}$ and $\sqrt{\alpha t}$, where D = mass diffusivity, a = thermal diffusivity (m²/s), and t = time (s). Mass diffusivities (D in m²/s) were obtained for selected examples from the D_0 and Q values compiled by Freer (1981) and values of $\log_{10}D$ at 500°C are plotted from the solution of the equation $D = D_0 \exp(-Q/RT)$. The values of D decrease with decreasing T but no account is taken of possible change from intrinsic to extrinsic diffusion. The number code + F corresponds to the references quoted by Freer (1981). Thermal diffusivity values of $\log_{10}\alpha$ at 20°C are taken from Stacey (1977, p. 334). These values change by less than an order of magnitude for temperatures up to about 600°C and pressures up to at least 3 kbar. The other abbreviations refer to estimated mass diffusivities for a range of processes at metamorphic conditions of about 600°C and 1 kbar from (F + H) Fletcher and Hofmann (1974); (R) Rutter (1976); (G + N) Giletti and Nagy (1981); (B) Brady (1983); (J) Joestein (1983).

number and involve fewer phases than the possible equilibria that may be written among the components of the chemically complex minerals. Ignoring for the moment the fact that the evolved fluid phase is not always preserved in the assemblage, it is necessary to consider the circumstances that permit both reactant and product to be present in order to recognize an equilibrium. Such a coexistence is usually ascribed to the P-T reaction intervals over which crystalline solutions are reacting but in many instances may also be due to incomplete reaction controlled by kinetics.

In the following sections we examine a whole range of proposed rate equations (Table 2), beginning with a review of the kinetics of nucleation and growth and then proceeding to examine overall transformation kinetics. Processes of grain growth and coarsening are also reviewed.

Kinetics of Nucleation and Growth

Nucleation

In solid-state reactions, homogeneous nucleation (occurring without the benefit of preexisting defects) is almost always subordinate to heterogeneous nucleation (occurring on grain boundaries, stacking faults, dislocations, etc.) (Christian, 1975). Provided that there is a sufficient density of suitable sites, heterogeneous nucleation is the general rule in solids because energy associated with defects contributes to the energy required to form a critically sized nucleus (Hanneman 1969; Putnis and McConnell, 1980, p. 147). Although examples of homogeneous nucleation in solids are rare (see Nicholson, 1970, p. 273), theory is best developed for this process (Volmer and Weber, 1926; Becker and Döring, 1935; Becker, 1940) and has led to an expression for steady-state nucleation rate (Eq. (1), Table 2) derived by Turnbull and Fisher (1949). At present, theories of heterogeneous nucleation are not quantitatively well developed (Russell, 1970; Nicholson, 1970). A theory of nucleation on grain boundaries (Christian, 1975, p. 448) is based on the reduction of interfacial or surface energy. The theory of nucleation on dislocations (Cahn, 1957) probably fails to account for nucleus-dislocation interaction (Nicholson, 1970, p. 287). None of the theories of nucleation appear to take into account the high strain energies associated with solid–solid reactions involving a large volume change (see below). Only in very few cases have experimental studies of metallic systems so far supported theory (e.g., Servi and Turnbull, 1966; Tanner and Servi, 1966).

Qualitatively, as shown in Fig. 2, nucleation theory predicts that (1) finite P and/or T overstepping of the equilibrium boundary is required for significant nucleation rates (see also Carpenter and Putnis, this volume); (2) the amount of overstepping is large for reactions involving high interfacial free energy (σ), high strain energy (ε), and small free energy of reaction (ΔG_r);

(3) the amount of overstepping to induce nucleation is less for high-energy defects than for low-energy defects and is greatest for homogeneous nucleation. The magnitude of the overstepping necessary for nucleation is generally unknown, but Nicholson (1970, p. 274) reports that for homogeneous nucleation in Cu–Co, 40°C undercooling is required. (4) Beyond the required overstepping, $\dot{N}_v$ increases very rapidly with further change in T or P (Fig. 2; Hanneman, 1969), especially according to Cahn's (1957) theory for nucleation on dislocations (see Rubie, 1983).

Useful objectives for experimental studies would be to determine (1) the required overstepping ($\Delta P, \Delta T$) for nucleation, (2) dominant sites of nucleation, (3) estimates of rates of nucleation by electron microscope (TEM and/ or SEM) methods (e.g., Tanner and Servi, 1966; see also Yund and Hall, 1970), and (4) effects of impurities on nucleation rates (see Yund and Hall, 1970).

The stage in a reaction at which all potential nucleation sites are consumed is termed site saturation (Cahn, 1956a), and following this the reaction proceeds by growth alone. Cahn (1956a) has shown theoretically that the time at which site saturation is reached is very sensitive to changes in nucleation rate ($\dot{N}_v$) and, thus, is likely to occur either very early or very late in a transformation. The widely used Avrami theory for overall transformation rate (Eqs. (7), Table 2) assumes either early site saturation or constant $\dot{N}_v$ throughout the transformation (see below). In experiments carried out at a great distance from equilibrium (large ΔT or ΔP), large numbers of nuclei are observed to form in a very short time, and Carlson (1983, p. 64) inferred that early site saturation occurs during the aragonite to calcite transformation. However, for reactions with a large volume change (ΔV), the reactant phase may become strained during the growth of the product phase (transformation strain, see below) with the consequent creation of *new* nucleation sites. For example, Carlson and Rosenfeld (1981, p. 624) reported fracturing of aragonite during the growth of calcite grains and the generation of new nuclei on the cracks (at $P = 1$ bar, $T = 375$–455°C). The continuous creation of new nucleation sites during a transformation by such a mechanism would invalidate assumptions of early site saturation and requires both further experimental and theoretical investigation.

Growth

Depending on the rate-controlling step, growth rate (velocity of interface migration) may be *interface-controlled* or *diffusion-controlled* (Christian, 1975, p. 6). For both types of solid-state reactions the rate is often observed to be exponentially temperature-dependent (Arrhenius relation; e.g., Eq. (2), Table 2). It is also dependent on the free energy of reaction (ΔG_r) and decreases to zero as equilibrium is approached (Fig. 2). For devolatilization

Table 2. Rate equations.

	Theoretical aspects/source	Applications (see Table 1)
Nucleation		
1. $\dot{N}_v = n_v \dfrac{kT}{h} \exp\left(\dfrac{-\Delta G^*}{kT}\right) \exp\left(\dfrac{-Q}{RT}\right)$	Turnbull and Fisher (1949), Cahn (1957), Christian (1975).	no application in geological reactions; precipitation in Cu–Co (Servi and Turnbull, 1966)
where $\Delta G^* = \dfrac{16\pi}{3} \dfrac{\sigma^3}{(\Delta G_v + \varepsilon)^2}$		
for a spherical nucleus		
Growth		
interface-controlled growth		
2. $\dot{x} = \dfrac{\delta kT}{h} \exp\left(\dfrac{-Q}{RT}\right)\left[1 - \exp\left(\dfrac{-\Delta G_r}{RT}\right)\right]$	Turnbull (1956), see also Fisher (1978, Eq. (11))	aragonite $\rightarrow$ calcite (A.1.9)
diffusion-controlled growth		
3. $x = (2K_3 t)^{1/2}$ $\dot{x} = K_3/x$	growth of a planar layer, Fisher (1978)	quartz + periclase $\rightarrow$ forsterite (Brady, 1983)
4. $r = (2K_4 t)^{1/2}$ $\dot{r} = K_4/r$	growth of a spherical particle, Fisher (1978)	growth of biotite–andalusite segregations (Fisher, 1978)
heat flow-controlled growth		
5. $x = K_5 t$ $\dot{x} = K_5$	growth of a planar layer, Fisher (1978)	theoretical only (Fisher, 1978)
6. $r = (K_6 t)^{1/3}$ $\dot{r} = \frac{1}{3} K_6^{1/3} t^{-2/3}$	growth of a spherical particle, Fisher (1978)	growth of biotite–andalusite segregations (Fisher, 1978)

Overall transformation kinetics
nucleation and interface-controlled
growth

7.1. $-\ln(1-\xi) = \frac{4}{3}\pi\int_0^t \dot N_v \dot x(t-\tau)\,d\tau$

7.2. $\xi = 1 - \exp(-K_7 t^n)$

or $\dot\xi = nK_7 t^{n-1}(1-\xi)$

Eqs. (7.1–7.6) allow for grain impingement

n = 4 nucleation and 3-d growth
n = 3 3-d growth (spheres)
n = 2 2-d growth (plates)
n = 1 1-d growth (needles)

Johnson and Mehl (1939), Avrami (1939, 1940, 1941)

7.3. $\xi = 1 - \exp(-\frac{1}{3}\pi \dot N_v \dot x^3 t^4)$

Eqs. (7.3–7.6) from Cahn (1956a):
n = 4 during nucleation stage

7.4. $\xi = 1 - \exp(-\frac{4}{3}\pi C_g \dot x^3 t^3)$

n = 3 nucleation on grain corners

7.5. $\xi = 1 - \exp(-\pi L_g \dot x^2 t^2)$

n = 2 nucleation on grain edges

7.6. $\xi = 1 - \exp(-2S_g \dot x t)$

n = 1 nucleation on grain surfaces

after site saturation

$(C_g$ = corners/unit vol., L_g = length of grain edge/unit vol., S_g = area of grain boundary/unit vol.)

applications of Eq. (7.2):

aragonite → calcite
A.1.6: n = 0.6–4.0
A.1.7: n = 1.3
A.1.8: n = 1
calcite → aragonite
A.2.2: n = 0.57–1.03
quartz → cristobalite
A.3.1: n = 1 assumed
Olivine → spinel
A.6.1: n = 1 assumed
A.7.2: n = 1.8
quartz → rutile (GeO$_2$)
A.8.1: n = 0.2–0.5
dehydration reactions
C.2.1: n = 2
hydration reactions
D.1.2: n = 1

Table 2. (Continued)

	Theoretical aspects/source	Applications (see Table 1)
diffusion-controlled growth		
8. $\xi = 1 - \exp(-K_8 t^n)$ $\dot{\xi} = nK_8 t^{n-1}(1 - \xi)$	$n = 0.5 \rightarrow 2.5$. No theoretical basis for diffusion-controlled reactions (see Ham, 1958, 1959)	exsolution (Yund and Hall, 1970; Yund *et al.*, 1972)
9. $\dfrac{2K_9 t}{r_0^2} = [1 - (1 - \xi)^{1/3}]^2$	$A + B \rightarrow C$; growth of C around spherical grains of either A or B when volume of products = volume of reactants (Jander, 1927)	decarbonation reactions C.3.1 hydration reactions D.1.1
10. $[1 + (z - 1)\xi]^{2/3} + (z - 1)(1 - \xi)^{2/3}$ $= z + 2(1 - z)K_{10} t/r_0^2$	as for Eq. (9) except $vol_{products}/vol_{reactants} = z$ (Carter, 1961), $(z \neq 1)$; see Hulbert *et al.* (1969) for similar eqs.	kinetics of oxidation of uniformly sized nickel spheres (Carter, 1961); no applications to geological reactions
martensitic transformations		
11. $\xi = 1 - \exp(-K_{11} t)$ $\dot{\xi} = K_{11}(1 - \xi)$	Poirier (1981b, 1982a) for olivine $\rightarrow$ spinel; equivalent to Eq. (7.2) with $n = 1$	no experimental applications
power law		
12. $\xi = (K_{12} t)^n$ $\dot{\xi} = nK_{12}^n t^{n-1}$	empirical equation; general form of Eqs. (3), (4), (5), (6); approximation of Eqs. (7.2), (8), and (11) for small ξ	aragonite $\rightarrow$ calcite, A.1.4, A.1.5 $n = 2$ assumed jadeite + quartz $\rightarrow$ albite B.1.1: $n = 0.51$–0.61 analcime + quartz $\rightarrow$ albite + H_2O C.8.2: $n = 1.4$–1.7 calcite + quartz $\rightarrow$ wollastonite + CO_2 C.3.3: $n = 0.17$–0.29

order of reaction equation

13. $\dot{\xi} = K_{13}(1 - \xi)^n$

based on theory of simple homogeneous reactions; no theoretical basis for solid-state reactions (Kingery et al., 1976; Gomes, 1961); equivalent to Eqs. (7.2), (8), (11) when $n = 1$

aragonite → calcite
A.1.3: $n = 1.2$–10.0
calcite → aragonite
A.2.1: $n = 3$
quartz → cristobalite
A.3.2: $n = 1$ assumed
olivine → spinel
A.7.1: $n = 1$ assumed.
dehydration reactions–C.1.1,
   C.4.1, C.5.1, C.6.1, C.6.2:
   $n = 1$ assumed.

constant rate equation

14. $\xi = K_{14}t \quad \dot{\xi} = K_{14}$

growth of planar layers with no impingement (Carlson, 1983); heat-flow or interface-controlled growth; equivalent to Eq. (5).

aragonite → calcite A.1.2
brucite → periclase + H_2O
C.3.1

Grain growth

15. $\bar{d}^2 - \bar{d}_0^2 = K_{15}t$

Martin and Doherty (1976)

grain growth of quartz and calcite aggregates (Tullis and Yund, 1982; Joesten, 1983)

16. $\bar{d}_e^{0.5} = K_{16}(T - T_0)$

Andrade and Aboav (1966)

no geological applications

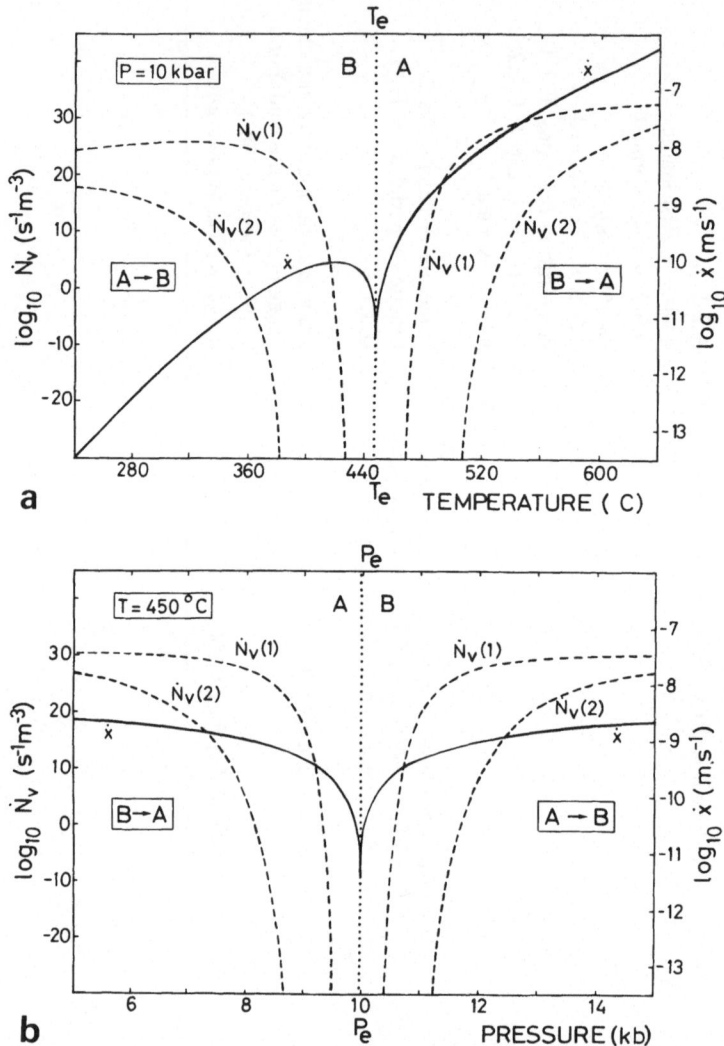

Fig. 2. Variation of nucleation rate $\dot{N}_v$ (broken lines) and growth rate $\dot{x}$ (solid lines) for hypothetical reactions $A \rightarrow B$ and $B \rightarrow A$, with temperature (a), and pressure (b), using Eqs. (1) and (2) (Table 2). The form of the curves is supported qualitatively by experimental data (e.g., Yund *et al.*, 1972). The model reactions are based, as far as possible, on the aragonite–calcite transformations, and free energy of reaction (ΔG_r) is calculated from data in Helgeson *et al.* (1978). Nucleation rates have been calculated ignoring strain energy (ε) and by assuming the preexponential factor $n_v k/h = 10^{40}$ (Christian, 1975, p. 441) and $Q = 163$ kJ/mol (Carlson and Rosenfeld, 1981). Two values of interfacial free energy have been used: $\sigma = 0.01$ J/m^2 (curves 1) and $\sigma = 0.02$ J/m^2 (curves 2). Parameters in the growth rate equation (Eq. 2) are those estimated by Carlson and Rosenfeld (1981) for the aragonite $\rightarrow$ calcite reaction and it is assumed that the same parameters approximately describe the reverse reaction ($Q = 163$ kJ/mol, $\delta = 5$ Å). Activation volume is assumed to be negligible.

reactions (larger ΔG_r than for solid–solid reactions) the growth rate may also be *heat flow-controlled* (Eqs. (5) and (6) in Table 2; Fisher, 1978; Ridley, this volume).

Growth rate has been measured experimentally at 1 bar for the reaction aragonite to calcite using a microscope heating stage (Carlson and Rosenfeld, 1981). Their results appear to validate Eq. (2) (Table 2). Measurements of growth rate can also be made at high pressure from a series of experiments of variable duration, assuming that nuclei form very early (see Yund and Hall, 1970; Yund *et al.*, 1972; Sung, 1979).

Overall Transformation Kinetics and Rate Equations

There are two basic approaches to the formulation of rate equations to describe the kinetics of solid-state reactions. *Empirical* rate equations relate experimental observation of amount of transformed material to temperature and duration of the experiments. Because any systematics among the variables are not considered in terms of specific reaction mechanisms, the empirical rate constants may permit interpolation among the data set but are not necessarily capable of being extrapolated (see below). Likewise, such empirical rate equations may require substantial modification if the experimental configuration is changed and therefore cannot necessarily be applied with confidence to geological processes. *Mechanism-based* or *model* rate equations are derived for specific reaction mechanisms and may be extrapolated beyond the range of the data set provided the specific mechanism still applies. We have attempted to review some of the available data on solid-state mineral reaction kinetics in terms of empirical or mechanism-based rate equations.

Current theories of overall transformation kinetics are generally based on the assumption of either early site saturation (Eqs. (7.4–7.6) and (9–11), Table 2) or of constant N_v (Eq. (7.3), Table 2).

Nucleation with Interface-Controlled Growth

Equations (7) (Table 2) were derived to describe the kinetics of polymorphic, discontinuous precipitation, and eutectoid reactions (Christian, 1975, p. 542). The original Johnson–Mehl equation (Eq. (7.2), Table 2) assumed constant N_v and $\dot{x}$ throughout the transformation, random nucleation, and impingement of growing spherical grains. Because constant N_v is rarely the case, Avrami (1939, 1940, 1941) modified the theory to incorporate the possibility of very early site saturation or decreasing N_v during the transformation, without in fact changing the form of the equation. The theory was further developed by Cahn (1956a) to describe transformations involving nucleation on different types of site (grain boundaries, edges, and corners). In an application to the aragonite–calcite transformation, Carlson (1983) has discussed how the theory can be modified to describe kinetics of powder reactions.

The derivation of Eqs. (7.2–7.6) (Table 2) (Christian 1975, p. 16 and Ch. 12) involves the integration of Eq. (7.1). To simplify this step, it is generally assumed that $\dot{N}_v$ is either constant or that $\dot{N}_v = 0$ after early site saturation and that $\dot{x}$ is constant. As discussed elsewhere, neither of these assumptions is necessarily valid, in which case the application of the Johnson–Mehl/Avrami theory can only be made by estimating $\dot{N}_v(t)$ and $\dot{x}(t)$ and then integrating Eq. (7.1) (Table 2) by numerical methods (see Carlson, 1983).

From Eq. (7) (Table 2), a plot of volume fraction transformed (ξ) against t gives a *sigmoidal* curve (see Christian, 1975; Kunzler and Goodall, 1970; Carlson, 1983). The constants in Eq. (7.2) (Table 2) can be found by plotting lnln $[1/(1 - \xi)]$ against ln t (Carlson, 1983, Figs. 1 and 5); the slope gives n and the intercept on the ordinate axis gives ln K_7. Although, in principle, the determination of "n" gives an indication of the dominant nucleation sites (following Cahn, 1956a; Eqs. (7.4–7.6), Table 2) or the morphology of growing nuclei (Avrami, 1939, 1940, 1941; see Table 2), in practice (Kunzler and Goodall, 1970) such an interpretation is highly unreliable. Frequently, plots of lnln $[1/(1 - \xi)]$ against ln t for experimental data give curved lines instead of the ideally straight line (Carlson, 1983, Figs. 1 and 5). One interpretation of the curvature is that the value of n is changing (Carlson, 1983, p. 64). There are at least two other interpretations. (1) The grain size is nonuniform (see below); in this case the overall-transformation kinetics will be dominated initially by fine-grained reactants and later by remaining unreacted coarse-grained particles. Thus, the *effective* grain size of untransformed material increases, and the value of K_7 decreases with time. (2) Transformation stress develops as a result of volume change and causes a decrease in growth rate ($\dot{x}$) and therefore a decrease in K_7 with time (see below; Kunzler and Goodall, 1970). If $K_7(t)$ does vary, a unique value of n cannot be obtained from a lnln $[1/(1 - \xi)]$ against ln t plot (Fig. 3). Thus, little faith can be placed on the values of n so obtained to indicate reaction mechanisms, nucleation sites, or growth morphologies, etc. (see Christian, 1975, p. 542). Because there is no way to differentiate between changing n and changing K_7 (see Fig. 3), $\dot{N}_v(t)$, $\dot{x}(t)$, and grain size distribution of unreacted materials must be assessed before Eqs. (7) can be applied.

Nucleation with Diffusion-Controlled Growth

Provided all nucleation occurs at a very early stage of reaction and that growing grains or layers are widely separated, then overall transformation kinetics can be derived from the parabolic law (Eqs. (3) and (4), Table 2). These equations were originally formulated to describe the growth of single

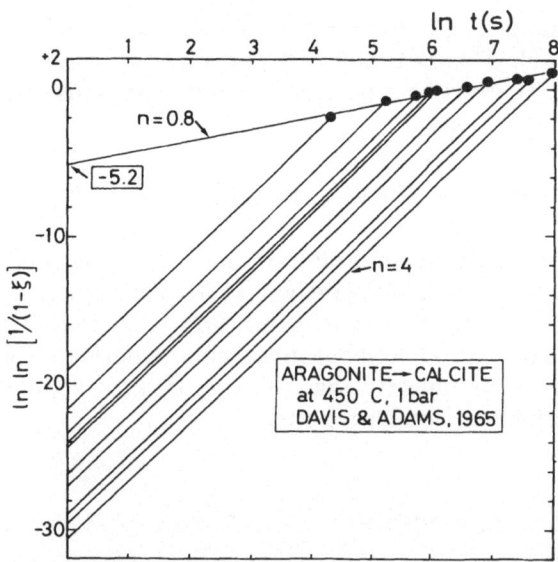

Fig. 3. Two possible interpretations of kinetic data for the aragonite to calcite transformation at 450°C and 1 bar (data from Davis and Adams, 1965, Fig. 4c) using Eq. (7.2) (Table 2). The usual interpretation is a straight line fit, which in this case gives $n = 0.8$ and $\ln K_7 = -5.2$. An alternative possibility is that K_7 decreases as extent of reaction (ξ) increases (see text) such that the data points lie on a continuous series of lines perhaps, but not necessarily, of constant slope (arbitrarily $n = 4$ in this diagram).

grains or layers. However, as noted by Christian (1975, p. 20 and Ch. 12), if there is impingement of growing grains or of diffusion domains it becomes increasingly difficult to predict overall transformation kinetics. The Avrami equation (Eq. (8), Table 2) has been used to describe rates of obviously diffusion-controlled reactions, with n values ranging from 0.5 to >2.5 for various growth morphologies and nucleation rate models (Christian 1975, p. 542). The lack of a sound theoretical basis limits its application to the early stages of reaction (e.g., $\xi < 0.5$) (Ham, 1958, 1959; Christian, 1975, p. 20; see also Yund and Hall, 1970, p. 396; Yund *et al.,* 1972, p. 262).

Equations (9) and (10) in Table 2 and other similar equations (see Hulbert *et al.,* 1969) have been derived to model the kinetics of powder reactions of the type $A + B \rightarrow C$, where C forms a growing layer around uniformly sized spherical grains of either A or B. Following initial nucleation, the reaction is subsequently controlled by diffusion of species through the growing layer of C. Equation (10) (Table 2) has been successfully fitted to experimental data where particles were spherical and initially of uniform size (Carter, 1961). However, for mineral reactions of this type (e.g., calcite + quartz to wollastonite + CO_2, Kridelbaugh, 1973, and albite to jadeite + quartz, Rubie, unpublished data) attempts to fit Eqs. (9) and (10) to the experimental data were unsuccessful. This is because if particles are not spherical and of a uniform size, $\xi(t)$ cannot be predicted by such equations (see Carter, 1961).

Nucleation with Heat Flow-Controlled Growth

For high enthalpy reactions in metamorphic rocks, the rate of heat supply may be the rate-controlling step (Fisher, 1978; Yardley, 1977, 1983; Ridley, this volume). Assuming early site saturation, the overall transformation rates for particular textural configurations can be derived from Eqs. (5) and (6) (Table 2). It seems unlikely that the rate of heat supply could control kinetics in experimental configurations.

Martensitic Transformations

Martensitic transformations change crystals into a crystallographically different structure by a shear process similar to mechanical twinning. An increase in temperature usually causes a martensitic transformation to occur instantaneously (Haasen, 1978, Fig. 9.23), but it can also proceed isothermally if nucleation or interface migration are assisted by a shear stress. The occurrence of such "military" transformations (Christian, 1975, p. 6) in minerals is poorly known. The mechanism has been suggested for the Mg_2SiO_4 olivine to spinel transformation (Poirier, 1981a), and for transformations between calcite and aragonite (Gillet and Madon, 1982) and between Al_2SiO_5 polymorphs (Doukhan and Christie, 1982). A rate equation derived theoretically by Poirier (1981b) (Eq. (11), Table 2) is identical in form to Eq. (7.6) for nucleation and growth. This again emphasizes that a rate equation, fitted to experimental data, may provide *no* information on reaction mechanisms.

Other Forms of Rate Equations

Additional rate equations in common use include power law, order of reaction, and constant-rate equations.

The power law equation (Eq. (12), Table 2) is a generalized form of the equation for diffusion and heat flow-controlled growth (Eqs. (3) to (6), Table 2). It is also an *approximation* of the Avrami equation (Eqs. (7), Table 2) for small values of ξ and t, because it is the first term of a power series expansion of this equation. For values of n other than 1 (spherical grains, diffusion-controlled), $\frac{2}{3}$ (spherical grains, heat flow-controlled), and 0.5 (planar layers), Eq. (12) is purely empirical. It has been used by Kridelbaugh (1973) for calcite + quartz to wollastonite + CO_2 ($n = 0.17-0.29$) and by Matthews (1980) for analcite + quartz to albite + H_2O ($n = 1.4-1.7$) reactions.

The order of reaction equation (Eq. (13), Table 2) was initially derived for very simple *homogeneous* reactions involving one elementary step. Under these circumstances the rate of a reaction is given by

$$-dC/dt = KC_1^\alpha C_2^\beta C_3^\gamma\ldots\ldots\ldots$$

where order of reaction is determined from the sum of the stoichiometric reaction exponents ($\alpha + \beta + \gamma + \cdot \cdot \cdot$), C is the concentration of reactants, and order and molecularity (numbers of molecules or atoms participating in a reaction) are the same (Kingery *et al.*, 1976, p. 381). For complex *heterogeneous* reactions, order and molecularity are quite different and "in solid-state reactions the concepts of concentration and order of reaction generally have no significance" (Gomes, 1961). For solid-state reactions, Eq. (13) can be regarded as empirical (see Kingery *et al.*, 1976, p. 382). This has been the most widely used rate equation in experimental studies of a whole variety of both polymorphic and dehydration reactions (Table 2).

Overall transformation rates may be constant (Eq. (14), Table 2) for the case of growing planar layers with no growth impingement when the reaction is controlled by heat flow (Eq. (5), Table 2) or by interface processes (Eq. (7) with $n = 1$). The constant reaction rate for aragonite to calcite obtained by Brown *et al.* (1962) has been explained by Carlson (1983, p. 68) as reflecting nucleation of calcite on aragonite cleavage surfaces and rapid coalescence of grains to form planar layers. Subsequent advance of the pseudoplanar fronts of calcite into the aragonite leads theoretically to a first-order time dependence. Why constant reaction rates have not been found in any other studies of the aragonite to calcite transformation is unclear (see Table 1).

Application and Extrapolation of Rate Equations

We distinguished above between *empirical* rate equations and *model-based* rate equations. However, even if a particular mechanism is specified in the choice of a rate equation, the fitted equation may still be inadequate to describe the transformation, unless all the assumptions used to derive the equation are valid. Furthermore, it is possible that *any* of the listed rate equations (Table 2) can be fitted to a given set of kinetic data provided that they contain at least two coefficients (see Rao, 1973; and discussion in Hulbert *et al.*, 1969, p. 583). The difficulties in fitting Eqs. (10) and (11) to some data sets may be because these equations contain only a single coefficient. This point is emphasized in Fig. 4, which shows data of Davis and Adams (1965, Fig. 3b) for the aragonite to calcite transformation at $T = 450°C$ and $P = 8$ kbar. Davis and Adams (1965, Table 1) originally fitted their data to Eq. (13) (Table 2). Using this equation to estimate the time for $\xi =$

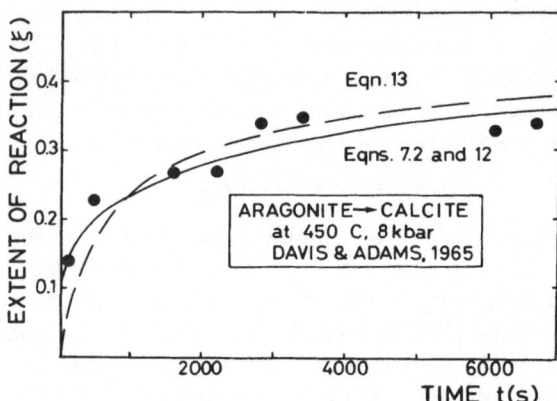

Fig. 4. Experimentally determined data points ($\xi(t)$) for the aragonite $\rightarrow$ calcite reaction at 450°C and 8 kbar (from Davis and Adams, 1965, Fig. 3b). Three rate equations have been fitted to these points. The broken line shows the fit of Eq. (13) (Table 2) where $n = 10.0$ and $K_{13} = 8.1 \times 10^{-20}$ (%)$^{-9}$/min (Davis and Adams, 1965). The solid line shows the fit of Eq. (7.2) (Table 2) with $n = 0.25$ and $K_7 = 5.06 \times 10^{-2}$ s$^{-0.25}$, and also the fit of Eq. (12) (Table 2) with $n = 0.22$ and $K_{12} = 1.39 \times 10^{-6}$/s. Over the range of ξ shown, Eqs. (7.2) and (12) give almost identical curves.

0.99, a duration of 8×10^{19} *s* is obtained. In Fig. 4 the same data have been fitted to Eqs. (7.2) and (12) (Table 2), and both equations fit equally well with similar correlation coefficients of 0.96 on $\ln\ln(1/(1 - \xi))$ against $\ln t$ and $\ln \xi$ against $\ln t$ plots, respectively. Equations (7.2) and (12) give estimated times to reach $\xi = 0.99$ of 8.5×10^7 *s* and 6.9×10^5 *s,* respectively. These fits demonstrate the flexibility of two-coefficient empirical rate equations and the large uncertainties encountered in data extrapolation. Furthermore, this example emphasizes the problem that even if a particular rate equation fits a given kinetic data set, it cannot be used to identify a reaction mechanism on the basis of rate equation coefficients. In all cases the reaction mechanism must be deduced if possible from microstructural observations using microscopic (optical, SEM, TEM) techniques.

A principle aim of collecting experimental data on reaction kinetics is to be able to extrapolate, via a rate equation, to a geological situation (e.g., Matthews, 1980; Carlson and Rosenfeld, 1981). As discussed in the previous paragraph, unless a model-based equation is used, extrapolation to geological conditions can be extremely uncertain. If a model-based equation is used, a very careful assessment of the relevance of experimental parameters to the geological situation is first required. Even then there is a further problem that requires careful consideration. Fisher (1978, p. 1043) has suggested that the rate law for a metamorphic process can change with reaction progress (ξ). For example, a reaction can initially be interface-controlled, then change to being diffusion-controlled, and then to being heat flow-controlled (Fisher, 1978, Fig. 7; see also Martin and Fyfe, 1970; Schmalzried, 1978, p. 286). Thus the applicability of rate-controlling steps in the experimental configuration to a geological process requires assessment before extrapolations are attempted.

Temperature Dependence of Reaction Rates

It is normally found that rates of solid-state reactions obey an Arrhenius relationship with temperature, although the physical basis of this relationship is less sound here than it is for reactions in liquids and gases (Gomes, 1961). There are a number of difficulties in measuring the activation energy (Q) for an overall transformation rate. (1) Measurements of $\xi(t)$, amount transformed at successive times, are difficult to make for many reactions and may be subject to large uncertainties. (2) Reaction mechanisms and therefore activation energies may change over the temperature range of the experimental study (e.g., the change from intrinsic to extrinsic controlled diffusion), but the data resolution may be inadequate to recognize this fact. (3) The relative importance of nucleation and growth (with respective activation energies) may change over the T range of study and so change the activation energy for the overall transformation rate (see also Wegner and Ernst, 1983, p. 172). (4) There are a number of factors, reviewed below, that may have a significant effect on Q. These problems are apparent for the aragonite to calcite transformation, where determined activation energies range from 160 to 410 kJ mol^{-1} (Kunzler and Goodall, 1970). Consequently, there is a huge uncertainty in reaction rates estimated by extrapolation. For example, extrapolating from 400 to 200°C using the lower value of Q reduces the rate by a factor of 10^5, whereas using the upper value the rate is reduced by a factor of 10^{13}! In spite of these problems, accurate determination of activation energy and an understanding of factors that control this parameter are necessary if reliable extrapolations are to be made outside the temperature range of an experimental study.

Determination of Activation Energy from Experimental Data

Common experimental methods employed to evaluate activation energy include (1) direct determination of growth rate, (2) change of rate method, (3) extent of reaction method, and (4) determining the temperature dependence of a rate constant.

1. Carlson and Rosenfeld (1981) measured the rate of growth of calcite from aragonite at $P = 1$ bar, at a series of fixed temperatures with a microscope heating stage, by observing the displacement of an interface. Activation energy for growth was obtained from the slope of a ln $\dot{x}$ against $1/T$ plot.

2. A variation of the growth rate method, described by Kittl and Cabo (1969), ascertains whether the activation energy for growth is dependent upon the amount of transformation (ξ). After measuring the velocity of interface movement ($\dot{x}_1$) at temperature T_1 until a certain amount of transformation is achieved, the temperature is changed to T_2 and the new interface velocity $\dot{x}_2$ is measured. The activation energy is given by

$$Q = \frac{R \ln(\dot{x}_1/\dot{x}_2)}{(1/T_1 - 1/T_2)}$$

assuming that the pre-exponential factor in the Arrhenius equation is constant over the temperature range of interest.

3. The time (t_{ξ_1}) for a given extent of reaction (ξ_1) at constant T is measured at various temperatures and the activation energy is measured from the slope of a $\ln(t_{\xi_1})$ against $1/T$ plot as shown in Fig. 6 (e.g., Davis and Adams, 1965; Kunzler and Goodall, 1970; Brar and Schloessin, 1979).

4. For homogeneous reactions in gases and liquids, the *rate constant* (K) is the proportionality factor between reaction rate and the concentration of the reactants. For such a reaction, the activation energy is derived from the temperature dependence of the rate constant through the Arrhenius relation (Kingery *et al.*, 1976, p. 382). In solid-state reactions, because concentration and order generally have no significance, the rate constant must be differently defined. According to Gomes (1961), K must be independent of time and have a factor s^{-1} in its dimensional formula. Many of the constants (K) in the rate equations of Table 2 do not conform to this rule, and consequently the temperature dependence of K may lead to a false or *apparent* activation energy (see Matthews, 1980, p. 395).

A problem with all of the methods used to evaluate activation energy is that a change of reaction rate with temperature depends not only on the activation energy Q, but also on the change in driving force (ΔG_r) of the reaction with T. As the amount of temperature overstepping of an equilibrium boundary is reduced, the rate decreases towards zero, whatever the reaction mechanism, because ΔG_r decreases to zero. From Eq. (2) (Table 2), to obtain a value of Q, $\ln[\dot{x}/T[1 - \exp(\Delta G_r/RT)]]$ plotted against $1/T$, will compensate for the variations of ΔG_r with T. For reactions with a moderate or small ΔG_r, plots of $\ln \dot{x}$ against $1/T$, $\ln(t_{\xi_1})$ against $1/T$, and $\ln K$ against $1/T$ will all be nonlinear, especially within 100–200°C of equilibrium (Figs. 5 and 6).

Factors Affecting Activation Energies

Crystallographic orientation of an interface, chemistry of the reacting system, the reaction mechanism, and the presence and chemistry of a fluid may all affect the values of activation energies.

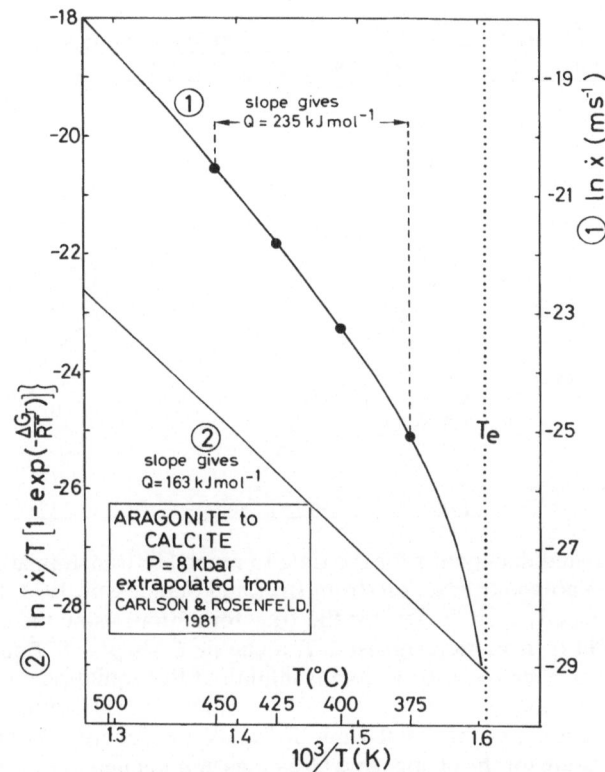

Fig. 5. Arrhenius plots for growth rate of calcite from aragonite using Eq. (2) (Table 2) and parameters of Carlson and Rosenfeld (1981). Their value of activation energy for growth of 163 kJ/mol was obtained from a ln $\dot{x}$ against $1/T$ plot at conditions distant from equilibrium ($P = 1$ bar, $T = 375$–$455°C$). Here $\dot{x}$ has been extrapolated to $P = 8$ kbar using free energy data of Helgeson *et al.* (1978). Curve 1 (right-hand scale) is a ln $\dot{x}$ against $1/T$ plot and is strongly nonlinear owing to the proximity of equilibrium conditions. Measuring the average slope of this curve, from the four indicated construction points, over the above T range gives a false value for Q of 235 kJ/mol. Curve 2 (left-hand scale) is a plot of ln $\{\dot{x}/T[1 - \exp(\Delta G_r/RT)]\}$ against $1/T$, the slope of which gives the correct value for Q. T_e is the equilibrium temperature.

It has been shown by several studies (e.g., Carlson and Rosenfeld, 1981) that growth rates of a product phase are dependent on the crystallographic direction of growth. Although not well documented for mineralogical reactions, variation of activation energy for growth with crystallographic direction has been established in some metallic systems. For example, for transformation in AgZn, Kittl and Cabo (1969) deduced that the activation energy for growth changed by up to 42 kJ mol^{-1} from an average of 147 kJ mol^{-1} depending upon the orientation of the product grain within the matrix. Consequently, the observed grain shape depended upon temperature.

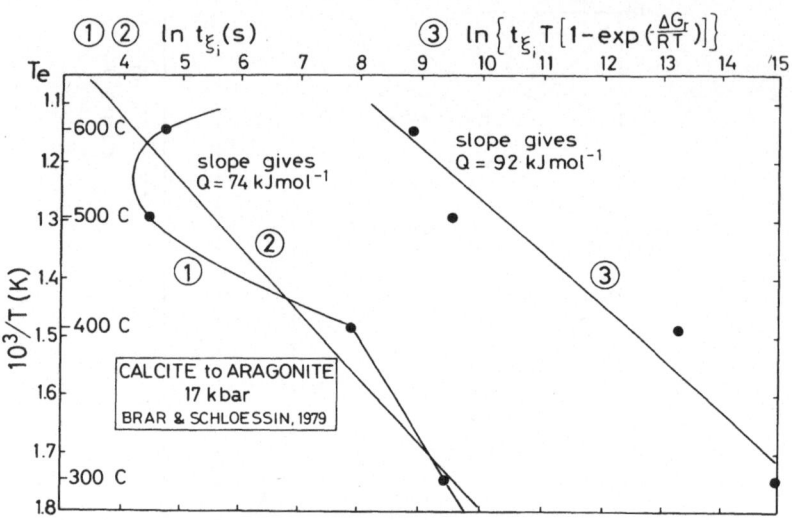

Fig. 6. Extrapolated data points for the time to reach 5% transformation ($\xi_1 = 0.05$) for the calcite $\rightarrow$ aragonite reaction (from Brar and Schloessin, 1979, Fig. 5) plotted to determine the activation energy for the transformation, using the extent of reaction method. If $\ln t_{\xi_1}$ is plotted against $1/T$, a classic C-shaped TTT curve should be obtained (curve 1) because $\ln t_{\xi_1}$ becomes infinite at the equilibrium temperature T_e (see Putnis and McConnell, 1980, p. 144). If a straight line (2) is fitted to data points defining such a curve, an incorrect value for activation energy is obtained (74 kJ/mol). To compensate for the change in ΔG_r as T_e is approached, it is necessary to plot $\ln \{t_{\xi_1} T[1 - \exp(-\Delta G_r/RT)]\}$ against $1/T$ (line 3) which in this case gives $Q = 92$ kJ/mol. The exact form of curve 1 is poorly defined owing to the small number of data points. The apparent change in slope below 400°C could be the result of poor data, and this is assumed to the case when constructing curve 3. (ΔG_r calculated from $\Delta P \Delta V$ where $\Delta V = 0.2784$ J/bar/mol using the equilibrium curves of Irving and Wyllie (1975) and Johannes and Puhan (1971).)

Activation energies for nucleation and for growth may be strongly dependent on variations in chemical composition, at least for some metallic systems. Bell and Farnell (1969) have shown that the activation energy for growth during the decomposition of iron–nitrogen austenite to bainite rises uniformly from 83 kJ mol^{-1} at 1.1 wt% N_2 to 125 kJ mol^{-1} at 2.6 wt% N_2. The different impurity content of materials used by Kunzler and Goodall (1970) in their study of the aragonite to calcite reaction may in part explain the large range of activation energies obtained.

The addition of H_2O to a reacting system can affect reaction mechanisms and activation energies, as is well documented in some diffusion studies (Yund, 1983). As shown by Yund and Tullis (1980), Q for Si–Al disordering in albite varies from 364 kJ mol^{-1} for samples disordered in air to 280 kJ mol^{-1} for samples disordered with $P_{H_2O} = 10$ kbar. The hydrolytic weakening that affects quartz (Blacic and Christie, 1984) is recognized in other silicate systems (e.g., Tullis *et al.*, 1979), but the roles of H^+, OH^-, H_3O^+, or H_2 are not well understood in their interaction with defects (see also Hobbs, 1981). The addition of H_2O may completely change the reaction mechanism and therefore the activation energy. In the absence of H_2O, the aragonite–calcite transformation apparently proceeds by a nucleation and growth process (Carlson and Rosenfeld, 1981), but Brown *et al.* (1962) found that H_2O had a pronounced catalytic effect, presumably by permitting a faster dissolution/precipitation mechanism to operate with a *reduced* activation energy. However, Yund *et al.* (1972, p. 266) found that although the rate of nepheline–kalsilite exsolution is enhanced by the presence of H_2O, the activation energy is unaffected. The fluid composition can also change the activation energy. Matthews (1980), in his study of the kinetics of the analcite + quartz to albite + H_2O reaction, found $Q = 186$ kJ mol^{-1} with 0.1 M NaCl and 255 kJ mol^{-1} with 0.05 M Na_2SiO_5 solutions measured from values of a rate constant defined by a power law equation ($\xi = Kt^n$).

Pressure Dependence of Reaction Rates

The pressure dependence of, for example, the growth rate equation (Eq. (2), Table 2) is contained in the two distinct quantities ΔG_r (free energy of the reaction) and Q (the activation energy) (see Hanneman, 1969, p. 792). The magnitude of the pressure dependence of ΔG_r is determined by the molar volume change of reaction (ΔV). At constant temperature, ΔG_r in Eq. (5) (Table 2) can be replaced by $\Delta V\Delta P$ where ΔP is the amount of pressure overstepping relative to the equilibrium boundary. The activation volume (ΔV^*) can be considered theoretically in terms of the volume of formation of a hypothetical activated state. Defining $\Delta V^* = (\partial Q/\partial P)_T$, then we may redefine Q as

$$Q = H^* + P\Delta V^*$$

where H^* is the activation enthalpy that is independent of pressure. These definitions permit a rewriting of a rate equation into activation and equilibrium terms (cf. Eq. (2), Table 2; Hanneman, 1969). ΔV^* is often assumed to be small (see Yund *et al.*, 1972; Carlson and Rosenfeld, 1981) and its measurement is extremely difficult. Ideally, measurements are made as far from the equilibrium boundary as possible, so that the effect of changing ΔP in the equilibrium term is small; then to determine ΔV^* it must be assumed that the reaction mechanism is constant over a large pressure range. An example of the possible difficulties is demonstrated by the results of Zeto and Roy (1969) who calculated apparent activation volumes (ΔV^*) for the GeO_2 quartz to rutile transformation in the range -15 to -25 cm^3 mol^{-1} (the molar volumes are about 25 and 16 cm^3 mol^{-1} respectively). Zeto and Roy (1969) therefore deduced an extremely dense activated state. However, they also discussed experimental factors, such as the effects of pressure on grain size and intra-crystalline strain energy, that were probably really responsible for these *deduced* values of ΔV^*. Other methods for estimating ΔV^* have included determining the effect of pressure on the rate of dislocation climb in olivine (Kohlstedt *et al.*, 1980). Empirical equations that relate diffusion coeffi-cients, activation energy, and activation volume to melting point diffusivity, melting temperature (T_m), and the variation of T_m with pressure have been discussed by Brown and Ashby (1980) and Sammis *et al.* (1981).

Grain-Size Dependence of Reaction Rates

The extent to which overall transformation kinetics will be a function of grain size depends on whether nucleation occurs on grain surfaces and edges as opposed to intracrystalline defects. For a nucleation and interface-con-trolled growth reaction (described by Eqs. (7), Table 2), the time for com-plete reaction is predicted to be proportional to grain diameter (d) (see Cahn, 1956a). For diffusion-controlled reactions, this time is proportional to d^2. Thus, in experiments where surface or grain boundary nucleation domi-nates, the grain size must be carefully characterized. In some experimental studies the reaction rate has been observed to increase with decreasing grain size (Kunzler and Goodall, 1970, p. 324; Wegner and Ernst, 1983). However, because of a large *range* of grain sizes in each experiment, the grain-size *distribution* must have been an important factor and the variation of rate with grain size cannot be quantified from their published data.

For the transformation calcite to aragonite at 17 kbar and 450°C, Brar and Schloessin (1979, Fig. 9) determined that the volume fraction transformed after 60 minutes for grain size fractions between 2 μm and 36 μm was a function of $d^{1/2}$, contrary to theoretical predictions. However, in experi-ments at high pressure, the actual grain-size distribution may bear no simple relationship to the prepared grain size of a powdered sample. In their investi-gation of the transformation kinetics of GeO_2 (quartz to rutile) at high pres-sure, Zeto and Roy (1969) found that transformation rates were *independent* of the *prepared* grain size. They concluded that *application of pressure* had reduced all grains to a size less than 44 μm diameter. A similar result was obtained by Rubie (unpublished data) in kinetic experiments on the jadeite + quartz to albite reaction. Such considerations make it difficult to interpret results such as those of Brar and Schloessin (1979) mentioned above.

The effect of grain-size distribution on overall transformation kinetics has been discussed by Gallagher (1965) who calculated $\xi(t)$ for known spherical particle-size distributions during diffusion-controlled powder reactions (Fig. 7; see also Kapur, 1973). For large standard deviations the initial reaction rate is extremely rapid as very fine-grained material reacts. In later stages, the rate decreases because the unreacted material is mostly coarse grained.

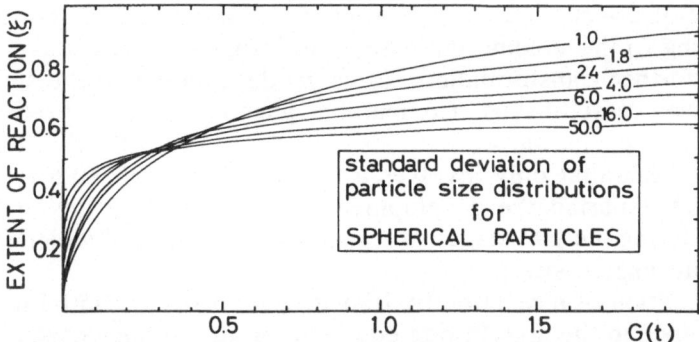

Fig. 7. Theoretically derived curves showing extent of reaction ξ against G, where G is a linear function of time, calculated for particle-size distributions varying from a standard deviation of 1.0 to 50.0. Diffusion-controlled growth and spherical particles are assumed. (From Gallagher, 1965.)

Experimentally determined rate curves for powders at high pressure (e.g., Kridelbaugh, 1973, Figs. 2 and 3; Zeto and Roy, 1969, Figs. 1 and 2) often have forms qualitatively similar to the high standard deviation curves of Gallagher (1965), and thus if such kinetic data are extrapolated, the results must be extremely uncertain if grain size and grain-size distribution have not been carefully characterized.

Transformation Stress

For solid-state reactions of geological interest the volume change is often large, e.g., −17% for albite to jadeite + quartz and +8.15% for aragonite to calcite. There is evidence that high transformation stresses develop in reacting materials as a consequence of such volume changes. Carlson and Rosenfeld (1981) observed the development of transformation *strain* during the growth of calcite grains in aragonite at $P = 1$ bar. The formation of strain haloes and lamellar twins that propagated out from calcite grains into the aragonite was followed by the development of radial and circumferential fractures in the surrounding aragonite. The nucleation of "second generation" calcite grains occurred on these fractures. These phenomena developed extensively when calcite grains reached a diameter of 40 to 50 μm. Boettcher and Wyllie (1967) observed that calcite grown rapidly from aragonite is biaxial, perhaps indicating stored transformation strain energy. During experiments on the transformation of Ni_2SiO_4 (olivine to spinel), Boland and Liebermann (1982) reported the development of a high dislocation density in the residual olivine phase, possibly as a result of the volume change. In a study of the transformation of quartz to cristobalite ($\Delta V = +13.7\%$), Chaklader and Roberts (1961) found that for quartz samples that were relatively free of defects and imperfections, reaction was restricted to the surface. Owing to the volume increase, transformation stress developed and inhibited further transformation. After gentle grinding of the sample, the reaction was reestablished. Comparable phenomena have been observed during some metallic transformations. For example, in the β' to ζ° transformation in AgZn that only involves a volume change of $+0.6\%$, Kittl and Cabo (1969) reported the development of deformation twins and surface rumpling during the late stages of transformation, both of which indicate a high transformation stress.

The inhibition of a reaction by transformation stress can be understood with reference to the growth rate equation for an interface-controlled reaction (Eq. (2), Table 2). A reaction involving a volume increase and proceeding at pressures below the equilibrium boundary will be characterized by localized stress development around a growing nucleus. This may result in a localized pressure increase back towards equilibrium. Then, because ΔG_r changes towards zero, the growth rate slows. As shown by data of Davis and Adams (1965) and as discussed by Kunzler and Goodall (1970), the reaction of aragonite to calcite is sometimes greatly inhibited after 25% to 50% reaction. According to the theoretical calculation of Kunzler and Goodall (1970) using compressibilities of $CaCO_3$, transformation stress should completely inhibit growth rate after 19% reaction at 400°C, considering only *elastic* strain. Honda and Sato (1954) have also analysed the problem by considering elastic strain.

The common observation (e.g., Fig. 5; Schmalzried, 1974, p. 167; see also Davis and Adams, 1965) that transformation curves flatten out before complete reaction ($\xi < 1$) means that growth rate may not remain constant for the duration of the transformation (which makes the evaluation of n and K_7 in Eqs. (7), Table 2 difficult; see Fig. 3). The problem may be considerably greater in a porous powder aggregate than in an annealed aggregate of low porosity (see below).

Transformation stress is expected to produce initially elastic and later plastic strain in the reacting system. Thus, under certain circumstances, the rate of stress relaxation (i.e., accommodation of volume change) may control the growth rate of product grains and consequently the overall transformation kinetics. Mechanisms of stress relaxation in solid-state reactions require careful experimental and transmission electron microscope investigations.

Effect of a Fluid Phase

Experimental studies of heterogeneous reactions indicate that the presence of a hydrous fluid has a huge effect on reaction rates. For example, the presence of H_2O increases the rate of reaction of $SiO_2 + 2\,MgO \rightarrow Mg_2SiO_4$

by 8 to 10 orders of magnitude in comparison with the dry system (see Fyfe *et al.*, 1958, p. 84–85). Brown *et al.* (1962, Fig. 4) found a similar difference for aragonite to calcite, although according to extrapolations from their data the difference decreases with increasing temperature. As pointed out by Yund (personal communication), water can enhance the kinetics of a transformation by changing the path or lowering the energy of the activated state, but the mechanism is still solid-state. On the other hand, water may provide a new faster path that is not solid-state, such as dissolution and reprecipitation. Apart from aragonite to calcite, studies of most reactions have been made with water present, and in some cases with large quantities (e.g., Matthews, 1980). It is to be anticipated that the results from such studies are not simply applicable to reactions in metamorphic rocks in which only very small traces of H_2O may be available. At present there are few experimental data to enable us to predict how rates may vary with *quantity* of fluid or with activity of H_2O, etc. (but see Wegner and Ernst, 1983). It has been suggested that the rheological properties of rocks may be sensitive to small changes in fluid content (Tullis and Yund, 1980). If the same is true for reaction kinetics, characterization of fluid content in rocks may be the largest problem in the application of experimental kinetic data.

The composition of the fluid phase has been found to affect reaction rates. The systematic studies by Fyfe and Bischoff (1965) on the aragonite to calcite transformation in aqueous solutions showed that CO_2, $NaCl$, $FeCl_2$, $NiCl_2$, $FeSO_4$, and $MgCl_2$ were all accelerators of rate relative to pure H_2O, whereas $NaOH$ and $SrCl_2$ were transformation inhibitors. However, Wood and Walther (1983) concluded that fluid composition has little effect on rates above 400°C.

Effect of Dislocations and Strain on Reaction Rates

Data exist on effects of dislocation density and of simultaneous plastic deformation (strain) on diffusion rates in albite (Yund and Tullis, 1980; Yund *et al.*, 1981). A high dislocation density (5×10^9 cm^{-2}) in albite was found to increase the rate of oxygen diffusion by only 0.5–0.7 orders of magnitude, which is in accordance with the theoretical model of Hart (1957). Yund and Tullis (1980) also found that a high density of static dislocations ($\sim 10^{12}$ cm^{-2}) had a relatively small effect on Al–Si order–disorder kinetics in albite. They found that simultaneous plastic deformation increased rates of order–disorder and inferred that the movement of dislocations through a crystal is necessary for the greater atomic mobility in the region of dislocation cores to affect all regions of the crystal.

In theoretical models in which polymorphic reactions are considered as martensitic transformation (Hornstra, 1960; Poirier, 1981a; Gillet and

Madon, 1982), a high dislocation density and a high shear stress should accelerate reaction rates. Little experimental information is available on these effects in heterogeneous reactions. Wenk *et al.* (1973) reacted Solenhofen limestone to aragonite during deformation experiments. The limited data on $\xi(T, \dot{\epsilon}, t)$ suggest that reaction rates were not drastically different from those measured by Brar and Schloessin (1979) for the same reaction in powders and single crystals under supposed hydrostatic conditions. Davis and Adams (1965) presented data on $\xi(t)$ for the reaction of single crystal aragonite to calcite at 450°C and 1 bar under two types of experimental conditions (Fig. 8, curves A and B). The data defining curve A was obtained without stressing the single crystal, whereas that defining curve B was obtained after initially subjecting the sample to high shear stresses. The second procedure produced a much faster initial reaction rate (Fig. 8) presumably because of the development of a high density of dislocations and/or microfractures.

Strain resulting from very high shear stresses produced by mechanical grinding can increase reaction rates by several orders of magnitude. Under such conditions, the kinetics of transformations between calcite and aragonite are fast at *room temperature* (Dachille and Roy, 1961; Dandurand *et al.*, 1982). Also, metastable phases can form in regions of high dislocation density; for example, Green (1972) was able to produce coesite in highly strained quartz samples at pressures 5–20 kbar below the stability field of coesite, in the temperature range 450–900°C. Such results probably have limited appli-

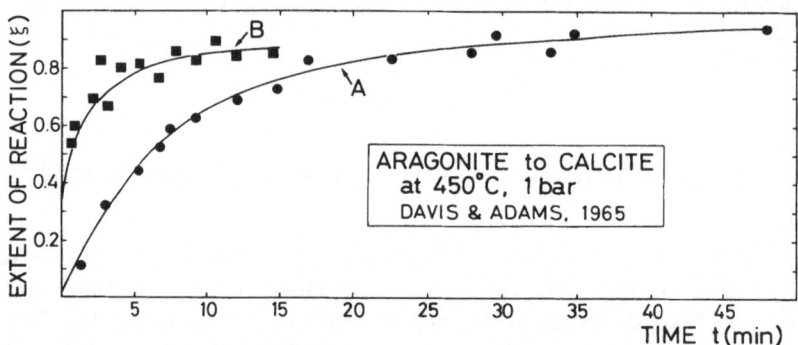

Fig. 8. Two sets of kinetic data for the aragonite → calcite reaction at $T = 450$°C and $P = 1$ bar, using single crystal Kamsdorf aragonite as starting material (from Davis and Adams, 1965, Figs. 2c and 4c). The data points (solid circles) defining curve A were collected after rapidly raising T to 450°C at 1 bar. The data (solid squares) defining curve B were obtained using a beryllium pressure vessel (Davis and Adams, 1964) in which high shear stresses are to be expected; pressure was increased to 10 kbar, the temperature was then raised slowly to 450°C, and P was reduced rapidly to 1 bar to initiate the reaction. This second procedure produced a much faster initial reaction rate, presumably because of the inducement of a high density of defects and/ or microfractures during application of high P.

cability to geological processes, but they may be of importance in interpreting results of kinetic experiments using crushed powders. Subjecting a powder to high pressure is expected to create plastic deformation in localized regions of high shear stress (see Petrovich, 1981a, 1981b). Nucleation rates and perhaps subsequent growth rates are likely to be *enhanced* in such damaged regions (Christian, 1975, p. 12). The importance of the effect on overall transformation kinetics has yet to be assessed.

The effect of dislocations and strain on reaction rates in rocks is very difficult to evaluate. This is because when deformation has enhanced metamorphic reactions it can often be shown that the deformation allowed fluid access to the reacting system (Beach, 1980; Rutter and Brodie, this volume). Because the addition of H_2O to an initially dry system may accelerate reaction rates by 8–10 orders of magnitude (see above), any contribution of strain energy may be completely swamped!

Time–Temperature–Transformation Diagrams, Isokines, and Isochrons

Time–temperature–transformation (*TTT*) diagrams (Putnis and McConnell, 1980; Ganguly, 1982) are usually constructed from isothermally derived experimental data and are also referred to as isothermal transformation (*I–T*) diagrams. Although such a diagram may be useful in qualitative discussions of nonisothermal temperature–time paths (Putnis and McConnell, 1980, p. 149), strictly speaking they cannot be applied *quantitatively* without modification (Porter and Easterling, 1981, p. 346; Yund, in press, Fig. 10). Diagrams have been constructed for reactions in metallic systems for nonisothermal cooling paths (e.g., Porter and Easterling, 1981, p. 344) and are referred to as continuous cooling transformation (*CCT*) or cooling transformation (*C–T*) diagrams. In these diagrams the *C*-shaped curves are shifted downwards to lower temperatures and to the right to longer times compared with *I–T* diagrams. Such diagrams are difficult to adapt to geological use because the equilibrium temperature on which the diagram is based is *pressure* dependent for most reactions, and in general, processes of heating or cooling in rocks are not isobaric. The addition of pressure to such a diagram makes it three-dimensional, and its graphical simplicity-value is lost.

Isokines, i.e., lines of constant kinetics (e.g., constant $\dot{\xi}$, $\dot{x}$, or $\dot{N}_v$), or isochrons, i.e., lines of constant time for a given extent of reaction (e.g., $t_{\xi=0.99}$), can be plotted on a $P–T$ diagram. Isokines or isochrons can be constructed for overall transformation kinetics from a rate equation (e.g., Davis and Adams, 1965, Fig. 6) or for the growth rate of the product phase (e.g., Carlson and Rosenfeld, 1981, Fig. 14). The general assumption in these plots is that the derived rate equation can be extrapolated to all $P–T$ conditions at which the reaction products are stable. This is very misleading because, as shown above, nucleation rates will be negligible within a $P-T$ region adjacent to the equilibrium boundary (Fig. 2). Because the reaction cannot start in this region, the overall transformation kinetics must be effectively zero. Although the constant growth rate curves constructed by Carlson and Rosenfeld (1981) are theoretically correct, they are also open to misinterpretation because, in the absence of nucleation rate data, they provide limited information about overall transformation kinetics at conditions fairly close to equilibrium.

Kinetics and Microstructural Development in Mineral Experiments and in Rocks

Even if we could be satisfied with experimental kinetic data and their interpretation, it is no small problem to know how such data can be used to decipher mineral microstructures in rocks to obtain, for example, pressure–temperature–time information. It is useful to consider each of the items discussed above in terms of how they could influence mineral textures and microstructures in rocks.

The Role of Nucleation Sites in Microstructural Development

Nuclei preferentially form on sites that require the lowest activation energy for nucleation. Although the range of nucleation sites is partly controlled by chemical potential gradients, the kinetics of nucleation are largely controlled by crystalline imperfections. Even when site saturation occurs early and growth is the dominant process throughout a transformation, "nucleation kinetics control the initial disposition of microstructures and thus play a vital role in determining the evolutionary pattern of (precipitation) structures" (Aaronson *et al.*, 1971, p. 16).

Clearly the type of nucleation site may affect overall transformation kinetics, which for polymorphic transformations, for example, may be different if nucleation occurs on grain boundaries or on intragranular defects. In the first case, grain size will be an important factor for overall kinetics. In the second case, defect density is important and the kinetics will be independent of grain size. Depending on sample preparation, different experiments could indicate the dominance of different kinds of nucleation sites and, therefore, of different kinetics. As a further example it is clear that phase transformations among the Al_2SiO_5 polymorphs may be *interface*-controlled if sillimanite nucleates on kyanite, but *diffusion*-controlled if sillimanite nucleates on another phase such as biotite.

Powder Experiments

Mineral powders, frequently used in kinetic experiments (see Table 1), have some properties that make it unlikely that reaction mechanisms, kinetics, and microstructural development in such materials are easily comparable to such processes in rocks. Some of the problems associated with predicting rates of powder reactions have been emphasized by Schmalzried (1974, p. 102; 1978, p. 291). Mechanical grinding during the preparation of powdered samples results in grains with considerable subsurface damage and with a high dislocation density in a thin surface layer (Petrovich, 1981a, 1981b; Nitkiewicz *et al.*, 1983). In quartz, the proliferation of dislocations resulting from localised plastic deformation during grinding may become so excessive that amorphous silica can develop locally. Placing a powdered aggregate under high pressure will further the development of plastic deformation in localized regions of high shear stress (observed by a broadening of X-ray diffraction peaks by Zeto and Roy, 1969; see also Yund *et al.*, 1972, p. 263). Subsurface structural damage (i.e., a high density of high-energy defects) ensures that *grain surface* nucleation rates are greatly dominant (as observed experimentally by Zeto and Roy, 1969; Sung, 1979; Brar and Schloessin, 1980; Gordon, 1971; Matthews, 1980; Carlson and Rosenfeld, 1981; Carlson, 1983; Wegner and Ernst, 1983). Nucleation rates on grain surfaces and on grain boundaries may differ by orders of magnitude because surface free energy and interfacial free energy (which largely control these respective nucleation rates) are different (Martin and Doherty, 1976, p. 159). Likewise, these rates must differ because the strain energies arising from the volume change upon nucleation on grain surfaces versus grain boundaries are expected to be different, because only in the latter case is a nucleus totally constrained by the matrix (Yund and Hall, 1970, p. 390). For these reasons alone, the kinetics and microstructural development of reactions in powders may be quite different from those in rocks.

A powder aggregate is porous even at high pressures, unless extremely fine grained. Many grain surfaces are separated by voids, and in the presence of H_2O or CO_2 these will be fluid filled. Products forming on such surfaces are able to grow into the fluid-filled voids and idomorphic crystals may result; see scanning electron micrographs of reaction products shown by Matthews (1980, Fig. 6) and Gordon (1971). Thus, it is probably necessary to assume that reaction mechanisms, and therefore growth kinetics, are different in a (porous) powder aggregate compared to a nonporous aggregate in which nucleation has occurred on grain boundaries and growth is constrained by the matrix. Furthermore, Wood and Walther (1983) concluded that the rate-limiting step in experiments using crushed powders with several percent H_2O present is the rate of surface reaction and noted that "the presence of a fluid provides an abundant medium for diffusion." In rocks, grain boundary diffusion is generally thought to be rate-controlling and may be slower by 5 to 10 orders of magnitude (see Fig. 1) than diffusion in bulk water (Fletcher and Hofmann, 1974; Ildefonse and Gabis, 1976; Rutter, 1976, 1983; Giletti and Nagy, 1981; Brady, 1983; Joesten, 1983). Thus kinetic data from such experiments are not likely to be directly applicable to rocks undergoing metamorphism.

As discussed above, the grain size of a powder aggregate is impossible to control at high pressure, and the broad grain-size distribution obtained makes characterization of this variable difficult (but see Kapur, 1973).

The effects of transformation stress must differ in powders and in nonporous crystalline aggregates, both at low and high pressure. At low pressure, fracturing of fragments will be the dominant stress relaxation mechanism. Even at elevated pressures, owing to the porosity of powders, fracturing and spalling to produce new small fragments is likely, and stress relaxation will then only partly be accommodated by plastic flow. Furthermore, as a result of fracturing and the formation of new surfaces during the transformation, new nucleation sites will develop to modify the kinetics further. If readjustment of the pore space can accommodate the volume change of a transformation, then initial growth may not be controlled by the accumulation of transformation stress.

Single Crystal Experiments

Kinetic studies have been made on some reactions at high pressure using a single crystal reactant starting material (e.g., calcite to aragonite; Davis and Adams, 1965; Brar and Schloessin, 1979, 1980). This approach suffers from some of the problems described for powder reactions. In particular if *surface nucleation* of the product dominates, then the results are difficult to compare with a *grain boundary nucleation* mechanism that is more applicable to rocks. For surface nucleation, the *size* of the crystal is critical to the overall transformation kinetics, whereas for nucleation on intracrystalline defects the kinetics are independent of crystal size. Such intracrystalline defects may proliferate in a single crystal during the application of high pressure, especially in solid-media high-pressure devices where pressure is usually applied before heating the sample (see Fig. 8). It is even possible that the single crystal becomes polycrystalline during application of high pressure.

Grain Size of Reaction Products

It is sometimes regarded that the grain size of a metamorphic rock is an important microstructural property because it may preserve a record of nucleation and growth kinetics during mineral reactions (Kretz, 1966; Jones and Galwey, 1966) and consequently may even be related to the kinematics of tectonic processes (Rubie, 1983). In other cases it is considered that the grain size records part of the deformational history of a rock and attempts have been made to correlate grain size and differential stress during dynamic recrystallization of monominerallic dunites, quartzites, and limestones (e.g., Christie and Ord, 1980). Alternatively, if it can be shown that *grain growth* or *grain coarsening* has occurred under static conditions, then grain size may record part of the thermal history of a rock (e.g., Jones *et al.*, 1975; Tullis and Yund, 1982; Joesten, 1983). In addition, grain size can be important in determining flow regimes for a deforming rock (Elliott, 1973; Schmid *et al.*, 1977), and it is sometimes necessary to understand the change in grain size with time to interpret the deformational history of a rock (Rubie, 1983, 1984).

As soon as a mineral reaction has started by the initial nucleation of the product phases, nucleation and growth can be regarded as competing processes in that nucleation sites may be consumed by either nucleation or by growth of early-formed grains. A fine-grained product can result if nuclei form at all possible nucleation sites before these sites are consumed by the growth of early-formed nuclei. This condition is favored by a high ratio of $\dot{N}_v/\dot{x}$. Conversely, if $\dot{N}_v/\dot{x}$ is small, then rapid growth of only a small number of early-formed nuclei can consume the reactant together with the majority of potential nucleation sites, thus resulting in a coarse-grained product. Most reactions during regional metamorphism occur in response to tectonically (kinematically) induced changes in P and/or T. Along a PTt path crossing an equilibrium boundary, a reaction will only start when the critical degree of ΔP or ΔT overstepping required for nucleation is reached, at which stage $\dot{N}_v/\dot{x}$ is very low. To achieve a high value of $\dot{N}_v/\dot{x}$, a further $P-T$ interval must be crossed (Fig. 2) before the reaction is complete. This is most likely to happen if rates of change of P and/or T are fast (Rubie, 1983). In addition to the ratio $\dot{N}_v/\dot{x}$, the relation between the crystallographic orientation of the reactant and product phases is important. For example, if a large crystal of aragonite reacts to calcite topotactically with all new calcite grains in the same orientation, their coalescence must lead to a coarse-grained product, even if $\dot{N}_v/\dot{x}$ was high (see Carlson and Rosenfeld, 1981, Figs. 2 and 3). For a fine-grained product to develop, calcite nuclei of *varying* orientation would have to form.

Two additional factors that can also control the development of grain size are the production or consumption of heat during a reaction (heat of reaction plus advected heat, radiogenic heat, or frictional heat) and the development of transformation stress. During all reactions, heat is either produced or consumed at a rate governed by the reaction kinetics and the enthalpy of reaction (Ridley, this volume). Reactions can become heat flow-controlled depending on the transport rate of this heat away from or to the reaction site. The direction of temperature change at the reaction site is to induce a shift back towards equilibrium, resulting in a decrease in the free energy for the reaction. For devolatilization reactions in particular, with a large heat of reaction, the rate of heat input into the metamorphic pile is considered by Yardley (1977, 1983) to be the ultimate rate-controlling kinetic step. Because of the contrasting variation of nucleation and growth rates with temperature (Fig. 2), the reaction enthalpy and rate of heat conduction can influence the microstructural evolution and the resulting grain size of a metamorphic rock (Ridley, this volume). Because of the small amount of reacting material in experimental assemblies, and the proximity of the heat source, reaction enthalpy will have negligible effect on the thermal history and consequent microstructural evolution of the sample.

Transformation stress, occurring as a result of volume change during a reaction, has an effect somewhat analogous to heat of reaction. As discussed above, unless stress relaxation occurs by deformation of the reacting system, internal pressure may return the system towards equilibrium with a consequent decrease in growth rate. However, complexities with regard to overall transformation kinetics must arise if defects, created by the deformation accompanying stress relaxation, consequently act as new nucleation sites.

Grain Growth and Particle Coarsening

In general the grain size of a metamorphic rock derived from a fine-grained protolith increases with increasing grade of metamorphism. Although this trend is partly due to the growth of porphyroblasts during prograde reactions, there is evidence that the grain size of a stable assemblage increases at high temperature with time (e.g., Griggs *et al.*, 1960; Tullis and Yund, 1982; Rubie, 1983; Joesten, 1983).

The reduction of the total grain boundary energy of a system produces the driving force for grain growth. This driving force decreases with increasing grain size (Martin and Doherty, 1976). Although theoretically a single phase aggregate will ultimately tend to coarsen into a single crystal whose shape minimizes the surface free energy, the low driving force results in an infinitesimally slow approach towards this state. The kinetics of grain growth have been predicted to follow a law of the form $\bar{d}^2 - \bar{d}_0^2 = K_{15}t$ (Eq. (15), Table 2) where $\bar{d}_0$ is the initial mean grain size, $\bar{d}$ is the final mean grain size, K_{15} is a temperature-dependent constant, and t is time. However, experimental data have generally been fitted to a relationship $\bar{d} = Kt^n$, which is only theoretically valid if $\bar{d}_0 << \bar{d}$. Although the exponent n is theoretically expected to be 0.5, experimentally determined values are often much lower, and there is evidence that n *decreases* with increasing grain size (Martin and Doherty, 1976, Fig. 4.44). This has led to the suggestion that the indefinite increase of $\bar{d}$ with time is incorrect and that, owing to a velocity-independent drag on grain boundaries, there is a *stable* average grain diameter $\bar{d}_e$ that is a function of temperature (Martin and Doherty, 1976). Andrade and Aboav (1966) have proposed that this stable grain diameter is given by

$$\bar{d}_e^{0.5} = K_{16}(T - T_0)$$

(Eq. (16), Table 2) where K_{16} and T_0 are material constants. The concept of a *stable* grain size for a given mineral as a function of temperature has several geological consequences. For example, in a thermal aureole, even if grain growth kinetics are adequately described by $\bar{d}^2 - \bar{d}_0^2 = Kt$, it is certainly not straightforward to relate grain size to thermal history or maximum temperature achieved at various locations in the aureole (see Spry, 1969, pp. 125–127; Joesten, 1983). However, if the concept of a *stable* grain size applies to monomineralic rocks (quartzites and marbles), then a grain size–distance variation might preserve a record of maximum temperature, independent of time. Relations between grain size and thermal history are further complicated by variations in P_{H_2O} and of nonhydrostatic stress. For example, Tullis and Yund (1982) have shown that addition of 1–2 wt% H_2O to quartz aggregates greatly increases the rate of grain growth. In metals, a small external shear stress is known to promote slow grain growth, when in the unstressed state the grains are of a stable size (Andrade and Aboav, 1966, p. 32).

Studies of Ostwald ripening in metals and ceramics (Greenwood, 1969) may have some application to the coarsening of a minor (second) phase in a rock, such as the coarsening of micas contained in a quartz-rich matrix. Reduction of the total grain boundary energy provides the driving force, and the coarsening process operates because of the greater solubility in the matrix or grain boundary network of small particles (which dissolve) than of the larger particles (which grow), relative to particles of the mean radius which have a zero growth rate. It has been found experimentally that the structure of the matrix affects the coarsening kinetics. In the case of a well-developed recovery substructure in which low-angle boundaries link all particles, the coarsening rate is increased compared to the case in which the recovery substructure has been eliminated by recrystallization (Stumpf and Sellars, 1969). Also, simultaneous deformation has been found to accelerate coarsening by up to three orders of magnitude, probably because of dislocation movement (Mukherjee and Sellars, 1969).

Although grain-size distributions of metamorphic minerals are commonly interpreted in terms of nucleation and growth models (e.g., Kretz, 1966; Spry, 1969) many size distributions could result from a postreaction coarsening process. Grain growth in a two-phase aggregate (e.g., quartz + mica) is much more difficult than in a single-phase aggregate because crystals of the second phase exert a retarding force on the migrating boundaries of the first phase. Second-phase particles can effectively pin grain boundaries and inhibit growth (Martin and Doherty, 1976, p. 234; Vernon, 1976, Fig. 5.2). However, if grains of the second (minor) phase undergo coarsening, they will decrease in number. Grain growth of the first (major) phase then becomes easier because first-phase boundaries become increasingly *unpinned* and therefore free to migrate.

Some Suggestions for Objectives and Methodology of Future Experiments

Objectives

Ideally the objective of experimental kinetic studies would be to determine overall transformation rates, nucleation rates, and growth rates as a function of all the variables discussed above. In addition, the microstructural evolution of the reacting material, the importance of which we have tried to emphasize in this paper, could be studied as a function of the above variables, using optical and electron microscope techniques. Some fundamental questions to be answered include:

1. What is the microstructure (e.g., distribution and density of defects, grain size, and grain-size distribution) of the starting material and how does it affect kinetics and microstructural development during the reaction?
2. How much overstepping of an equilibrium boundary is required to obtain a detectable nucleation rate? Carpenter and Putnis (this volume) suggest that the required overstepping could be large, but quantitative estimates are lacking.
3. Where are nucleation sites located and do the dominant sites vary with *P, T,* and fluid content, etc.?
4. Does nucleation continue throughout the duration of reaction (late-site saturation) or does it dominantly occur at a very early stage (early-site saturation)? This should be assessed from microstructural observations rather than from values of parameters in rate equations.
5. Is there crystallographic control over the orientation of reaction products, in which case can conclusions be drawn concerning reaction mechanisms (e.g., nucleation and growth or martensitic)?
6. How are volume changes accommodated during reaction? By what mechanism does transformation stress relaxation occur, and does the development of transformation strain, such as that apparently observed by Boland and Liebermann (1982), create new nucleation sites?
7. Is growth rate approximately isotropic or is it strongly anisotropic, and does this vary with temperature?
8. To what extent are developing microstructures modified by grain growth during reaction and after a reaction reaches completion? Can the grain-size distribution of reaction products be interpreted in terms of nucle-

ation and growth alone or must grain growth (coarsening) be considered as well?

9. Are rates strongly dependent on the *quantity* of fluid present or is only a very small amount required to achieve a maximum catalytic effect? Also, how does the *activity* of components of the fluid affect the kinetics? The role of a fluid requires detailed study; i.e., how it affects nucleation rates, growth rates, transformation stress relaxation, grain growth processes, and reaction mechanisms generally.

10. What information does the microstructure of a partially or completely reacted system give about possible sets of *PTt* conditions and about the role of fluids? The extent to which the microstructures of reaction products are a function of the variables that control the kinetics is at present unknown. There would appear to be great potential for comparing experimentally produced microstructures with those of metamorphic rocks.

11. How does deformation affect reaction kinetics and microstructural development, and what is the effect of the reaction on deformation? Although it has frequently been proposed that reactions can enhance the deformability of rocks (e.g., Sammis and Dein, 1974; White and Knipe, 1978; Poirier, 1982b; Rubie 1983, 1984; Brodie and Rutter, this volume), only limited experimental evidence is currently available (e.g., Vaughan and Coe, 1981).

12. Kinetic experiments are generally carried out under isothermal and isobaric conditions by suddenly placing reactants out of equilibrium often at large values of $(\Delta T$ and $\Delta P)$ overstepping, whereas during regional metamorphism reactions occur in response to extremely slow changes in T and/or P (see Rubie, 1983; England and Thompson, 1984). A major problem for the future will be to assess how kinetics and microstructural development differ under these contrasting reaction conditions (see Yund *et al.*, 1972, p. 268). Some of the problems of understanding transformation kinetics under nonisothermal conditions have been discussed by Cahn (1956b) and by Christian (1975, pp. 542–548).

Methodology of Experiments

It is clearly important to use starting materials with microstructures that are as closely comparable to those of natural rocks as possible. A crystalline aggregate of near-constant grain size and characterized by grain boundaries would be ideal for polymorphic reactions and for reactions of the type $A \rightarrow B + C + \ldots$. Three possible methods for achieving this configuration under high pressure would be to use (1) natural rock samples, (2) hot-pressed and annealed powders, and (3) an aggregate crystallized from a melt at high pressure.

Natural rock samples have been used frequently in deformation experiments (see Nicolas and Poirier, 1976) and in grain growth experiments (Tullis and Yund, 1982) but apparently not in experimental studies of heterogeneous reactions. The main problem must be the risk of considerable microstructural damage during the application of high pressure, especially in a solid-media apparatus. The possibility of repairing such damage by annealing under high-pressure conditions within the stability field of the reactant(s) requires investigation. To initiate the reaction, P and/or T are changed to the stability field of the product assemblage after annealing, without terminating the experimental run.

If crushed powders are annealed at high P and T, within the stability field of the reactant(s), there should be a reduction in porosity and surface strain energy. Fine-grained material recrystallizes, and there is an approach towards a relatively uniform grain size (see Griggs et al., 1960). As the microstructure changes from that of a crushed powder towards that of a nonporous crystalline aggregate, the reaction rate should decrease. However, in a study of the kinetics of calcite + quartz to wollastonite + CO_2, the annealed powder was found to react faster than the nonannealed material (Gordon, 1971), possibly reflecting that during "annealing," dislocations originally concentrated on grain surfaces migrated further into the grains. Evidently the time required for effective annealing requires careful investigation in conjunction with TEM observations. Following an annealing period, $P-T$ conditions can be changed to start the reaction without terminating the experiment.

For some starting materials, an aggregate can be crystallized from a melt at high pressure within the stability field of the reactant, followed by a change in $P-T$ conditions to initiate the reaction. A careful check for metastable glass would be required because such material could form nucleation sites not comparable to those in natural rocks.

For experimental investigations of reactions involving two starting phases, it may be possible to use single crystals of each phase in contact with each other. The growth rate of the product layer between the two single crystals could then be measured in a series of experiments of varying duration. Initial annealing may also be required to repair subsurface damage along the interface. It may be difficult to use this configuration in a solid media apparatus because of the possibility of producing intracrystalline damage during the application of high pressure (see Fig. 8). An alternative possibility, involving a single crystal of one reactant surrounded by a powder aggregate of the other reactant, has been described by Brady (1983). Provided the product phase nucleates on the single crystal to form a coherent layer, diffusion through this layer is rate controlling, and the problems usually associated with powdered samples only apply during the early stages of reaction.

Concluding Remarks

We have tried to emphasize that experimental studies of mineral kinetics should reflect the situations and problems that we are trying to understand in rocks. Even crude kinetic experiments are useful, because they provide constraining parameters that we do not yet have and obviously guide more systematic experimentation. In order that we may apply the results of kinetic experiments to natural processes, it is clearly important that observations on rocks be made with many possible models in mind. We appreciate that our coverage of the literature is incomplete and that our emphasis here strongly reflects our prejudices.

Acknowledgments

We thank M. A. Carpenter for many interesting discussions on the subject of kinetics; A. Putnis, R. H. Vernon, and R. A. Yund for reviewing the manuscript; S. Girsperger, W. D. Gunter, P. O. Koons, and R. Schmid for help in carrying out experiments that led to some of the ideas formulated in the paper; K. Malmström for typing the manuscript and drafting the figures; and H. Boedecker for much help in locating literature. Financial support from E. T. H. (grants #0.330.080.31/8 and 5.501.330.752/4) and the Schweizerische National Funds (grant #2.7 2010 167) is gratefully appreciated.

References

Aaronson, H. I., Aaron, H. B., and Kinsman, K. R. (1971) Origins of microstructures resulting from precipitation. *Metallography* **4**, 1–42.

Andrade, E. N. da C., and Aboav, D. A. (1966) Grain growth in metals of close-packed hexagonal structure. *Proc. Roy Soc. London* **291A**, 18–40.

Avrami, M. (1939) Kinetics of phase change. I. General theory. *J. Chem. Phys.* **7**, 1103–1112.

Avrami, M. (1940) Kinetics of phase change. II. Transformation–time relations for random distribution of nuclei. *J. Chem. Phys.* **8**, 212–224.

Avrami, M. (1941) Kinetics of phase change. III. Granulations, phase change, and microstructure. *J. Chem. Phys.* **9**, 177–184.

Beach, A. (1980) Retrogressive metamorphic processes in shear zones with special reference to the Lewisian complex. *J. Struct. Geol.* **2**, 257–263.

Becker, R. (1940) On the formation of nuclei during precipitation. *Proc. Phys. Soc.* **52**, 71–76.

Becker, R., and Döring, W. (1935) Kinetische Behandlung der Keimbildung in übersättingten Dämpfen. *Ann. Phys. Lpz.* **24**, 719–752.

Bell, T., and Farnell, B. C. (1969) The isothermal decomposition of nitrogen austenite to bainite, in *The Mechanism of Phase Transformations in Crystalline Solids*. Monograph and Rept. Series No. 33, pp. 282–287. Institute of Metals, London.

Bischoff, J. L. (1969) Temperature controls on aragonite–calcite transformation in aqueous solution. *Amer. Mineral.* **54,** 149–155.

Bischoff, J. L., and Fyfe, W. S. (1968) Catalysis, inhibition, and the calcite–aragonite problem: I. The aragonite–calcite transformation. *Amer. J. Sci.* **266,** 65–79.

Blacic, J. D., and Christie, J. M. (1984) Plasticity and hydrolitic weakening of quartz single crystals. *J. Geophys. Res.* **89,** 4223–4239.

Boehm, H. (1983) Modulated structures at phase transitions. *Amer. Mineral.* **68,** 11–17.

Boettcher, A. L., and Wyllie, P. J. (1967) Biaxial calcite inverted from aragonite. *Amer. Mineral.* **52,** 1527–1529.

Boland, J. N., and Liebermann, R. C. (1982) Mechanism of the olivine to spinel phase transformation in Ni_2SiO_4. *Trans. Amer. Geophys. Union,* **63,** 431.

Brady, J. B. (1983) Intergranular diffusion in metamorphic rocks. *Amer. J. Sci.* **283A,** 181–200.

Brar, N. S., and Schloessin, H. H. (1979) Effects of pressure, temperature, and grain size on the kinetics of the calcite → aragonite transformation. *Can. J. Earth Sci.* **16,** 1402–1418.

Brar, N. S., and Schloessin, H. H. (1980) Nucleation and growth of aragonite in a calcite single crystal, *Phase Transitions* **1,** 299–324.

Brodie, K. H., and Rutter, E. H. (1985) On the relationship between rock deformation and metamorphism, with special reference to the behavior of basic rocks, in *Advances in Physical Geochemistry,* Vol. 4, edited by A. B. Thompson and D. C. Rubie, pp. 138–179. Springer-Verlag, New York.

Brown, A. M., and Ashby, M. F. (1980) Correlations for diffusion constants. *Acta Metall.* **28,** 1085–1101.

Brown, W. H., Fyfe, W. S., and Turner, F. J. (1962) Aragonite in California glaucophane schists, and the kinetics of the aragonite–calcite transformation. *J. Petrol.* **3,** 566–582.

Cahn, J. W. (1956a) The kinetics of grain boundary nucleated reactions. *Acta Metall.* **4,** 449–459.

Cahn, J. W. (1956b) Transformation kinetics during continuous cooling. *Acta Metall.* **4,** 572–575.

Cahn, J. W. (1957) Nucleation on dislocations. *Acta Metall.* **5,** 169–172.

Carlson, W. D. (1983) Aragonite–calcite nucleation kinetics: An application and extension of Avrami transformation theory. *J. Geol.* **91,** 57–71.

Carlson, W. D., and Rosenfeld, J. L. (1981) Optical determination of topotactic aragonite–calcite growth kinetics: Metamorphic implications. *J. Geol.* **89,** 615–638.

Carpenter, M. A., and Putnis, A. (1985) Cation order and disorder during crystal growth: Some implications for natural mineral assemblages, in *Advances in Physical Geochemistry,* Vol. 4, edited by A. B. Thompson and D. C. Rubie, pp. 1–26. Springer-Verlag, New York.

Carter, R. E. (1961) Kinetic model for solid-state reactions. *J. Chem. Phys.* **34**, 2010–2015.

Chaklader, A. C. D., and Roberts, A. L. (1961) Transformation of quartz to cristobalite. *J. Amer. Ceram. Soc.* **44**, 35–41.

Chaudron, G. (1954) Contribution à l'étude des réactions dans l'état solide cinétique de la transformation aragonite–calcite, in *Reactivity of Solids*, 2nd. Int. Symp. on the Reactivity of Solids, pp. 9–20. Gothenburg.

Christian, J. W. (1975) *Transformations in Metals and Alloys. I. Equilibrium and General Kinetic Theory.* Pergamon Press, Oxford.

Christie, J. M., and Ord, A. (1980) Flow stress from microstructures of mylonites: Example and current assessment. *J. Geophys. Res.* **85**, 6253–6262.

Dachille, F., and Roy, R. (1961) Influence of "displacive-shearing" stresses on the kinetics of reconstructive transformations effected by pressure in the range 0–100,000 bars, in *Reactivity of Solids*, Proc. 4th Int. Symp. on the Reactivity of Solids, edited by J. H. de Boer, pp. 502–511. Elsevier, Amsterdam.

Dandurand, J. L., Gout, R., and Schott, J. (1982) Experiments on phase transformation and chemical reactions of mechanically activated minerals by grinding: Petrogenetic implications. *Tectonophysics* **83**, 365–386.

Davis, B. L., and Adams, L. H. (1964) X-ray diffraction evidence for a critical end point for cerium I and cerium II. *J. Phys. Chem. Solids* **25**, 379–388.

Davis, B. L., and Adams, L. H. (1965) Kinetics of the calcite–aragonite transformation. *J. Geophys. Res.* **70**, 433–441.

Doukhan, J. C., and Christie, J. M. (1982) Plastic deformation of sillimanite Al_2SiO_5 single crystals under confining pressure and TEM investigation of the induced defect structure. *Bull. Minéral.* **105**, 583–589.

Elliott, D. (1973) Diffusion flow laws in metamorphic rocks. *Geol. Soc. Amer. Bull.* **84**, 2645–2664.

England, P. C., and Thomson, A. B. (1984) Pressure–temperature–time paths of regional metamorphism. I. Heat transfer during the evolution of regions of thickened continental crust. *J. Petrol.* **25**, 894–928.

Evans, B. W. (1965) Application of a reaction-rate method to the breakdown equilibria of muscovite and muscovite plus quartz. *Amer. J. Sci.* **263**, 647–667.

Fisher, G. W. (1970) The application of ionic equilibria to metamorphic differentiation: An example. *Contrib. Mineral. Petrol.* **29**, 91–103.

Fisher, G. W. (1978) Rate laws in metamorphism. *Geochim. Cosmochim. Acta* **42**, 1035–1050.

Fletcher, R. C., and Hofmann, A. W. (1974) Simple models of diffusion and combined diffusion-infiltration metasomatism, in *Geochemical Transport and Kinetics*, edited by A. W. Hofmann, B. J. Giletti, H. S. Yoder, Jr., and R. A. Yund, pp. 243–259. Carnegie Institute, Washington.

Freer, R. (1981) Diffusion in silicate minerals and glasses: A data digest and guide to the literature. *Contrib. Mineral. Petrol.* **76**, 440–454.

Fyfe, W. S., and Bischoff, J. L. (1965) The calcite–aragonite problem, in *Dolomitization and Limestone Diagenesis*, Soc. Econ. Paleontol. Mineral. Spec. Publ. No. 13, edited by L. C. Pray, and R. C. Murray, pp. 3–13.

Fyfe, W. S., Price, N. J., and Thompson, A. B. (1978) *Fluids in the Earth's Crust.* Elsevier, Amsterdam.

Fyfe, W. S., Turner, F. J., and Verhoogen, J. (1958) Metamorphic reactions and metamorphic facies. *Geol. Soc. Amer. Mem.* **73.**

Gallagher, K. J. (1965) The effect of particle size distribution on the kinetics of diffusion reactions in powders, in *Reactivity of Solids,* Proc. 5th Int. Symp. on the Reactivity of Solids, edited by G. M. Schwab, pp. 192–203. Elsevier, Amsterdam.

Ganguly, J. (1982) Mg-Fe order–disorder in ferromagnesian silicates. II. Thermodynamics, kinetics, and geological applications, in *Advances in Physical Geochemistry* 2, edited by S. K. Saxena, pp. 58–99. Springer-Verlag, New York.

Giletti, B. J., and Nagy, K. L. (1981) Grain boundary diffusion of oxygen along lamellar boundaries in perthitic feldspars (abstract). *Trans. Amer. Geophys. Union* **62,** 428.

Gillet, P., and Madon, M. (1982) Un modèle de dislocations pour la transition aragonite ↔ calcite. *Bull. Minéral,* **105,** 590–597.

Gomes, W. (1961) Definition of rate constant and activation energy in solid state reactions. *Nature* **192,** 865–866.

Gordon, T. M. (1971) Some observations on the formation of wollastonite from calcite and quartz. *Can. J. Earth Sci.* **8,** 844–851.

Green, H. W. (1972) Metastable growth of coesite in highly strained quartz. *J. Geophys. Res.* **77,** 2478–2482.

Greenwood, G. W. (1969) Particle coarsening, in *The Mechanism of Phase Transformations in Crystalline Solids.* Monograph and Rept. Series No. 33, pp. 103–110. Institute of Metals, London.

Greenwood, H. J. (1963) The synthesis and stability of anthophyllite. *J. Petrol.* **4,** 317–351.

Griggs, D. T., Paterson, M. S., Heard, H. C., and Turner, F. J. (1960) Annealing recrystallization in calcite crystals and aggregates. *Geol. Soc. Amer. Mem.* **79,** 21–37.

Grimshaw, R. W., Hargreaves, J., and Roberts, A. L. (1956) Kinetics of the quartz transformation. *Trans. Brit. Ceram. Soc.* **55,** 36–56.

Haasen, P. (1978) *Physical Metallurgy.* Cambridge University Press, Cambridge.

Ham, F. S. (1958) Theory of diffusion-limited precipitation. *J. Phys. Chem. Solids* **6,** 335–351.

Ham, F. S. (1959) Stress-assisted precipitation on dislocations. *J. Appl. Phys.* **30,** 915–926.

Hamaya, N., and Akimoto, S. (1982) Experimental investigation on the mechanism of olivine → spinel transformation: Growth of single crystal spinel from single crystal olivine in Ni_2SiO_4, in *High Pressure Research in Geophysics,* edited by S. Akimoto and M. H. Manghnani, pp. 373–389. Reidel, Dordrecht.

Hanneman, R. E. (1969) Effect of high pressure on phase transformation rates, in *Reactivity of Solids,* Proc. 6th Int. Symp. on the Reactivity of Solids, edited by J. W. Mitchell, R. C. de Vries, R. W. Roberts, and P. Cannon, pp. 789–802. Wiley, New York.

Hart, E. W. (1957) On the role of dislocations in bulk diffusion. *Acta Metall.* **5,** 597.

Helgeson, H. C., Delany, J. M., Nesbitt, H. W., and Bird, D. K. (1978) Summary and critique of the thermodynamics of rock forming minerals. *Amer. J. Sci.* **278A**, 1–229.

Hobbs, B. E. (1981) The influence of metamorphic environment upon the deformation of minerals. *Tectonophysics* **78**, 335–383.

Holdaway, M. J. (1971) Stability of andalusite and the aluminum silicate phase diagram. *Amer. J. Sci.* **271**, 97–131.

Honda, K., and Sato, M. (1954) On the theory of transformation stress, in *Reactivity of Solids*, Proc. 2nd Int. Symp. on the Reactivity of Solids, pp. 847–857. Gothenburg.

Hornstra, J. (1960) Dislocations, stacking faults and twins in the spinel structure. *J. Phys. Chem. Solids* **15**, 311–323.

Hulbert, S. F., Brosnan, D. A., and Smoak, R. H. (1969) Kinetics and mechanism of the reaction between MgO and Cr_2O_3, in *Reactivity of Solids*, Proc. 6th Int. Symp. on the Reactivity of Solids, edited by J. W. Mitchell, R. C. de Vries, R. W. Roberts, and P. Cannon, pp. 573–584. Wiley, New York.

Ildefonse, J. P., and Gabis, V. (1976) Experimental study of silica diffusion during metasomatic reactions in the presence of water at 550°C and 1000 bars. *Geochim. Cosmochim. Acta* **40**, 297–303.

Irving, A. J., and Wyllie, P. J. (1975) Subsolidus and melting relationships for calcite, magnesite and the join $CaCO_3$–$MgCO_3$ to 36 kb. *Geochim. Cosmochim. Acta* **39**, 35–53.

Jander, W. (1927) Reaktionen im festen Zustande bei höheren Temperaturen. *Z. Anorg. Allg. Chem.* **163**, 1–30.

Joesten, R. (1983) Grain growth and grain-boundary diffusion in quartz from the Christmas Mountains (Texas) contact aureole. *Amer. J. Sci.* **283A**, 233–254.

Johannes, W., and Puhan, D. (1971) The calcite–aragonite transition, reinvestigated. *Contrib. Mineral. Petrol.* **31**, 28–38.

Johnson, W. A., and Mehl, R. F. (1939) Reaction kinetics in processes of nucleation and growth. *Trans. Amer. Inst. Min. Metall. Engrs.* **135**, 416–458.

Jones, K. A., and Galwey, A. K. (1966) Size distribution, composition and growth kinetics of garnet crystals in some metamorphic rocks from the west of Ireland. *J. Geol. Soc. London* **122**, 29–44.

Jones, K. A., Wolfe, M. J., and Galwey, A. K. (1975) A theoretical consideration of the kinetics of calcite recrystallization produced by two basalt dykes in Co. Antrim, Northern Ireland. *Contrib. Mineral. Petrol.* **51**, 283–296.

Kapur, P. C. (1973) Kinetics of solid state reactions of particulate ensembles with size distributions. *J. Amer. Ceram. Soc.* **56**, 79–81.

Kasahara, J., and Tsukahara, H. (1971) Experimental measurements of reaction rate at the phase change of nickel olivine to nickel spinel. *J. Phys. Earth* **19**, 79–88.

Kerrick, D. M. (1972) Experimental determination of muscovite + quartz stability with $P_{H_2O} < P_{total}$. *Amer. J. Sci.* **272**, 946–958.

Kingery, W. D., Bowen, H. K., and Uhlmann, D. R. (1976) *Introduction to Ceramics.* Wiley, New York.

Kittl, J. E., and Cabo, A. (1969) The $\beta' \rightarrow \zeta^0$ transformation in the AgZn system, in

The Mechanism of Phase Transformations in Crystalline Solids, Monograph and Rept. Series No. 33, pp. 260–265. Institute of Metals, London.

Kohlstedt, D. L., Nichols, H. P. K., and Hornack, P. (1980) The effect of pressure on the rate of dislocation recovery in olivine. *J. Geophys. Res.* **85**, 3122–3130.

Koons, P. O., and Rubie, D. C. (1983) The effect of deformation on the metamorphic evolution of quartz diorite in the eclogite facies (abstract). *Terra Cognita* **3**, 186.

Kretz, R. (1966) Grain size distributions for certain metamorphic minerals in relation to nucleation and growth. *J. Geol.* **74**, 147–174.

Kridelbaugh, S. J. (1973) The kinetics of the reaction: calcite + quartz = wollastonite + carbon dioxide at elevated temperatures and pressures. *Amer. J. Sci.* **273**, 757–777.

Kunzler, R. H., and Goodall, H. G. (1970) The aragonite–calcite transformation: a problem in the kinetics of a solid–solid reaction. *Amer. J. Sci.* **269**, 360–391.

Lasaga, A. C. (1981) Rate laws of chemical reactions, in *Kinetics of Geochemical Processes. Reviews in Mineralogy, 8,* edited by A. C. Lasaga and R. J. Kirkpatrick. Mineral. Soc. America, 1–68.

Lasaga, A. C., and Kirkpatrick, R. J. (editors) (1981) *Kinetics of Geochemical Processes. Reviews in Mineralogy, 8.* Mineral Soc. America.

Loomis, T. P. (1979) A natural example of metastable reactions involving garnet and sillimanite. *J. Petrol.* **20**, 271–292.

Martin, B., and Fyfe, W. S. (1970) Some experimental and theoretical observations on the kinetics of hydration reactions with particular reference to serpentinization. *Chem. Geol.* **6**, 185–202.

Martin, J. W., and Doherty, R. D. (1976) *Stability of Microstructure in Metallic Systems.* Cambridge University Press, Cambridge.

Matthews, A. (1980) Influences of kinetics and mechanism in metamorphism: A study of albite crystallization. *Geochim. Cosmochim. Acta* **44**, 387–402.

Mukherjee, T., and Sellars, C. M. (1969) Influence of concurrent deformation on coarsening of carbides, in *The Mechanism of Phase Transformation in Crystalline Solids,* Monograph and Rept. Series No. 33, pp. 122–124. Institute of Metals, London.

Nicholson, R. B. (1970) Nucleation at imperfections, in *Phase Transformations,* pp. 269–312. American Society for Metals, Metals Park, OH.

Nicolas, A., and Poirier, J. P. (1976) *Crystalline Plasticity and Solid State Flow in Metamorphic Rocks.* Wiley, London.

Nitkiewicz, A. M., Kerrick, D. M., and Hemingway, B. S. (1983) The effect of particle size on the enthalpy of solution of quartz: Implications for phase equilibria and solution calorimetry. *Geol. Soc. Amer. Abstr. Progs.* **15**, 653.

Petrovich, R. (1981a) Kinetics of dissolution of mechanically comminuted rock forming oxides and silicates. I. Deformation and dissolution of quartz under laboratory conditions. *Geochim. Cosmochim. Acta* **45**, 1665–1674.

Petrovich, R. (1981b) Kinetics of dissolution of mechanically comminuted rock forming oxides and silicates. II. Deformation and dissolution of oxides and silicates in the laboratory and at the earth's surface. *Geochim. Cosmochim. Acta* **45**, 1675–1686.

Poirier, J. P. (1981a) Martensitic olivine–spinel transformation and plasticity of the mantle transition zone, in *Anelasticity in the Earth,* edited by F. D. Stacey, M. S. Paterson, and A. Nicolas, pp. 113–117. Geodynamics Series **4.** American Geophysics Union and Geological Society of America.

Poirier, J. P. (1981b) On the kinetics of the olivine–spinel transition. *Phys. Earth. Planet. Int.* **26,** 179–187.

Poirier, J. P. (1982a) The kinetics of martensitic olivine γ-spinel transition and its dependence on material and experimental parameters, in *High Pressure Research in Geophysics,* edited by S. Akimoto and M. H. Manghnani, pp. 361–371. Reidel, Dordrecht.

Poirier, J. P. (1982b) On transformation plasticity. *J. Geophys Res.* **87,** 6791–6797.

Porter, D. A., and Easterling, K. E. (1981) *Phase Transformations in Metals and Alloys.* Van Nostrand Reinhold, New York.

Putnis, A., and McConnell, J. D. C. (1980) *Principles of Mineral Behaviour.* Blackwell, Oxford.

Rao, M. S. (1973) Kinetics and mechanism of transformation of aragonite to calcite. *Indian J. Chem.* **11,** 280–283.

Ridley, J. (1985) The effect of reaction enthalpy on the progress of a metamorphic reaction, in *Advances in Physical Geochemistry,* Vol. 4, edited by A. B. Thompson and D. C. Rubie, pp. 80–97. Springer-Verlag, New York.

Rubie, D. C. (1983) Reaction-enhanced ductility: The role of solid–solid univariant reactions in deformation of the crust and mantle. *Tectonophysics* **96,** 331–352.

Rubie, D. C. (1984) The olivine → spinel transformation and the rheology of subducting lithosphere. *Nature,* **308,** 505–508.

Russell, K. C. (1970) Nucleation in solids, in *Phase Transformations,* pp. 219–268. American Society for Metals, Metals Park, OH.

Rutter, E. H. (1976) The kinetics of rock deformation by pressure solution. *Phil. Trans. Roy. Soc. London* **283A,** 203–219.

Rutter, E. H. (1983) Pressure solution in nature, theory and experiment. *J. Geol. Soc. London* **140,** 725–740.

Rutter, E. H. and Brodie, K. H. (1985) The permeation of water into hydrating shear zones, in *Advances in Physical Geochemistry,* Vol. 4, edited by A. B. Thompson and D. C. Rubie, pp. 242–250. Springer-Verlag, New York.

Sammis, C. G., and Dein, J. L. (1974) On the possibility of transformation superplasticity in the Earth's mantle. *J. Geophys. Res.* **79,** 2961–2965.

Sammis, C. G., Smith, J. C., and Schubert, G. (1981) A critical assessment of estimation methods for activation volume. *J. Geophys. Res.* **86,** 10707–10718.

Schmalzried, H. (1974) *Solid State Reactions.* Academic Press, New York.

Schmalzried, H. (1978) Reactivity and point defects of double oxides with emphasis on simple silicates. *Phys. Chem. Minerals* **2,** 279–294.

Schmid, S. M., Boland, J. N., and Paterson, M. S. (1977) Superplastic flow in finegrained limestone. *Tectonophysics* **43,** 257–291.

Schramke, J. A., Kerrick, D. M., and Lasaga, A. C. (1983) Irreversible thermodynamics applied to the reaction: muscovite + quartz ↔ andalusite + sanidine + water. *Geol. Soc. Amer. Abstr. Progs.* **15,** 681.

Servi, I. S., and Turnbull, D. (1966) Thermodynamics and kinetics of precipitation in the copper–cobalt system. *Acta Metall.* **14,** 161–169.

Spry, A. (1969) *Metamorphic Textures.* Pergamon Press, Oxford.

Stacey, F. D. (1977) *Physics of the Earth.* Wiley, New York.

Stumpf, W. E., and Sellars, C. M. (1969) Effect of matrix structure on carbide coarsening and transformations, in *The Mechanism of Phase Transformations in Crystalline Solids, Monograph and Rept. Series* No. 33, pp. 120–122. Institute of Metals, London.

Sung, C. M. (1979) Kinetics of the olivine → spinel transition under high pressure and temperature: Experimental results and geophysical implications, in *High Pressure Science and Technology,* Vol. 2, edited by K. D. Timmerhaus and M. S. Barber, pp. 31–42. Plenum Press, New York.

Tanner, L. E., and Servi, I. S. (1966) Direct observations of size, distribution and morphology of precipitates in copper–cobalt alloys. *Acta Metall.* **14,** 231–234.

Tanner, S. B., Kerrick, D. M., and Lasaga, A. C. (1983) The kinetics and mechanisms of the reaction: calcite + quartz $\rightleftharpoons$ wollastonite + carbon dioxide. *Geol. Soc. Amer. Abstr. Progs.* **15,** 704.

Thompson, A. B. (1970) A note on the kaolinite–pyrophyllite equilibrium. *Amer. J. Sci.* **268,** 454–458.

Thompson, A. B. (1971) Analcite–albite equilibria at low temperatures. *Amer. J. Sci.* **271,** 79–92.

Tolokonnikova, L. I., Topor, N. D., Kadenatsi, B. M., and Solov'yeva, T. B. (1974) Kinetics of the aragonite–calcite transition studied by a dilatometric method on a derivatograph. *Zh. Fiz. Khim.* **48,** 2616.

Tsuzuki, Y., and Mizutani, S. (1971) A study of rock alteration process based on kinetics of hydrothermal alteration. *Contrib. Mineral. Petrol.* **30,** 15–33.

Tullis, J. A., Shelton, G. L., and Yund, R. A. (1979) Pressure dependences of rock strength: Implications for hydrolytic weakening. *Bull. Minéral.* **102,** 110–114.

Tullis, J. A., and Yund, R. A. (1980) Hydrolytic weakening of experimentally deformed Westerly granite and Hale albite rock. *J. Struct. Geol.* **2,** 439–451.

Tullis, J. A., and Yund, R. A. (1982) Grain growth kinetics of quartz and calcite aggregates. *J. Geol.* **90,** 301–318.

Turnbull, D. (1956) Phase changes. *Solid State Phys.* **3,** 225–306.

Turnbull, D., and Fisher, J. C. (1949) Rate of nucleation in condensed systems. *J. Chem. Phys.* **17,** 71–73.

Vaughan, P. J., and Coe, R. S. (1981) Creep mechanisms in Mg_2GeO_4: Effects of a phase change. *J. Geophys. Res.* **86,** 389–404.

Vernon, R. H. (1976) *Metamorphic Processes.* George Allen and Unwin, London.

Volmer, M., and Weber, A. (1926) Keimbildung in ubersaettigten Gebilden. *Z. Phys. Chem.* **119,** 277–301.

Wegner, W. W., and Ernst, W. G. (1983) Experimentally determined hydration and dehydration reaction rates in the system MgO-SiO_2-H_2O. *Amer. J. Sci.* **283A,** 151–180.

Wenk, H. R., Venkitasubramanyan, C. S., and Baker, D. W. (1973) Preferred orientation in experimentally deformed limestone. *Contrib. Mineral. Petrol.* **38,** 81–114.

White, S. H., and Knipe, R. J. (1978) Transformation- and reaction-enhanced ductility in rocks. *J. Geol. Soc. London* **135**, 513–516.

Wood, B. J., and Walther, J. V. (1983) Rates of hydrothermal reactions. *Science* **222**, 413–415.

Yardley, B. W. D. (1977) The nature and significance of the mechanism of sillimanite growth in the Connemara schists, Ireland. *Contrib. Mineral. Petrol.* **65**, 53–58.

Yardley, B. W. D. (1983) Heat of reaction—the key to metamorphic kinetics? (abstract). *J. Geol. Soc. London* **140**, 162.

Yund, R. A. (1983) Diffusion in feldspars, in *Feldspar Mineralogy. Reviews in Mineralogy, 2,* edited by P. H. Ribbe. Mineral. Soc. America, 203–222.

Yund, R. A. (in press) Alkali feldspar exsolution: Kinetics and dependence on alkali diffusion, in *Feldspars and Feldspathoids; Structures, Properties and Occurrences,* edited by W. L. Brown. Reidel Publishing Co.

Yund, R. A., and Hall, H. T. (1970) Kinetics and mechanism of pyrite exsolution from pyrrhotite. *J. Petrol.* **11**, 381–404.

Yund, R. A., and Tullis, J. (1980) The effect of water, pressure, and strain on Al/Si order–disorder kinetics in feldspar. *Contrib. Mineral. Petrol.* **72**, 297–302.

Yund, R. A., McCallister, R. H., and Savin, S. M. (1972) An experimental study of nepheline–kalsilite exsolution. *J. Petrol.* **13**, 255–272.

Yund, R. A., Smith, B. M., and Tullis, J. (1981) Dislocation-assisted diffusion of oxygen in albite. *Phys. Chem. Minerals* **7**, 185–189.

Zeto, R. J., and Roy, R. (1969) Kinetics of the GeO_2 (quartz) $\rightarrow$ GeO_2 (rutile) transformation at pressures to 30 kbar, in *Reactivity of Solids,* Proc. 6th Int. Symp. on the Reactivity of Solids, edited by J. W. Mitchell, R. C. de Vries, R. W. Roberts, and P. Cannon, pp. 803–815. Wiley, New York.

Chapter 3
The Effect of Reaction Enthalpy on the Progress of a Metamorphic Reaction

J. Ridley

Introduction

Factors potentially controlling the rate of a metamorphic reaction are the rate of nucleation, the rate of incorporation of ions into the new mineral at an interface, the rate of diffusion to or away from the interface, and the heat flow required to counteract the reaction enthalpy.

This paper investigates one type of interrelationship between these factors, that between rock temperature and reaction progress. The problem can be summarized as follows.

Considerable energy input (up to a third of the total amount; Rice and Ferry, 1982) is required to produce the changes in mineral assemblage and phase chemistry during a typical prograde regional metamorphic reaction sequence. Because any "prograde" reaction will be endothermic and absorb heat from the surroundings, the rate of reaction is potentially controlled by the rate of energy input (e.g., Yardley, 1977; Fisher, 1978). Nucleation theory, however (McLean, 1965; Christian, 1975), suggests that there will always be a kinetic barrier to the nucleation of a new phase in a rock, and hence that reaction will only take place after a finite P- or T-overstepping of the theoretical equilibrium boundary in $P-T$ space. If a rock is heated above an equilibrium boundary before reaction begins, then some of the energy required for reaction might be supplied by lowering the temperature of the immediately surrounding rock. This temperature reduction will affect subsequent reaction behavior. It is this interrelationship, and the effects it may have on observable metamorphic textures such as grain size or the occurrence of porphyroblasts, that this paper specifically investigates.

The exact thermal and metamorphic evolution of a rock undergoing reaction in a system in which the rock temperature may change will be dependent on the rate laws for grain nucleation and growth. There are large uncertainties in the values of the many parameters of the published theoretical nucleation and growth rate laws (e.g., Russell, 1970). There are also uncertainties whether these laws, which have been developed for "simple" metallurgical systems, are applicable to "complex" geological silicate systems. Because of these uncertainties, the study here involves a mathematical simulation of reaction behavior, using only the simplest form of the rate laws and varying each parameter within reasonable limits to try to determine the types of reaction history possible, and which geological situations are most likely to give rise to each possible history. The first part of this paper briefly discusses the theory and nature of the generalized nucleation rate and growth rate laws. The second part discusses the types of thermal and reaction behavior predicted by the modelling, suggesting relations between these and the textural patterns seen in metamorphic rocks.

Nucleation Kinetics and the Overstepping of Reaction Equilibria

Nucleation theory as applied to metallurgical systems is discussed comprehensively by Russell (1970) and Christian (1975). The theory examines the energy balance on the formation of a single grain of a new phase through aggregation of ions or clusters of ions. For nucleation of a new phase within a uniform host, the basic equation for the rate of nucleation is:

$$\dot{n} = A_n \exp \left(\frac{16\pi\gamma^3}{3\Delta G^2 kT} \right) \tag{1}$$

where A_n is a constant related to the activation entropy (Fyfe et al., 1958, p. 72), γ is the interfacial energy per unit area, ΔG is the free energy change per unit volume of product, and k is Boltzmann's constant.

If one assumes that γ remains constant for a specific rock composition at a specific reaction, it can be seen from Eq. (1) that the nucleation rate will be critically dependent on the magnitude of ΔG, and hence the temperature or pressure overstepping of a reaction boundary. A 10% increase in the amount of overstepping may increase the nucleation rate by several orders of magnitude. It is this extremely potent dependence of nucleation rate on the value of ΔG that makes the overstepping required for nucleation effectively finite rather than infinitesimal (McLean, 1965).

The basic form of the nucleation rate equation is independent of whether nucleation is homogeneous (i.e., occurring without the influence of defects or interfaces) or along grain boundaries, edges, or at corners (heterogeneous in the sense of Cahn, 1956, or Russell, 1970). The mathematical formulation for nucleation kinetics along dislocation lines suggests an even stronger dependence of nucleation rate on the magnitude of temperature overstepping of the reaction boundary (Cahn, 1957; for details see Appendix).

Growth Laws and Accelerating Reaction Rates

If the growth rate is determined by the kinetics of accretion onto a crystal surface (interface control), then the growth rate of crystal linear dimensions should be constant. The overall reaction rate should accelerate unless the growing grains impinge.

The linear growth rate will be given approximately by

$$\dot{g} = A_g \exp\left(-\frac{\Delta G_a}{kT}\right)\left[1 - \exp\left(-\frac{\Delta G}{kT}\right)\right] \tag{2}$$

(e.g., Carlson and Rosenfeld, 1981) where A_g is a constant related to the activation entropy for growth, ΔG is the free energy change on reaction, ΔG_a is the kinetic barrier to growth (the activation energy).

If diffusion to the new surface is the rate-controlling factor, then the linear growth rate is given by

$$\dot{g} = \alpha_j(Dt)^{1/2} \tag{3}$$

where α_j is a geometry-specific constant related to the concentration of the rate controlling species outside and inside the new phase, and D is the effective, bulk diffusivity. Overall reaction rates will increase with time, though less steeply than for the interface controlled case.

Modelling: Principles and Parameters

In the general case of reaction behavior considered here, the reaction rate at any moment in time is a function of the temperature–time history up to that moment. (Isobaric conditions are assumed.) This T–t history is itself dependent on the enthalpy of reaction. Full details of the mathematical formulation and computation are given in the Appendix.

The results reported here are for a model "prograde" reaction, involving the growth of a new phase of different composition to those already present in the rock, and with an equilibrium temperature of 500°C. Reaction completion is taken arbitrarily after 20% by volume of the rock has transformed. Nucleation is assumed to be either at grain edges and corners, or along dislocations. Growth is assumed to be controlled by either diffusion or interface kinetics. A rate is calculated for both, and the slower is taken to be rate controlling. Grain impingement is ignored because the reaction involves only a small fraction of the rock. The modelling ignores lateral conduction of heat. It is assumed that there is a constant external energy source causing a general rise in rock temperature, and that there are no significant temperature gradients within the body of rock considered. The effects of this simplification are discussed below.

The effect of varying the magnitude of the following parameters has been investigated:

1. *The interfacial energy.* Brace and Walsh (1962) give a range of values for crystal surface energies of 0.2–2 J m^{-2}. It is assumed (e.g., Porter and Easterling, 1981) that interfacial energies in polycrystalline rocks are about one-third of these values (0.06–0.6 J m^{-2}). Modelling of grain-edge or grain-corner nucleation is undertaken by scaling down these values by 70–90% (see Christian, 1975, Fig. 10.14, after Cahn, 1956, assuming that ratios of interfacial energies approach unity, e.g., Vernon, 1968). The actual range of values used is therefore 0.01–0.2 J m^{-2}.

2. *The free energy released on reaction.* The variable on input is the reaction entropy. $\Delta G = \delta T \Delta S$, where δT is the temperature overstepping of the reaction boundary. Three representative values of ΔS have been taken; 20, 50, and 200 JK^{-1} mol^{-1}, corresponding respectively to likely values for a solid–solid reaction, and for reactions involving some or substantial loss of volatiles. The volume of one mole of reactants is taken as 450 cm^3.

3. *The pre-exponential constant in the nucleation rate law (A_n; Eq. (1)).* This is generally assumed to be $N \times v$ (N, Avogadro's number; v, atomic vibration frequency, $\approx 10^{13}$ s^{-1}) in metallurgical systems (e.g., McLean, 1965). Fyfe et al. (1958, p. 72) suggest that this should not be the case in geological systems. There is an extra term: the entropy of activation, which can be regarded as a measure of the probability of the statistical fluctuation in energy required for nucleation also resulting in an ionic configuration close enough to the structure of the new phase to promote nucleation. In geological systems this will in general be small. Values of the entropy of activation of 70–200 JK^{-1} mol^{-1} have been taken here (see Fyfe et al., 1958).

4. *The externally imposed heat input.* The variable on input is the temperature rise imposed in the case where there is no internal heat production. Two values have been taken, 20 and 200 K my^{-1}, representative, respectively, of conductive heating of a thickened crust (England and Richardson, 1977) and shear heating in a major crustal shear zone (Brun and Cobbold, 1980). The rock heat capacity is taken as 3.7×10^6 J kg^{-1} K^{-1}.

5. *The stored energy per unit length of dislocation and the dislocation density.* For most calculations the energy assumed is that given by approximate shear modulus and Burgers' vector data for quartz ($b = 4.8 \times 10^{-10}$ m, $u = 4.5 \times 10^{10}$ Nm^{-2}; Nicolas and Poirier, 1976, p. 201). A dislocation density of 10^{12} m^{-2} has been assumed. The rate of nucleation along dislocations is critically dependent on the magnitude of the stored energy (Cahn, 1957). Some calculations have therefore been undertaken assuming a rate control by a lower density of higher-energy dislocations (with $b = 6.4 \times 10^{-10}$ m), as might be characteristic of a mineral with large lattice repeats, or alternatively in intensely deformed grains.

6. *The activation energy for growth (ΔG_a; Eq. (2)).* This is theoretically

related to the activation energy for lattice diffusion (Christian, 1975, p. 480). Values have therefore been taken in the range 150–300 kJ mol^{-1} (Freer, 1981).

7. *The pre-exponential constant in the growth rate equation* (A_g; *Eq.* (2)). Similar considerations apply to this as for A_n. In the calculations reported, a constant value of $10^{-4} \times v$ has been taken.

8. *Bulk-rock diffusion* (*D*). Although attempts have been made to obtain sample experimental values for this (e.g., Brady, 1979), it remains effectively an unknown. Values must lie between those for diffusion in solutes ($10^{-8} - 10^{-9}$ m^2 s^{-1}) and those for diffusion within silicate lattices ($<10^{-16}$ m^2 s^{-1}). The range considered here is $10^{-11} - 10^{-17}$ m^2 s^{-1}.

General Features of the Predicted Temperature and Reaction Histories in an Isobaric System

It is found that there is a constant qualitative form to the predicted temperature–time (T–t) evolution, and hence reaction history, for any reaction with a large entropy change (Fig. 1).

Nucleation starts after a finite T-overstepping of the equilibrium boundary. The nucleation rate then increases rapidly with further heat input and temperature increase until the rate of grain growth is sufficient for the reaction enthalpy to affect the rock temperature. The rock is cooled sufficiently to first decrease the nucleation rate and then tŏ cause complete suppression of further nucleation. Growth, however, continues. In many model runs the temperature fell to a few degrees above equilibrium before completion of reaction. A steady-state temperature is reached as reaction rate slows down as the equilibrium boundary is approached.

This qualitative history holds for any set of "reasonable" values of the variables listed in the previous section. The only variable that significantly affects the form of the predicted T–t evolution is the entropy of reaction. For a model solid–solid, low ΔS reaction a temperature drop during reaction progress is in most instances predicted (Fig. 2). It is, however, rarely sufficient to suppress nucleation. The formation of new grains should continue throughout the progress of the reaction.

Changes in any of the other variables affect only the quantitative details of the predicted reaction progress, e.g., the exact extent of reaction before the cessation of nucleation or the total number of nuclei formed.

A lower interfacial energy, for instance, will decrease the temperature overstep of the equilibrium boundary required before the start of nucleation. However small the interfacial energy is, the temperature is still predicted to fall sufficiently to suppress nucleation (Fig. 3). If the interfacial energy is small, then the nucleation rate is more critically dependent on small absolute changes in temperature.

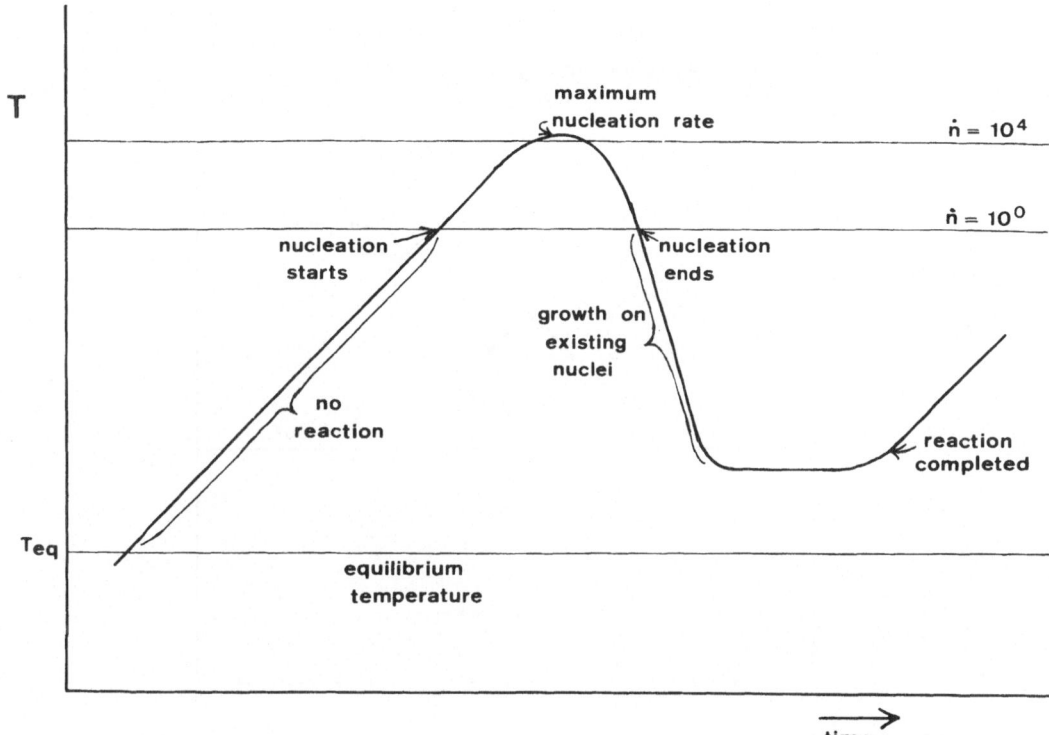

Fig. 1. Schematic Temperature–time $(T-t)$ history for a body of rock during the progress of a single, strongly endothermic prograde discontinuous reaction. For discussion see text. The predicted temperature drop after the onset of nucleation shown in this diagram and those following is much greater and steeper than would be the case in reality, because no account has been taken of increased heat flow into the reacting volume of rock ($\dot{n}$ = nucleation rate per unit volume).

The qualitative form of the $T-t$ history is independent of whether growth rate is controlled by interface kinetics or diffusion. If the growth rate is decreased (e.g., by decreasing the mass diffusivity), then the reaction rate will be initially reduced. A reduction in the initial growth rate, however, allows the temperature to rise to a level at which nucleation becomes more rapid. More nuclei form, hence counteracting the effect of reduced growth rates on the overall reaction rate. The reaction rate will always rise to a value great enough to give a rapid temperature drop (Fig. 4). A faster heating rate likewise allows more nuclei to form (Fig. 5), and hence also leads to a faster overall reaction rate.

The results presented in Figs. 2–5 are for nucleation along grain boundaries, edges, or corners. The mathematical laws for nucleation along dislocations predict an even more rapid fall-off of nucleation rate with decreasing

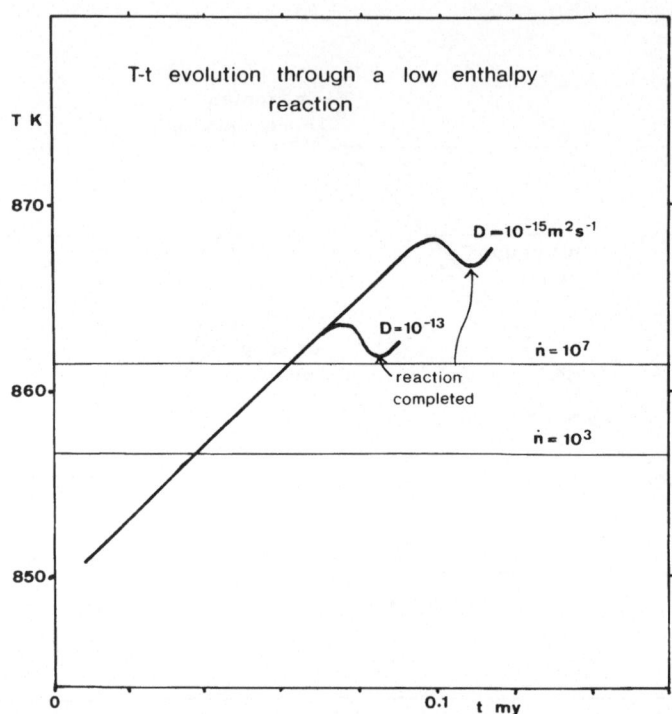

Fig. 2. Model *T–t* histories during a weakly endothermic reaction. A temperature drop during reaction is predicted, but this is not sufficient to reduce the nucleation rate significantly. The two curves show the effect of changing the model diffusivity. (Model parameters: equilibrium temperature, 773 K; surface free energy, 0.02 J m^{-2}; externally applied rate of temperature increase, 200 K/my. Units of nucleation rate in this and subsequent diagrams are m^{-3}/my).

temperature. For any set of conditions, nucleation will cease after a smaller proportion of the rock has reacted if it had taken place along dislocations (Fig. 6).

Grain-Size Distributions

Grain size and grain-size distributions are predicted in the models. An average grain diameter, for instance, can readily be calculated from the percentage volume reacted and the total number of nuclei predicted. Most of the models predict grain diameters in the range 0.05–25 mm. Whatever the grain size, however, the prediction is for grain-size distributions with certain constant qualitative characteristics.

If, in a metamorphic reaction, nucleation ceased significantly before the

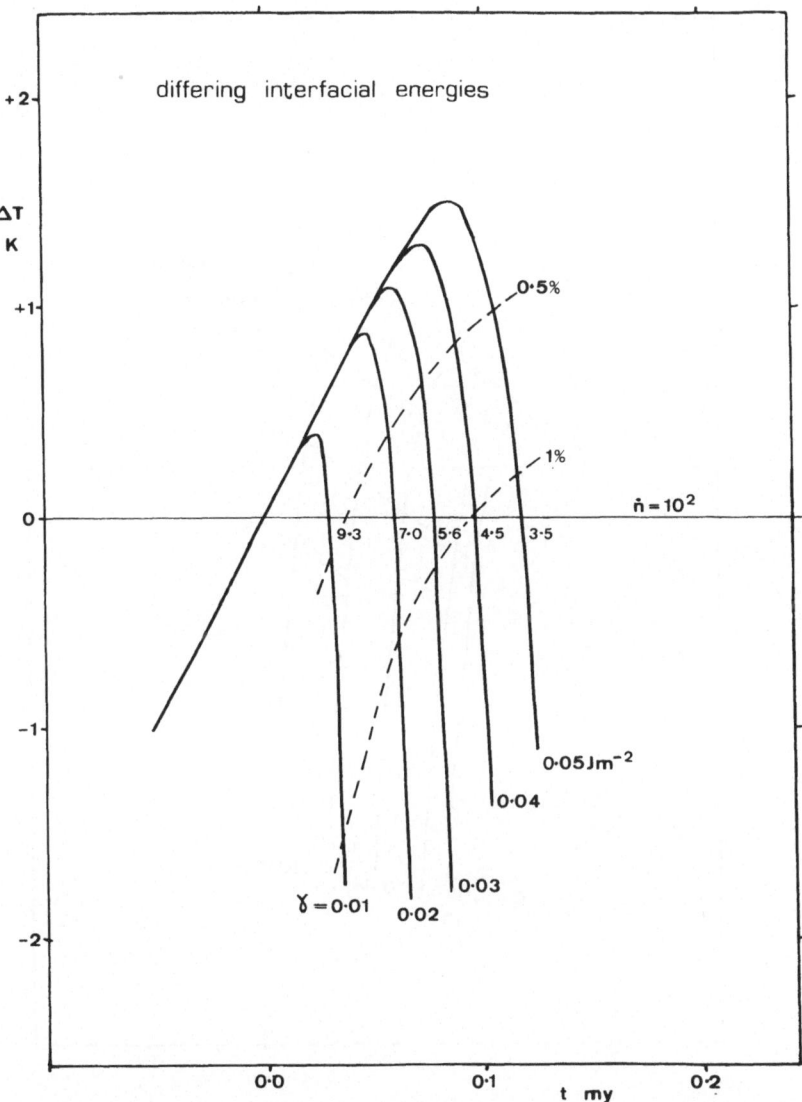

Fig. 3. Effect of differing interfacial energies on the T–t history during a reaction. The general form of all paths is the same. The diagram shows the effect on the rate and exact timing of the temperature fall after the onset of nucleation: taken arbitrarily as when the nucleation rate reaches 10^2 m^{-3}/my. This takes place at different absolute temperatures for each set of parameters. The temperature measure here is therefore ΔT, the difference between rock temperature and that at the onset of nucleation. The dotted lines contour the percentage of rock reacted at any time. The figures by the "onset of nucleation" line give $\log_{10}$ of the final total number of nuclei per cubic meter. A lower interfacial energy leads to a more rapid cut out of nucleation, although more nuclei form and the final grain size is smaller. (Model parameters: heat input, 20 K/my; diffusivity, 10^{-12} m^2/s.

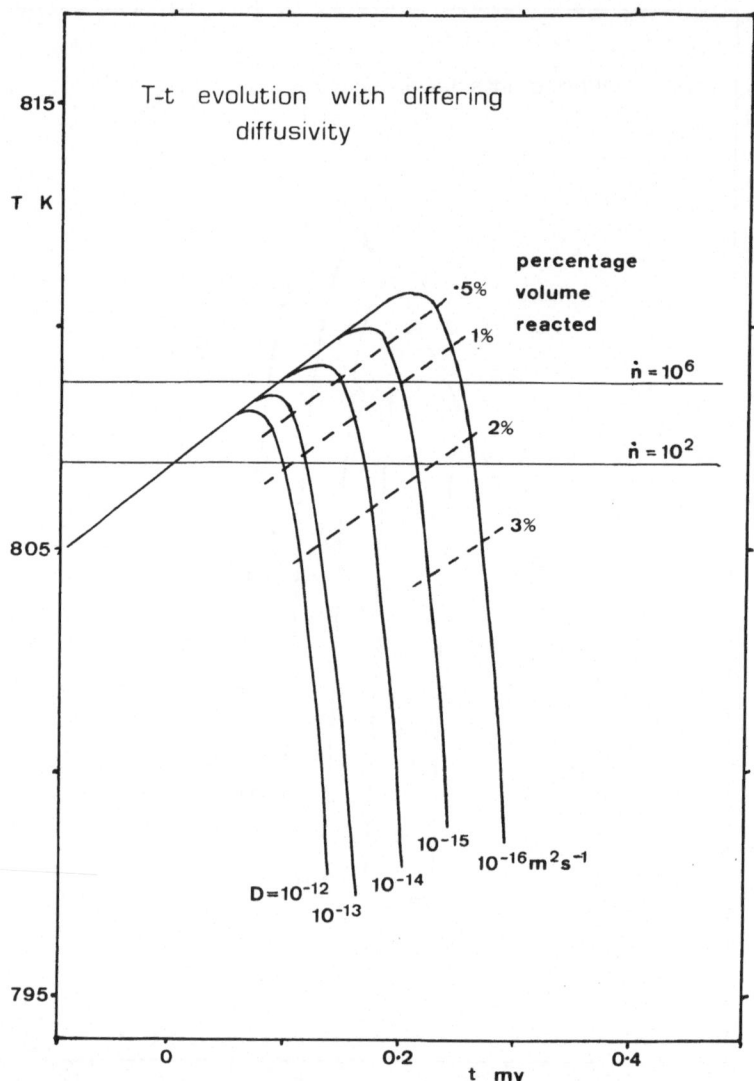

Fig. 4. Effect of differing diffusivities on the *T–t* history during a reaction. Notation as for Fig. 3. (Model parameters: heat input equivalent to a linear rate of temperature increase of 20 K/my; interfacial free energy, 0.05 J m⁻².)

reaction went to completion, then there should be no grains of the new phase smaller than a certain size; i.e., there should be a distinct minimum grain size. If the temperature evolution was such that the nucleation rate first increased, reached a maximum, then decreased (e.g., Fig. 1), one should expect a unimodal grain-size distribution, either symmetric or asymmetric

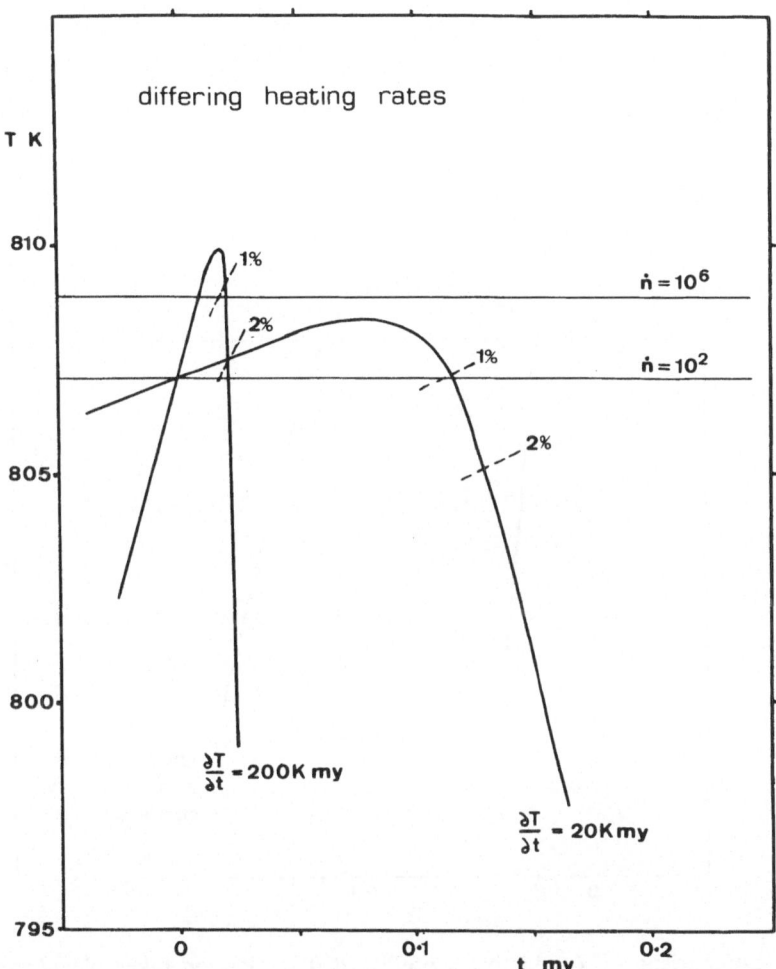

Fig. 5. Effect of differing heating rates on the *T–t* histories. A faster heating rate allows the rock temperature to reach a higher level before the effect of reaction enthalpy becomes important, consequently more nuclei are formed and subsequent reaction progress is faster. (Model parameters: diffusivity, 10^{-13} m²/s; interfacial free energy, 0.05 J m⁻².

(as would be given by the history of Fig. 1), but with a distinct modal grain size.

Both of these features are seen in most of the data presented in the literature on grain-size distributions of porphyroblasts (Jones and Galwey, 1964, 1966; Kretz, 1966, 1973). A distinct lower grain size cut off is seen in the data of Jones and Galwey (1966). Kretz (1966) carefully demonstrates its existence in one sample and concludes that there must have been some grain growth after nucleation ended.

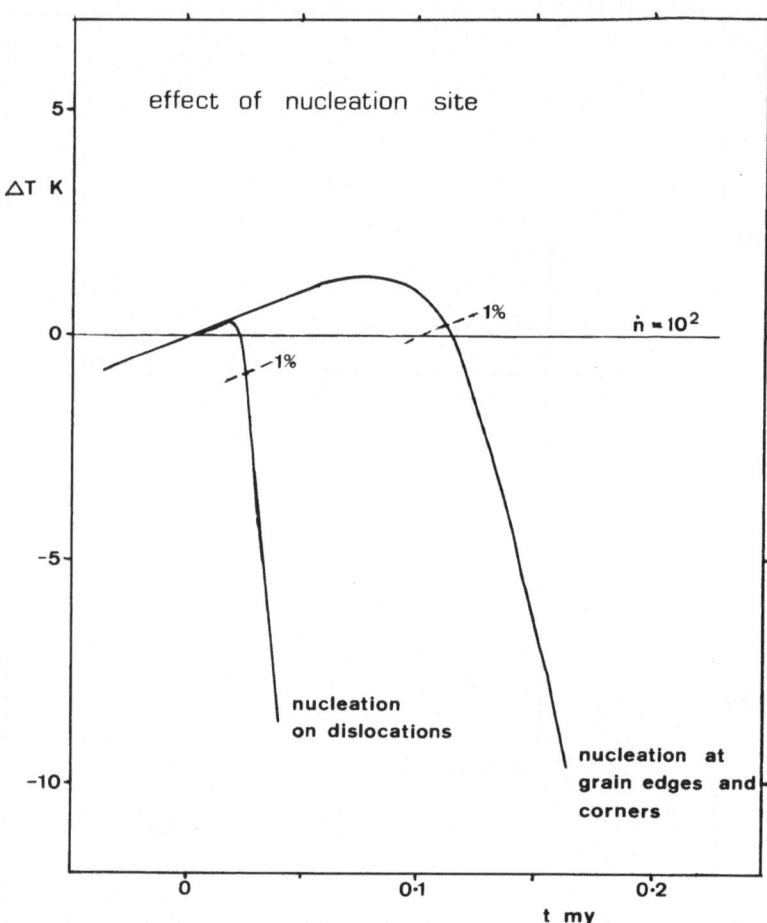

Fig. 6. Comparison of *T–t* histories predicted if nucleation takes place predominantly along dislocations or homogeneously. Model parameters as in preceeding diagram but with an interfacial free energy for nucleation along dislocations of 0.2 J m^{-2}. ΔT as in Fig. 3.

The exact ratio between the average and smallest grain size will depend on the proportion of rock reacting, the reaction enthalpy, and the exact growth and nucleation rate laws. The zero-dimensional modelling suggests that in a high enthalpy reaction nucleation should cease after only 0.5–2% of the total rock volume has reacted. If a reaction eventually involves 10% by volume of the rock, then the majority of growth would have taken place after all nuclei had formed, and the ratio of the smallest to average grain size should be much larger than that shown by any of the data of Jones and Galwey (1966). An incorporation of one-dimensional heat conduction in the modelling would decrease this ratio (see below).

Grain Size and Porphyroblast Growth

It is probably a general rule that porphyroblast phases have a crystal-lattice structure that contrasts with those of the overprinted phases. This has been suggested as a prerequisite for porphyroblast growth (Kretz, 1966). There would be no substrate for epitaxial growth, hence nucleation rates would be relatively slow.

It is also a general rule that the reactions forming porphyroblast phases, where they can be inferred, involve the loss of volatiles and will be strongly endothermic. (Exceptions exist where reactions involving substantial loss of volatiles do not give rise to porphyroblast growth, e.g., many sillimanite-forming reactions, but few examples have been reported of porphyroblasts growing from solid–solid reactions.) The modelling has shown that, for a reaction with a high negative enthalpy, nucleation will in general be suppressed after the reaction has gone to only a fraction of completion. The growth of porphyroblasts, in the sense of grains much coarser than average for the rock, will be favored by a suppression of nucleation, although it is not a necessary consequence of it. The consequence is rather that if conditions are such as to give a low ratio of nucleation rate to growth rate, then this suppression of nucleation will give a final texture in which there will be no small grains of a phase that appears as a porphyroblast.

This prediction is in contrast with that suggested for a low enthalpy reaction. In such a reaction, nucleation would continue until reaction completion. Even if the nucleation rate was relatively low there should still be a large number of small grains. The new phase would be unlikely to be recognized as a porphyroblast phase.

The factors found to promote a coarse grain size in a devolatilization reaction are the same as those suggested by Spry (1969, p. 138) as favoring porphyroblast growth: a high interfacial energy, a relatively high growth rate, and a relatively slow rate of heat input.

Interrelations Between Rate Controls in Metamorphism

Fisher (1978) examined the relationships between the various rate-controlling mechanisms in mineral growth. He suggested that, in general, a progressively growing mineral will pass through stages of interface, diffusion, and finally heat-flow controlled growth. The present study suggests superficially similar progressive histories. The relations between heat-flow and the other rate controls are, however, more involved and the two cannot be treated independently.

The late, flat portion of the T–t path shown in Fig. 1 is the result of an "equilibrium" being reached between reduced reaction rates with decreased reaction affinity as the equilibrium temperature is approached, and heat input. The exact reaction rate at this stage of the predicted history is a function of both the heat input and the reaction affinity–reaction rate relationship. In a one-dimensional model, this isothermal, "flat" portion of the T–t history would in many instances not be reached. Reaction would take place entirely within a regime in which reaction progress would control the rock temperature rather than the converse. This is, but for different reasons, the conclusion given by Fisher (1978).

Possible Restrictions on the Applicability of the Modelling

The qualitative form of the reaction history as shown in Fig. 1 is a result of certain direct predictions of the nucleation and growth rate equations, especially that reaction rates should increase with time. It is necessary to discuss whether there are any situations in which the assumptions behind these equations may not be valid.

The equations have been developed for metallurgical systems where there is in general only one reactant phase, and this forms the whole of the reacting system. Reactions in geological systems involve, in contrast, only a proportion of the rock, and both reactants and products may be dispersed. It is possible that interface kinetics or diffusion rates associated with the reactant phase control the reaction rate, in which case reaction rates will be constant or decrease with time. Such a rate control has been suggested by Carmichael (1979) and Aagaard and Helgeson (1982).

Whether, in any system, rates are controlled by processes associated with the products or reactants will be in part a function of the distribution of grains of the various phases present. The overall rate of diffusion from reactants to products will be controlled by diffusion to or from the phase with the largest average spacing. This will be, at least initially, in most reactions, a product phase. It seems reasonable, therefore, that reaction rates will accelerate at least over the first fraction of reaction progress, and hence it is unlikely that rate control by the reactants will change the basic form of the reaction histories proposed.

Reaction rates may also be significantly reduced if nuclei occupy all available sites before reaction has gone to completion (site saturation, Cahn, 1956; Rubie and Thompson, this volume). A restriction of nucleation to grain corners is most likely to lead to site saturation. A rock with an average grain size of 100 μm will have $\simeq 10^{12}$ grain corners per cubic meter, several orders of magnitude more than the predicted number of nuclei in any model. It seems reasonable therefore to ignore the possible effects of site saturation if it can be assumed that all grain corners are effectively identical. Yardley

(1977), however, suggested that nucleation may in some circumstances be restricted to very specific sites in a multiphase rock—certain two- or three-phase contacts that are activated by a relatively small overstepping of the reaction boundary. In such a situation, effective site saturation could be important. Whether the temperature will fall after such site saturation will depend on the exact density of possible sites and the growth rate at each site. Even if sites are at 1 cm intervals the temperature would still fall unless bulk diffusion rates were low ($<10^{-16}$ m^2 s^{-1}). Yardley further points out that an observation of a restriction of nucleation to specific sites may imply a flattening off or a lowering of temperature after the start of reaction. A further increase in temperature would allow nucleation on different types of sites.

Effects of One-Dimensional Heat Transport

The modelling discussed above has been for the thermal evolution of a single body of rock for which there is a constant externally imposed heat input. The temperature reductions predicted during reaction will therefore be maxima because no account has been taken of increased heat conduction into a reacting volume of rock. A more realistic treatment would involve the simultaneous modelling of the metamorphic and thermal evolution of a one-dimensional pile.

If reaction takes place within a restricted thickness of rock then conduction from the surroundings will be rapid enough to prevent a significant temperature drop. If reaction takes place in a thick pile of rocks, then the temperature difference between the two ends will be such that reaction is not synchronous throughout. The absorption of heat at one end of the pile may affect the thermal evolution at the other.

Figure 7 shows the ratio of the temperature drop predicted between the zero-dimensional calculations undertaken and a simple one-dimensional model allowing heat flow into a reacting column of rock. It is seen that the temperature in a 100 m thick reacting layer would fall by only 10–20% of the amount shown in Figs. 2–5. This would still be sufficient to give a suppression of nucleation in a high enthalpy reaction, though a greater volume of rock would have reacted before nucleation ceased.

Extension of Modelling to Multivariant Reactions

The discussion above has been based on a hypothetical discontinuous reaction. There are extra theoretical complications in a multivariant reaction. In such a reaction, the reaction affinity will be reduced as reaction proceeds. Certain general features can be suggested. If the lower temperature edge of a

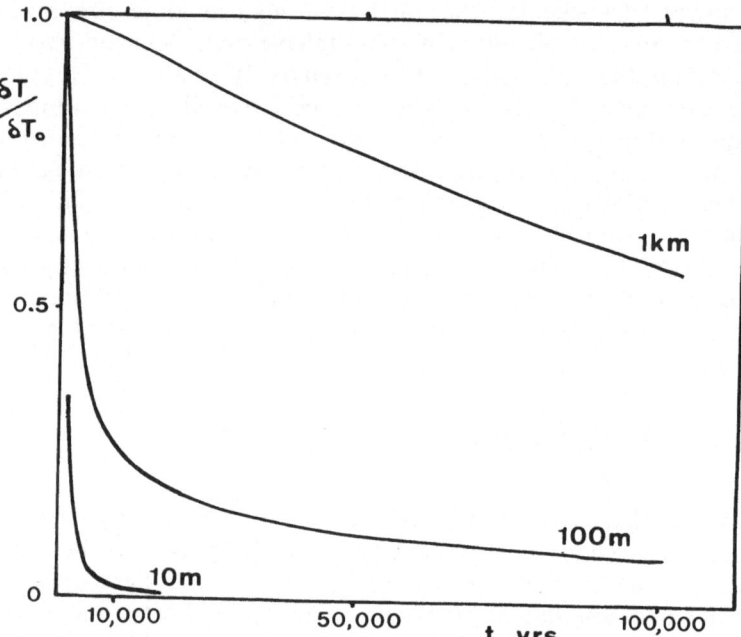

Fig. 7. Ratio of temperature fall during reaction in a one-dimensional model to that given by the zero-dimensional calculations described above, against length of time for the reaction. Curves are given for reacting piles of rock 10, 100, and 1000 m thick. The ratio is calculated from the analytical solution of the heat-conduction equation for a finite thickness heat source:

$$\frac{\delta T}{\delta T_0} = 1 - 4i^2\text{erfc}\left(\frac{l}{2\sqrt{(kt)}}\right)$$

(Carslaw and Jaeger, 1959, p. 80) where l is the thickness of the reacting pile, and k is the conductivity.

divariant field in $T-X$ space is overstepped, then once nucleation actually starts the first fraction of growth up to the then prevailing equilibrium proportions may be effectively univariant (Rubie, 1983). Behavior similar to that predicted by the modelling is therefore expected for the initial stages of, for instance, garnet growth over chlorite.

Mineral zonation in such a situation will be dependent on both growth and heating rates. "Reverse" zoning may form if the rock cools after nucleation. "Prograde" zoning may only start after the formation of an unzoned core growth if the temperature neither rose nor fell significantly during the initial, "univariant" period of growth.

Summary

Modelling of temperature–time histories of "prograde" metamorphic reactions suggests that, if there is a kinetic barrier to the nucleation of the product phases in a metamorphic reaction, and if this reaction involves a large positive reaction enthalpy, then in general the local rock temperature will fall after the onset of nucleation as a result of heat absorption in reaction. This fall in temperature will eventually be sufficient to suppress further nucleation, and over much of the reaction progress grain growth will take place without the concurrent formation of new nuclei. This qualitative history is predicted for any reaction with a large positive reaction enthalpy. It is only marginally affected by the exact growth rate and nucleation rate laws.

Different grain size distributions are predicted for high and low enthalpy reactions. The new phase produced in a high enthalpy reaction, assuming reaction has gone to completion, will show an approximately normal grain-size distribution and a definite minimum grain size. In a low enthalpy reaction there will be a dominance of small grains. The growth of a product as a porphyroblast phase is strongly favored, though is not inevitable, if the reaction involves a high enthalpy.

Although the analysis here has been for a nominally univariant reaction, there will be a similar effect in the history of a multivariant reaction because delayed nucleation through a P- or T-overstepping of the reaction boundary will give rise to an initial, effectively univariant period of reaction. Zonation patterns in the products of multivariant reactions may vary depending on the relative rates of growth and heating.

Acknowledgments

The work for this contribution was undertaken during the tenureship of a Royal Society European Programme Research Fellowship at ETH, Zürich. This and the help of the staff at ETH are gratefully acknowledged. I thank D. C. Rubie, A. B. Thompson, and B. W. D. Yardley for critical reviews of earlier versions of the manuscript, and the editors of the volume for painting the language in apposite shades.

Appendix: Computation of Temperature–Time Histories

The computation of the Temperature–time histories reported involves an explicit difference approximation to the zero-dimensional conservation of energy equation. The time step in each calculation is taken as small enough (generally 1000–5000 yr) so that a doubling of the interval does not change the predicted temperature at any point in the history by more than 1 K.

For each time step a nucleation rate is calculated from either the equation for homogeneous nucleation (1) (with scaled interfacial energies to model grain-corner or grain-edge nucleation), or that for nucleation along dislocations (see below). All nuclei formed within a single time step are assumed to grow at the same rate.

The volumetric growth rate for any grain is controlled both by the temperature and by the length of time since the grain first nucleated (Eqs. (2) and (3)). The total growth rate for each time step is therefore calculated by summing the products of growth rate and number of nuclei formed at each previous time step.

The rate of nucleation along dislocations is obtained by first calculating the parameter α_D, where

$$\alpha_D = \frac{\Delta G b^2 \mu}{2\pi^2 \gamma^2 (1 - \nu)} \qquad \text{(Cahn, 1957)}$$

(where ΔG is the free energy of reaction per unit volume, b is the magnitude of the Burgers vector, μ is the shear modulus, and ν is Poisson's ratio), and using the graphical relationship given by Cahn (1957) between α_D and the ratio between the free energy required for homogeneous nucleation and nucleation on dislocations.

References

Aagaard, P., and Helgeson, H. C. (1982) Thermodynamic and kinetic constraints of reaction rates among minerals and aqueous solutions. I. Theoretical considerations. *Amer. J. Sci.* **282**, 237–285.

Brace, W. F., and Walsh, J. B. (1962) Some direct measurements of the surface energy of quartz and orthoclase. *Amer. Mineral.* **47**, 1111–1122.

Brady, J. B. (1979) Intergranular diffusion in quartz–periclase reaction couples. *Carnegie Inst. Washington Yearbook* **78**, 577–581.

Brun, J. P., and Cobbold, P. R. (1980) Strain heating and thermal softening in continental shear zones: A review. *J. Struct. Geol.* **2**, 149–158.

Cahn, J. W. (1956) The kinetics of grain boundary nucleated reactions. *Acta Metall.* **4**, 449–459.

Cahn, J. W. (1957) Nucleation on dislocations. *Acta Metall.* **5**, 169–172.

Carlson, W. D., and Rosenfeld, J. L. (1981) Optical determination of topotactic aragonite–calcite growth kinetics: Metamorphic implications. *J. Geol.* **89**, 615–638.

Carmichael, D. M. (1979) Some implications of metamorphic reaction mechanisms for geothermobarometry based on solid–solution equilibria (abstract). *Geol. Soc. Amer. Abstr. Progs.* **11**, 398.

Carslaw, H. S., and Jaeger, J. C. (1959) *Conduction of Heat in Solids*. Clarendon Press, Oxford.

Christian, J. W. (1975) *The Theory of Transformations in Metals and Alloys, Part 1: Equilibrium and General Kinetic Theory*, 2nd ed. Pergamon Press, Oxford.

England, P. C., and Richardson, S. W. (1977) The influence of erosion upon the mineral facies of rocks from different metamorphic environments. *J. Geol. Soc. London* **134**, 201–214.

Fisher, G. W. (1978) Rate laws in metamorphism. *Geochim. Cosmochim. Acta* **42**, 1035–1050.

Freer, R. (1981) Diffusion in silicate minerals and glasses: A data digest guide to the literature. *Contr. Miner. Petrol.* **76**, 440–454.

Fyfe, W. S., Turner, F. J., and Verhoogen, J. (1958) Metamorphic reactions and metamorphic facies. *Geol. Soc. Amer. Mem.* **73**, 259 pp.

Jones, K. A., and Galwey, A. K. (1964) Study of possible factors concerning garnet formation in the rocks from Ardara, Co. Donegal. *Geol. Mag.* **101**, 76–93.

Jones, K. A., and Galwey, A. K. (1966) Size distribution, composition and growth kinetics of garnet crystals in some metamorphic rocks from the West of Ireland. *J. Geol. Soc. London* **12**, 29–44.

Kretz, R. (1966) Grain size distributions for certain metamorphic minerals in relation to nucleation and growth. *J. Geol.* **74**, 147–174.

Kretz, R. (1973) Kinetics of the crystallization of garnet at two localities near Yellowknife. *Can. Miner.* **12**, 1–20.

McLean, D. (1965) The science of metamorphism in metals, in *Controls of Metamorphism*, edited by W. S. Pitcher and G. W. Flinn, pp. 103–118. Oliver and Boyd, Edinburgh.

Nicolas, A., and Poirier, J. P. (1976) *Crystalline Plasticity and Solid State Flow in Metamorphic Rocks*. Wiley, London.

Porter, D. A., and Easterling, K. E. (1981) *Phase Relations in Metals and Alloys*. Van Nostrand Reinhold Co., New York.

Rice, J. M., and Ferry, J. M. (1982) Buffering, infiltration and the control of intensive variables during metamorphism. *Reviews in Mineralogy, 10, Characterization of Metamorphism through Mineral Equilibria*, Miner. Soc. America, 263–326.

Rubie, D. C. (1983) Reaction-enhanced ductility: The role of solid-solid invariant reactions in deformation of the crust and mantle. *Tectonophysics*, **96**, 331–352.

Russell, C. K. (1970) Nucleation in solids, in *Phase Transformations*, pp. 219–268. American Society for Metals, Metals Park, OH.

Spry, A. (1969) *Metamorphic Textures*. Pergamon Press, Oxford.

Vernon, R. (1968) Intergranular microstructures of high grade metamorphic rocks at Broken Hill, Australia. *J. Petrol.* **9**, 1–22.

Vernon, R. H. (1976) *Metamorphic Processes*. George Allen and Unwin Ltd., London.

Yardley, B. W. D. (1977) The nature and significance of the mechanism of sillimanite growth in the Connemara schists, Ireland. *Contr. Miner. Petrol.* **65**, 53–58.

Chapter 4
The Influence of Defect Crystallography on Some Properties of Orthosilicates

B. K. Smith

Introduction

Dislocations and other crystallographically controlled defects are now known to control both mechanical and geochemical behavior of minerals by a degree that is a function of the metamorphic and deformation history. Dislocation creep of some kind is undoubtedly the controlling mechanism for accommodating large plastic strains at sufficiently high temperatures and pressures (see Kirby, 1983, and references therein), and recent electron microscopy has led to an increased appreciation of the ability of defects to modify the chemical response of minerals to retrograde metamorphism. (Knipe and Wintsch, 1982). Despite this recent interest in dislocation mechanisms, a notable limitation of many of these applications is the lack of a realistic concept of the structure of a mineral dislocation and how the crystallography of mineral defects affects their transport, interaction, and chemical properties.

This paper describes several ways in which a complex dislocation structure may be modelled in a large unit cell, orthosilicate mineral (garnet), and then applies some similar arguments to the olivine–spinel system. Basically, the approach considers the dislocation core to be "extended" such that a perfect dislocation *dissociates* into partial dislocations joined by very thin ribbons of stacking fault. This changes the displacement per dislocation segment from a large, high-energy lattice translation into smaller, coupled translations on the subunit cell scale.

Physical Basis for Dislocation Dissociation

To a first approximation, dissociation reactions would appear to be likely in large unit cell silicates because the energy per unit length of a dislocation is a function of Burgers vector length; i.e., for an edge dislocation (Hirth and Lothe, 1968, p. 210):

$$E_e = \frac{\mu b^2}{4\pi(1 - \nu)} \left(\ln\left(\frac{R}{2\zeta}\right) \right) \qquad (1)$$

or, for a screw dislocation

$$E_s = E_e \cdot (1 - \nu) \qquad (1')$$

where E is the total strain energy of the dislocation per unit length, R is the effective elastic cutoff radius (generally taken as one-half the distance between two dislocations), ζ is one-half the interplanar spacing of the slip plane, b is the Burgers vector, μ the shear modulus, and ν Poisson's ratio. Although core energy (i.e., the nonelastic energy within the core) is not included in the formulas above, it is clear that a dissociation of b into partial dislocations ("partials") b_p such that the vector sum of b_p equals b and

$$\Sigma(b_p)^2 \leq b^2 \qquad (2)$$

will lower the energy per dislocation length. This is the so-called Franks criterion (Poirier, 1975), and one can note that any set of partials with b_p that are all mutually acute will satisfy this condition. This is even true for dissociations that force the partials out of the glide plane ("sessile" dissociation); furthermore, improvements on the Frank criterion can incorporate the increase in energy owing to increases in the edge character of the partials compared to the perfect dislocation (Fig. 1, see also Eq. (5), below).

In addition to the energy decrease provided by dissociation, the movement of an extended dislocation on a slip plane is more realistic on the atomic scale than motion of large perfect dislocations. Because the Burgers vectors in orthosilicates are in the range of 0.5–1.2 nm, it is energetically improbable that glide is accommodated by simultaneous rearrangement of one unit cell's worth of bonds (upwards of 50 bonds per displacement depending on slip plane width and orientation). From an atomistic viewpoint, the frequency of jumps of one species to the sublattice sites of nearly equivalent energies will be much higher than for jumps of one lattice dimension; because the velocity of a dislocation is a function of how rapidly the crystal can bring itself back into registry as the dislocation passes, those defects with the highest frequency jumps to near equivalent sites will have the highest velocities. Hence, for a dislocation density ρ caused by a nonhydrostatic stress, the Orowan formula

$$\dot{\varepsilon} = \upsilon b \rho \qquad (3)$$

shows that dissociation of dislocations with b into higher velocity partials

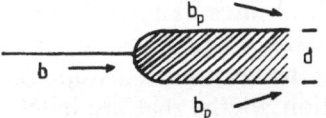

Fig. 1. Schematic dissociation reaction. Lined region is stacking fault; symbols from text.

leads to higher strain rates $\dot{\varepsilon}$ than would be possible for perfect, relatively low velocity dislocations. These arguments do not apply for immobile dislocations formed during growth or recovery; these latter defects have been observed to have very large b's and can be considered hollow (Frank, 1951; Van der Hoek *et al.*, 1982). I will limit my discussion here to those defects formed during deformation.

Because the displacement of a partial dislocation is a nonsymmetric translation of the unit cell, a plane of disregistry must exist between two parallel partials. Atoms displaced by the first partial are forced into sites that are either unoccupied or nonexistent in the undeformed crystal and are not returned to their normal position until another partial completes the lattice translation. Obviously, the greater the affinity of a particular atomic species for its site in a perfect lattice, the more difficult it becomes to move this species into another, less favorable site by a partial. The increase in energy resulting from this mismatch is defined as the stacking fault energy γ_o, which tends to limit the width by which partials can separate: While the elastic strain fields of two parallel partials of similar sign will force them apart, the increase in energy of atoms on unfavorable sites in the stacking fault tends to suppress this separation. Because there is no easy way (at present) to predict stacking fault energies from crystal energy calculations, stacking fault energies are usually derived by (e.g., Hirth and Lothe, 1968, p. 298):

$$\gamma_o = \frac{\mu}{2\pi d} (b_1 \cdot l_1)(b_2 \cdot l_2) + \frac{(b_1 \times l_1) \cdot (b_2 \times l_2)}{(1 - \nu)} \tag{4}$$

where d is the separation distance of the two repulsing partials, and l_i is the direction of the dislocation segments. Note that if this formula is used to calculate γ_o's from direct measurements of d (from TEM micrographs, for example), the energy required to move the partials apart through perfect crystal is neglected (Peierls forces, core drag, etc.). In any case, it is exceedingly difficult to measure d's for minerals, because they are usually in the range of 3.0–5.0 nm (Boland and Liu, 1983).

Lacking a good quantitative measure of γ_o, the best one can do is to consider the atomic mismatch across the stacking fault. This is easily justified by comparing minerals to metals. For closepacked fcc metals, a disruption of the stacking sequence can be introduced by insertion of an hcp layer; because the differences in the stacking energies for the two different sequences (ABCABC for fcc, ABABAB for hcp) are expected to be small, one could postulate that γ_o for fcc metals is also low (which, in fact, it is). Spinels, with a structure based on fcc stacking of oxygen sheets, may also contain stacking faults that retain the integrity of this structural element by altering the sequence of tetrahedral and octahedral sites. The mismatch caused by the cation sequences near the fault causes the stacking fault energy of spinel to be much higher than metals, but the very slight disruption of the oxygen substructure still permits dissociation widths that are much greater than in silicates (cf. Vaughan and Kohlstadt, 1981; Van der Biest and Thomas, 1974). For most silicate minerals, juxtaposition of various cation sites by a partial requires significant shifts in the oxygen topology, with an ensuing increase in γ_o. Even small structural modifications (such as the α–β transition in quartz) have demonstrably large mechanical effects during plastic deformation (Ross *et al.*, 1983), which again argues for large energy contributions from mismatch strains. Finally, the role of ordering in modifying site energies in silicates may introduce a substantial barrier for dissociation reactions that superimpose no longer equivalent sites.

Clearly, the simple application of the Frank criterion (Eq. (2), above) as an assessment of the proclivity for dissociation is not adequate unless it is combined with some measure of γ_o. Formally, a better criterion may be given as:

$$\hat{E}_{p,s+e} + \gamma_{o,i} \cdot d_i \leq \hat{E}_{u,s+e} \qquad (5)$$

where $\hat{E}_p, \hat{E}_u$ are the total energies of the screw and edge components of the partials and undissociated dislocation, respectively, and the i subscript refers to the stacking fault ribbons separating n pairs of partials.

Extended Dislocations in Garnet

Dissociated dislocations have been produced during experimental deformation of garnet, and TEM observations of these dislocations combined with those found in naturally deformed rocks have provided the best data on dissociation mechanics of any orthosilicate material. In this section I will present some models for the crystal structure of extended dislocations, followed by an overview of the types of dissociations that are observed for specific experimental deformation conditions or in natural samples.

Models

From the discussion above, the main control on the formation of extended dislocation segments in garnet is the minimization of the misfit of the faulted crystal layers. This is not necessarily a function of the crystal structure alone (thermodynamic and kinetic effects will be discussed at the end of this section), but the coherence of the fault with unfaulted structure is certainly a dominant factor in lowering γ_o. These low mismatch structures were modelled using two criteria. First, a specific structural element of the unit cell was chosen such that partial dislocations either displace it to an equivalent position without breaking any bonds or else displace this element by the smallest possible shift. In practice, this entails choosing the densest sublattice of the unit cell; in garnet, this is defined by the octahedral sites, because the repeat distance a_Y is shorter than that of the other cation polyhedra (Fig. 2). The second criterion was introduced to measure the strength of the bonds that must be broken across the fault. Initially, polyhedral compressibilities (Table 1) were used as a rough indicator of which polyhedra were easiest to distort or break. These data show that the SiO_4 tetrahedra are by far the strongest subunit, followed by the octahedra and dodecahedra in order of decreasing strength. Ideally, bond strength calculations could also be performed to measure this effect for specific fault models (Lasaga, 1980; Lasaga and Cygan, 1982); although this approach is somewhat subjective, it gives results for olivine that are reasonably consistent with the polyhedral compressibilities.

Rabier et al. (1976) proposed a series of dissociation reactions that satisfy the first criterion above, i.e., they leave the octahedral sublattice unaltered, even though the tetrahedral (or Z) site is apparently more difficult to break. For dislocations in bcc lattices, the shortest perfect Burgers vectors are $a_o/2\langle 111 \rangle$ and $a_o\langle 100 \rangle$. For their dissociation (Fig. 3), some possible reactions are:

Order Faults (OF): $a_o\langle 001 \rangle$ $a_o/2\langle 001 \rangle + a_o/2\langle 001 \rangle$
Electrostatic Faults (EF): $a_o/2\langle 111 \rangle$ $a_o/4\langle 111 \rangle + a_o/4\langle 111 \rangle$
                              $a_o\langle 001 \rangle$ $a_o/4\langle 111 \rangle + a_o/4\langle \bar{1}\bar{1}1 \rangle$

Note that reactions with perpendicular partials (e.g., [001] and [110]) are precluded by their lack of any mutual repulsion (Eq. (4)). Self energies for all of these dislocations are listed on Table 2.

The OF and EF dissociation schemes above effect the dodecahedral (X) and Z sites in markedly different ways. In Fig. 2, the X and Z sites are seen to alternate along the $\langle 100 \rangle$ directions. For the OF partials, this X–Z alternation is disrupted, but no new sites are formed (hence these are Order Faults, or OF). The EF partials necessitate both disruption of the sequence and occupancy of new sites by X and Z species. This is most easily seen by viewing down [111] (Fig. 3(b)), where the $a_o/4\langle 111 \rangle$ partials can be related to

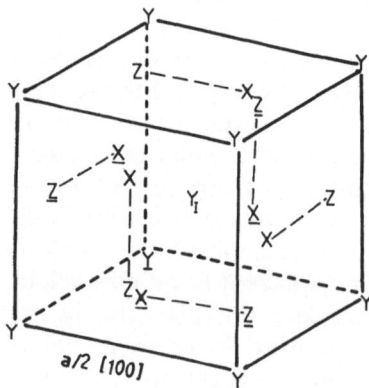

Fig. 2. An octant of the garnet unit cell, defined by Y sites on the corners of the cube. Underlined X and Z sites are on the hidden faces, and Y_I is at the center of the cell at (1/4,1/4,1/4).

Table 1. Polyhedral compressibilities.

Mineral	Site	(GPa)
Garnet	X	115–130 ($\pm$12.5)
	Y	220 ($\pm$50)
	Z	300 ($\pm$100)
Olivine	M (Fo)	150
	M (Fa)	180
	T	250

From Hazen and Finger (1976, 1978).

the undeformed structure by rotations of X and Z to new sites. These small shifts involve occupation of new crystallographic sites that are nearer to the undeformed sites than any X or Z site in the perfect crystal, and one would expect that the electrostatic repulsions of the faulted nearest neighbors would produce relatively high γ_{EF}.

For synthetic garnets (with X, Y, Z all trivalent), Rabier *et al.* (1976) postulated that dissociations of the OF type could be more stable (lower γ_{OF}), because they only require disruption of the X–Z order without the electrostatic repulsions of the EF faults. For silicate garnets, the presence of highly different charge densities on the X and Z sites does not lend itself to this argument. From Fig. 3(a), the now-joined tetrahedra must share edges previously shared with dodecahedra, which is a highly energetic state. Alternatively, the EF faults require that the small shifts about $\langle 111 \rangle$ bring highly charged Z sites within close proximity. However, the integrity of the tetrahedra are preserved for EF faults, which satisfies the second criterion established for these models. In conclusion, neither the OF nor the EF dissociations can be selected as more probable on the basis of the models.

Observation of Extended Dislocations

Transmission electron microscopy (TEM) of deformed garnets has provided some answers about the dissociation reactions that are unobtainable from the crystallographic models alone. Although interpretation of stacking fault and dislocation contrast of garnet defects suffers from the same problems that plague observations of most large unit cell silicates, several observations of extended dislocations have now been made for this mineral (Rabier *et al.*, 1981; Smith, 1982). These results offer convincing if not demonstrative evidence that EF faults are much more prevalent than OF faults.

For dislocations that are not sufficiently extended for observation of the stacking fault structure, the predominance of $a_o/2\langle 111 \rangle\{011\}$ slip indicates that $a_o/2\langle 111 \rangle$ dislocations are significantly less energetic than $a_o\langle 001 \rangle$ dislocations. Even if some component of this energy difference can be attributed

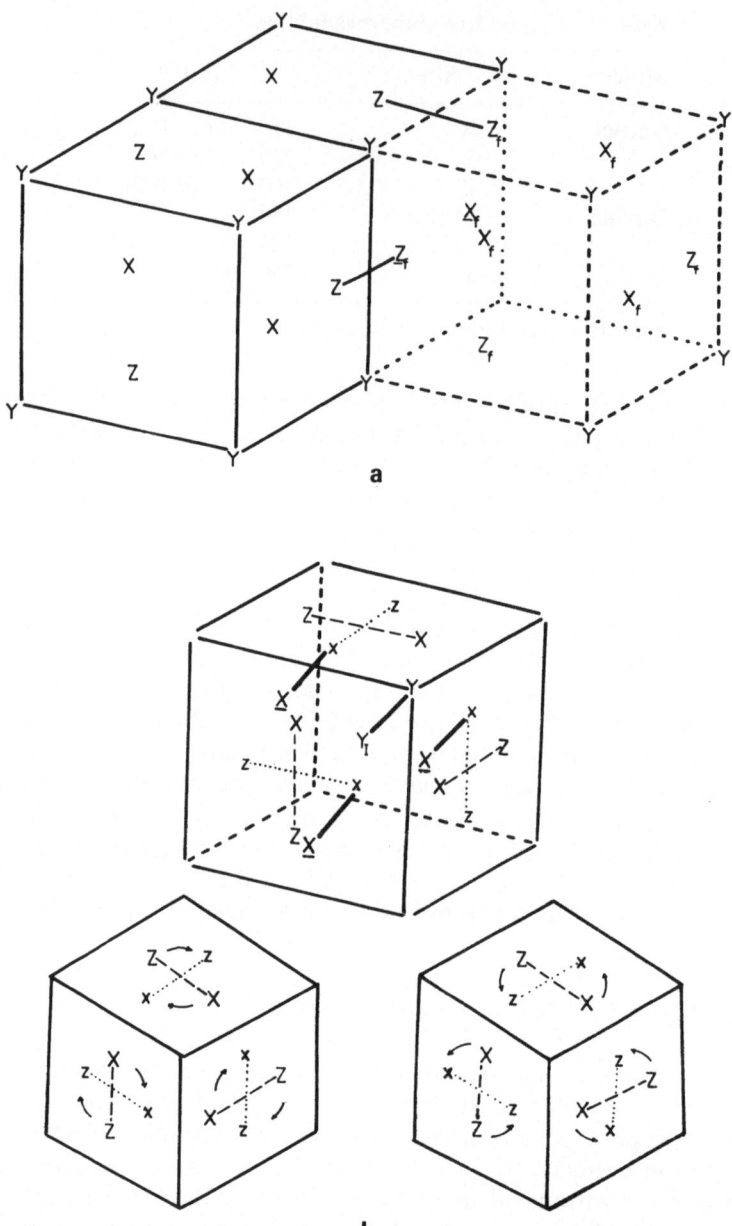

Fig. 3. (a) Structure of an OF partial ($a/2\langle001\rangle$). Faulted sites are labeled with f, and disrupt the Z–X alternation. (b) Crystal structure of an EF partial with $b = a/4\langle111\rangle$. This translation brings Y_I to Y on the corner and requires that new sites be formed on the cell faces (denoted by lower case x and z). Depending on the sense of the partial, the two different configurations shown on the bottom are possible for these new sites.

Table 2. Garnet dislocation self-energies.

Dislocation type	$\|b\|$ (nm)	E_s (joules $\times 10^{-8}$)	E_e (joules $\times 10^{-8}$)
Perfect:			
$a/2\langle111\rangle(011)$	0.993	4.123	5.648
$a\langle100\rangle(011)$	1.147	5.501	7.535
Partials:			
$a/4\langle111\rangle(011)$	0.497	1.033	1.415
$a/2\langle100\rangle(011)$	0.574	1.378	1.887
$a/2\langle011\rangle(011)$	0.822	2.825	3.870
OF $(= 2(a/4\langle100\rangle))$		2.756	
EF $(= 2(a/4\langle111\rangle))$		2.066	

Notes: $\nu = 0.270$; $\mu = 92.0$ GPa; R $= 250.0$ nm; $\xi = 0.822$ nm.

to the relative differences in self energies (which are not particularly large, cf. Table 2), extension of the $a/2\langle111\rangle$ dislocation cores might also contribute to the stability of these defects. In this case, EF faults would have lower stacking fault energies than OF's, because the latter can only stabilize $a_o\langle001\rangle$ dislocations. Note that if EF faults are present, γ_{EF} must still be very high, because there is no observable separation of partial dislocations on glide planes for specimens deformed in this temperature range (homologous temperature T/Tm 0.5–0.7).

Stacking fault fringes have been imaged in two different populations of garnet: (1) silicate garnets that were deformed under ostensibly hydrous conditions (but compare Kirby and Kronenberg, 1984), and (2) synthetic garnets deformed at high homologous temperatures (0.85–0.95) (Rabier, 1983, personal communication). For the synthetic garnets, extended segments separated by 10–50 nm have also been observed using weak beam dark field TEM techniques (Rabier *et al.*, 1981), but the high temperatures and sessile properties of their experiments do not seem especially relevant to the defect microstructures I have observed in these silicate garnets. However, it is important that for the stacking faults in both synthetic and silicate garnets, EF type dissociations best explain the contrast and crystallographic orientations of these defects.

For the example shown in Fig. 4(b), the imaging conditions precluded a rigorous solution of the displacement vectors of the fault; however, both the asymmetry of the stacking fault fringes and the alignment of the fringes on several *different* {110} slip planes are indicative of a glissile $a/4\langle111\rangle\{011\}$ fault. The curved nature of some of the fringes may be due to some component of climb in the EF partials but could also be caused by roughness of the specimen surface (note the discontinuous nature of the Bragg contours across the micrograph). The dissociation widths of several 100 nm are unu-

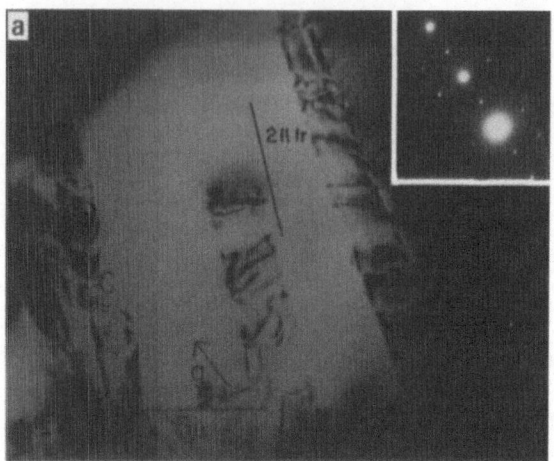

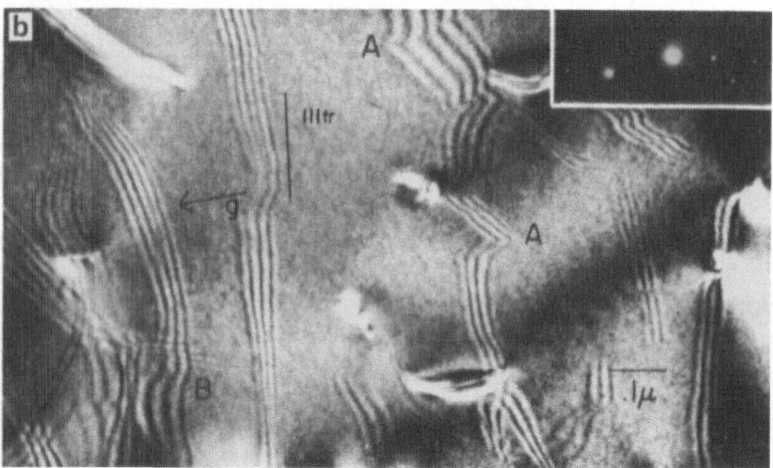

Fig. 4. Transmission electron micrographs of deformed garnet. (a) Low-temperature deformation with dislocations restricted to (211) glide planes and no obvious dissociation of the cores. Note residual strain contrast on the slip surface. (b) Widely dissociated segments connected by stacking fault ribbons in a hydrous, high-temperature deformation experiment.

sual and rarely observed, but they do point out the possibility of drastically reduced γ_{EF} in hydrous environments.

The γ_{EF} lowering may be best explained by the location of hydroxyl groups in the garnet structure. By analogy to the hydrogrossular structure (Meagher, 1980, and references therein), the location of hydrogen in garnet is restricted to the Z (tetrahedral) sites. This is equivalent to the replacement of H_4O_4 for SiO_4, or in Kroeger notation (Lasaga, 1981),

$$H_Z^{\cdot} = 4(Si_Z^{\cdots\cdots})$$

In other words, the number of positively charged (H˙) species is equal to four times the number of Si on the same (Z) site (with four positive charges, or Si$\cdots\cdot$). By replacing a rigid SiO_4 tetrahedra group in the fault region, water could promote dissociation by either destroying or distorting the Z site. The electrostatic repulsions across an extended EF defect may also be lowered by H_Z, as individual H or OH point defects could be removed from the nearest neighbor sites with a smaller residual charge imbalance than removal of a Si interstitial. For example, removal of a positive $H_Z^{\cdot}$ could be accommodated by oxidation of iron, such that,

$$H_Z^{\cdot} = 3(Si_Z^{\cdots\cdot}) + Fe_{X,Fe}^{\cdot}$$

with residual charge imbalance between the X and Z sites. Such locally charged defects can be expected in garnet as in other low conductance solids (Hobbs, 1981; Stocker and Smyth, 1978).

Although water on Z sites provides a crystallographic means for lowering γ_o, it does not explain why EF defects are much more favorable than OF. The predominance of EF faults may therefore be due to the ability of these faults to glide or cross-slip more efficiently than the OF defects: The small X–Z shifts combined with a higher cation site density on the EF fault plane may facilitate glide if this motion occurs as small atomic jumps into Peierls minima (Hirth and Lothe, 1968, p. 331). For OF faults, the number of cation sites on the fault is not changed (only their sequence), and hence there is little or no damping of the barriers to glide by an OF-extended defect.

Dissociated Dislocations and Metamorphic Reactions

Suzuki Effects

The preceding model for H concentrations on EF fault surfaces is a special case of a more general phenomenon of enhanced solubility on stacking faults (the Suzuki effect). If the stacking fault energy is again taken as a function of atomic or lattice mismatch, it is clear that diffusion of species that alleviate some of this misfit strain will be energetically favored. Conversely, diffusion of material that aids in relieving the stacking fault energy away from the fault could raise γ_o, such that the extended segments either contract or increase their intrinsic strain energy by bowing, cross-slipping, etc. This process can thus be stated as:

$$-\frac{(\partial\gamma_o)}{(\partial\mu_i)} = C_i - \frac{(x_i)}{(x_c)}C_c \tag{6}$$

where μ_i is the chemical potential of the ith diffusing species, C_i and C_c are the concentrations of the ith species and the bulk crystal within the fault volume, and x_i and x_c are the molar proportions of these components in the unfaulted crystal (Suzuki, 1957). In general, the ith species can be defined as any component that has an equilibrium concentration in unfaulted crystal (e.g., point defects, vacancies, and atoms on different sites).

The lower value of γ_o attendant with a flux of the ith component to the fault (increased C_i) can be accompanied by an increased dissociation width d (from Eq. (3); Haasen, 1979, p. 162). However, for moving dislocations, this increased width can only be maintained if the zone of nonstoichiometry (the "stacking fault atmosphere") diffuses at a rate equivalent to the movement of the partials (Fig. 5). If, on the other hand, the atmosphere diffuses more slowly, the excess concentrations responsible for γ_o-lowering must be accommodated in the perfect crystal structure behind the trailing partial. The segregation of i in unfaulted crystal results in an increase of energy:

$$dG_{f \to s} = \sum_{X_c}^{X_s} (\mu_{i,c} - \mu_c)\, dX_i \qquad (7)$$

where the subscripts f, s, c refer to fault, segregated, and perfect crystal, respectively. Setting this energy term equal to a stacking fault energy times a change in dissociation width,

$$dG_{f \to s} = \gamma_s\, d(d) \qquad (8)$$

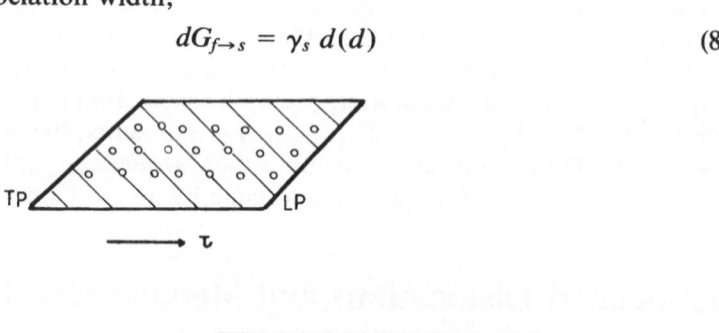

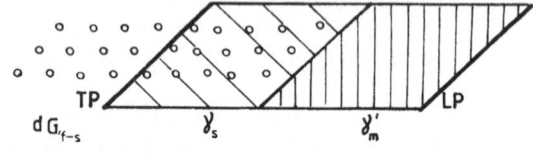

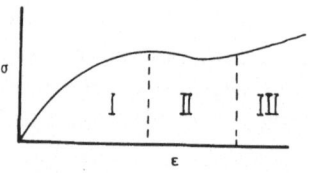

Fig. 5. Schematic representation of Suzuki locking. "o" are species that were once on the fault but that must be incorporated into perfect crystal if their diffusion cannot keep up with the movement of the partials. LP and TP are the leading and trailing partials, respectively, and energy terms below lower figure correspond to the forces that either impede or aid motion. Inset of stress/strain behavior correlative to this mechanism is from an almandine experiment with $\dot{\varepsilon} = 10^{-5}$/s, $T = 825°$ C, $P = 15$ kbar.

with γ_s the fault energy for segregation, then

$$\gamma'_m = \gamma_s - \gamma_o \tag{9}$$

with γ'_m the resulting stacking fault energy for a mobile fault in front of a segregated region (Hirth and Lothe, 1968, p. 631). Depending on the compatibility of the ith species in the bulk crystal, this term can either impede or enhance dissociation. For motion under a resolved stress τ, the driving force for dislocation movement

$$F_m = \tau b_p \tag{10}$$

must overcome the energy resulting from segregation of the atmosphere behind the trailing partial; if it cannot, this dislocation is held in place while the leading partial passes into crystal that lacks an atmosphere until a separation given by γ'_m is obtained. This "Suzuki locking" behavior qualitatively explains the wide segments of stacking fault shown in Fig. 4(b): With the relatively high strain rates and stresses used in this experiment ($\dot{\varepsilon} = 10^{-5}/$ sec, $\tau = 150$ MPa, $T/Tm = .80$), it is plausible that the atmosphere of H produces a drag on the trailing partial if it cannot diffuse at a rate sufficient to keep up with the lead partial.

At first glance, this mechanism apparently contradicts the arguments of the previous section that hydrogen (or "water") is instrumental in stabilizing EF faults through dampening electrostatic repulsions and other barriers to glide: this locking process requires that the leading partials pass through essentially unaltered garnet structure. However, the mechanical behavior of this specimen is consistent with both EF stabilization by H_Z and Suzuki locking (Fig. 5; also Smith and Wenk, in preparation). As the yield stress is approached, hydrogen diffuses to the nucleating dislocations and stabilizes EF defects that can then glide more efficiently. This results in the stress drop after yield in region II. As these faults continue to move, the segregation of the H cloud behind the trailing partial leads to strain hardening (region III), because the formation and motion of EF faults in crystal behind the leading partial is more difficult than in the "wet" regions (Fig. 5). At constant strain rate, new ribbons of stacking fault with $\gamma'_{EF,m} > \gamma_{EF,m}$ can only be formed by increasing the resolved shear stress, which causes the observed hardening. Similar models have been proposed for the same kind of behavior with nonextended dislocations in quartz (Griggs, 1974), but in the case of garnet it is more likely that dissociation exerts the fundamental control on the strain hardening behavior as opposed to dislocation core effects (cf. Haasen, 1979).

Retrograde Metamorphism in Naturally Deformed Garnets

A combination of both γ_o-lowering and Suzuki locking also explains the orientation and location of chlorite replacement bands within highly deformed almandines from amphibolites of the North Cascade Central Schist

Belt (Smith, unpublished observations). A schematic diagram of the progress of this reaction is shown in Fig. 7, which accounts for the ubiety of this replacement, as follows: (1) Chlorite is always found in association with slip bands (or occasionally at polygonized boundaries) within the deformed garnets (Fig. 6(a)). (2) The 001 planes of chlorite are parallel to the 110 form in garnet, which is garnet's preferred slip plane. This form has the highest density of octahedral sites in both chlorite and garnet, but the conformation of these sites in their *undeformed* crystal structures is quite different.

I believe that the growth of chlorite proceeds in the following way. First, passage of an EF defect during garnet deformation is accompanied by a flux of H_Z' (and probably other species) to the slip plane, by analogy to region I in

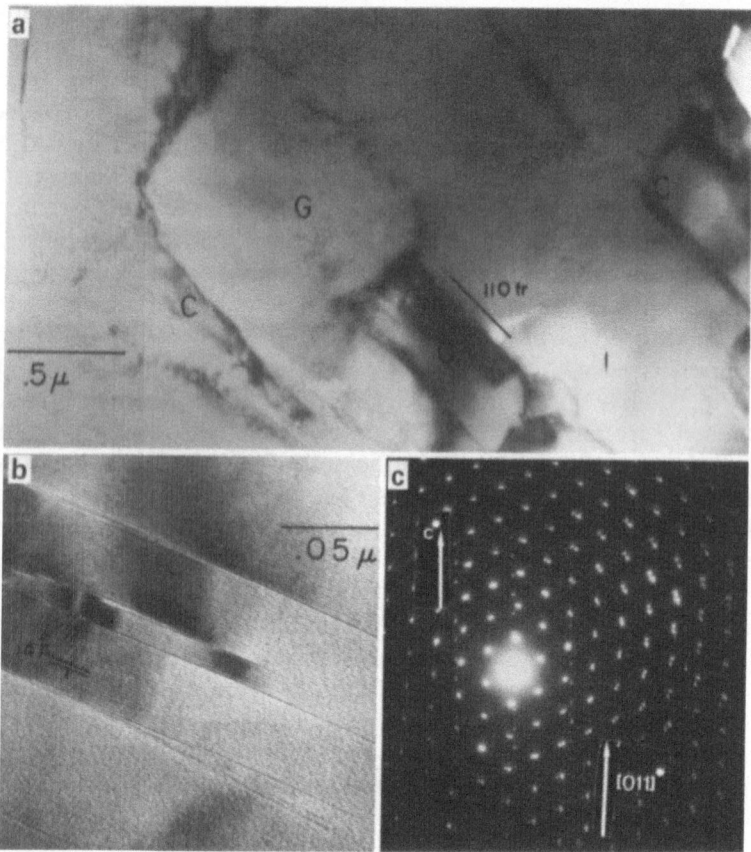

Fig. 6. Electron micrographs of the chlorite replacement bands in naturally deformed garnets. (a) Bands of chlorite (C) localized on the slip planes of the garnet (G). (b) enlargement of the chlorite bands; note that they are not especially deformed and are probably post-deformation growth. (c) SAD of the garnet–chlorite crystallographic relationship ($[110]_g$ parallel to c_{ch}^*). $[111]_g$ orientation.

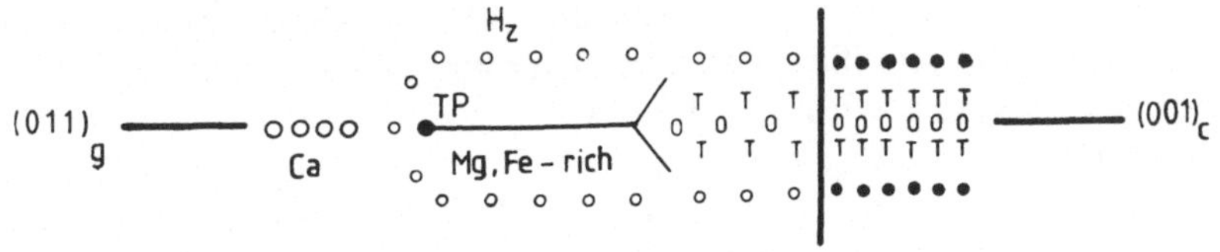

Fig. 7. Cartoon of the proposed chlorite replacement of garnet (compare with Fig. 5). Segregates of grossular rich components (Ca) behind a once-mobile stacking fault leaves a zone of Mg,Fe-enriched garnet near the fault. The combination of water segregation near the fault (small circles, H_Z) plus the increase in structure compatibilities (shown on right) provides the best site for chlorite nucleation on (011). Filled circles in chlorite correspond to the brucite layer, T and O are tetrahedral and octahedral sites (Z and Y in previous garnet figures).

the stress/strain curve in Fig. 5. As deformation proceeds, this is probably accompanied by Suzuki locking, such that segregation of the higher activity coefficient species occurs behind the trailing partials on the slip surface. In addition to the production of "water"-rich zones, it is likely that segregates could contain relatively high grossular (or Ca_I) concentrations, due to its large deviations from ideal behavior in almandines with the compositions found in this suite (pyralspite 80%, ugrandite 20%) (Cressey et al., 1978). This grossular segregation thus leaves relatively enriched Mg, Fe^{VIII} $a_o/4\langle 111 \rangle$ EF segments lying on $(01\bar{1})$, with the possibility of higher "water" activities within this region as well. By

$$(Mg,Fe)_3^{VIII}(Al)_2^{VI}Si_3^{IV}O_{12} + 8H_Z^{\cdot} + 2(Mg,Fe)_X^{x} + 6O_i''$$
$$= (Mg,Fe)_5^{VI}(Al)^{VI}(AlSi_3)^{IV}O_{10}(OH)_8 + 12V_O \quad (11)$$

where the 12 oxygen vacancies are in the *garnet* structure, hence

$$\Delta G = \Delta G^0 - RT \ln \frac{(a_{V_O})^{12}a_{ch}}{(a_{H_Z^{\cdot}})^8(a_{Mg,Fe_X^{\cdot\cdot}})^2(a_{O_i''})^6a_g} \quad (12)$$

assuming neutral oxygen vacancies and neglecting the probable speciation of the hydrogen with the oxygen interstitials. Although this reaction is not especially exact (e.g., all transition states of Fe are neglected, and there is assumed ideal mixing of cations on the various sites between garnet and chlorite), it does indicate that both increases in X site Mg or Fe (by removal of Ca) or displacement of an oxygen to an interstitial position promote this replacement. Furthermore, the vacancy producing steps of this reaction, viz.,

$$Al_Y = Al_Z + 2V_O$$

are structurally assisted by EF defects. From Fig. 3, the displaced halves of the garnet structure across the fault can juxtapose tetrahedral sites on either side of the octahedral layer (which lies in the fault), thereby introducing a distorted tetrahedral–octahedral–tetrahedral sequence normal to {110}. While there are still significant atomic rearrangements required to bring this sequence into registry with the sheet structure of chlorite, the presence of this sequence in faulted garnet certainly lowers the activation energy for nucleation compared to unfaulted garnet (which does not possess any analogous structure). In addition to the chemical drive, this reaction is certainly assisted kinetically, because dampening of distortion may also be provided by polytypism of the chlorite layers (Spinnler *et al.*, 1984) and by pipe diffusion of necessary structural components along the nearby partials (Yund *et al.*, 1981). It should be stressed that the growth of chlorite by a dissociated garnet defect mechanism is attributable to neither a topotactic nor martensitic mechanism (the disregistry and stoichiometric differences for these two phases are too large), but instead it demonstrates that extended defects in garnet provide energetically favorable ("activated") sites for chlorite growth in agreement with the TEM results.

Finally, it will be noted that the model described above does not include any explicit contribution from the stored elastic strain energy of the extended dislocations. Calculations of this energy effect for various dislocation arrays (Table 3) show that this "free energy drive" is rather trivial, especially in view of the very high free energies of formation (and corresponding large uncertainties) for most sheet silicates (Helgeson *et al.*, 1978). While some attempts to incorporate strain energy into thermodynamic formulations have been successful for pure phases (Holder and Granato, 1969), a more comprehensive assessment of both mechanical and chemical responses to deformation would seem to be essential in any model of strain-induced metamorphism of minerals.

Application of Dissociation Reactions to the Olivine–Spinel System

The large [100] Burgers vector of olivine led Poirier (1975; Poirier and Vergobbi, 1978) to propose a dissociation scheme based on small shifts of the oxygen sublattice. Observations of extended dislocations have been made with the TEM (Vander Sande and Kohlstedt, 1976; Zeuch and Green, 1978) but so far confirmation of any dissociation reaction has not been possible. In the high-pressure polymorphs of olivine (β and γ spinel), stacking fault structures are now well known (Madon and Poirier, 1983; Boland and Liu, 1983). While it is clear that dissociation reactions in olivine generate stacking fault energies that must be very high, some insights into possible roles for extended dislocations in assisting (or causing) the olivine-to-spinel transformation can be provided by comparing their defects to those in garnet.

Spinel stacking faults display several features that were presumed to be

Table 3. Excess free energy from defects in garnet.

Dislocation b	Type	ρ (cm^{-2})	ΔG/mol, STP (joules)	ΔG/mol, 300°C, 300 MPa (joules)
$a/2\langle 111 \rangle$	screw	1.4E9	53.5	53.5
	screw	3.5E10	873.6	882.0
	edge	1.4E9	73.3	73.1
	edge	3.5E10	1196.9	1364.3
$a\langle 100 \rangle$	screw	1.6E9	77.2	77.2
	screw	4.0E10	1224.8	1229.8
	edge	1.6E9	105.8	105.8
	edge	4.0E10	1677.8	1678.0

important both mechanically and chemically in garnet. In germanate spinel, the TEM analysis of Vaughan and Kohlstedt (1981) demonstrated that (1) excess concentrations of a specific species are localized within the fault, even though the crystal is very nearly stoichiometric, and (2) there is a strong stress dependence on dissociation width. For the last point, it was observed that wide stacking fault ribbons are caused only during stress relaxation experiments, whereas quenched specimens contained dislocations that were only slightly extended. Again, in order to stabilize these stacking faults, the ability of some species to diffuse either to or away from the fault ribbon appears to control the effective γ_o. This chemical drive may in fact be stronger than in garnet, because the spinel stacking faults can also be considered as antiphase boundaries with an attendant decrease in energy as the ordered domains coarsen.

The structure of extended dislocations in olivine is not well understood. The criteria applied by Poirier (1975) that SiO$_4$ tetrahedra retain their integrity for his dissociation reactions may be overcome in a manner similar to garnet: H$_Z$ replacement for Si$_Z$ was recently proposed from infrared spectroscopy of a natural hydrous olivine (Beran and Putnis, 1983). The common occurrence of CO$_2$-rich micro-inclusions along recovered dislocations in peridotites (Green and Radcliffe, 1975) may also be indicative that a carbon phase was localized near an extended core, if the bubbles are related to segregated zones behind locked segments behind extended dislocations. Lacking a good idea of the crystallography of olivine defects, one is limited in assessing which carbon or hydrogen species may promote dissociation, but the rapid diffusibility of these elements together with the large number of possible crystallographic configurations that they may have (Freund, 1981) could account in part for the observed oxygen fugacity dependence on creep.

Because olivine dissociation appears to be energetically unfavorable and no mechanism for lowering of these high γ_o has been experimentally observed with the introduction of volatile species, I think it unlikely that olivine can transform to spinel by a mechanical process alone (i.e., martensitic transformation). Alternatively, the preferred orientation of nuclei of spinel formed during this transformation under nonhydrostatic stress (cf. Hamaya and Akimoto, 1982; Vaughan *et al.*, 1982) may be attributed to the localization of chemical (and hence site energy) fluctuations on spinel stacking faults. If the spinel (or a transitional phase with similar stacking fault properties) is deformed by dislocation glide or creep synchronously with the transformation, those grains oriented such that the stacking faults are in a nearly epitaxial alignment with the olivine could grow preferentially as the misfitting olivine sites are incorporated as excess concentrations on the spinel stacking fault. A critical parameter here is the size of the newly formed spinel grains: if the grain size is sufficiently small, deformation by grain boundary processes will undoubtedly predominate, and high stacking fault densities on $\{011\}_{sp}$ (or possibly $\{111\}_{sp}$) will not be available to provide energetically favorable growth sites. This is clearly the case in the germanate spinels (Vaughan and Coe, 1981), where the small grain sizes of the spinel phase activates a superplastic deformation mechanism with no preferred orientation of the spinel; the relevance of experimental studies of germanate spinels as analogues of silicate spinel behavior have been recently questioned, however (Weidner and Hamaya, 1983). If the spinel crystals are sufficiently large to deform by dislocation processes, the stacking fault mechanism described above would still require diffusion across an olivine–spinel interface, but for the nearly epitaxial growth of these two phases the activation energy for this diffusion is much lower than that required for nucleation and growth of randomly oriented phases (Turnbull, 1956).

Conclusions

Very thin ribbons of stacking fault between partial dislocations are a more crystallographically and mechanically reasonable defect structure in large unit cell silicates than perfect, nonextended dislocations. For garnet, this defect structure explains both the type of defects that are observed and the increased separation of partials in ostensibly "wet" crystals. The geometry of the faulted structures and their ability to cause planar chemical anomalies also provides a conceptually simple framework for describing metamorphic replacement reactions in terms of site activities. This is a significant improvement over models that consider the static strain energy of perfect dislocation arrays as the sole driving force for retrogression.

More quantitative applications of the extended dislocation models will require better estimates of (1) the segregation and activities of solute species near the faults, and (2) stacking fault energies themselves. The chemical parameters are becoming available from published calorimetric investigations and the advent of TEM chemical analyses of very small volumes (with high-angle EDX and STEM optics). Estimates of the magnitudes of γ_o may be more difficult to obtain, but lacking better TEM data on dissociation widths, a possible approach may be calculation of the energy of the fault structure from atomic potentials. Such calculations have been demonstrated to predict within reasonable error the physical properties of the Mg olivine spinel system (Price and Parker, 1984; Miyamoto and Takeda, 1983), and calculations are underway for faulted crystal structures.

Finally, a better understanding of the crystal structure of defects in minerals must be forthcoming if we are to incorporate dislocations effectively into geological models. For the garnet chlorite reaction described here, it is clear that simple assessments of the stored strain energy within a perfect dislocation does not suffice to describe this reaction; similar difficulties can be expected for applications of mechanical models based on simple defect morphologies in metals to the complicated dislocation structures of deformed minerals. There remains an urgent need to correlate deformation microstructures on the unit cell scale to the structures found within thin sections, hand specimens, or in the field. Because there is no escaping the strong role exerted by dislocations during the formation of large-scale phenomena (retrograde metamorphism, fabric development, etc.), more exacting models of dislocations themselves can contribute greatly to a better understanding of more readily observed geologic features.

Acknowledgments

Electron microscopy and deformation experiments were conducted in the Department of Geology, University of California, Berkeley, with the support and encouragement of Prof. H-R. Wenk. W. M. Meier provided me with the opportunity to think about faulted structures during a stay at the Institute for Crystallography at the ETH. I am also grateful for Penrose grant funding from the GSA that subsidized the collection of the Cascade garnets as part of a broader field study.

References

Beran, A., and Putnis, A. (1983) A model for the OH positions in olivine, derived from infrared-spectroscopic investigations. *Phys. Chem. Minerals* **9**, 57–60.

Boland, J. N., and Liu, L. (1983) Olivine to spinel transformation in Mg_2SiO_4 via faulted structures. *Nature* **303**, 233–235.

Cressey, G., Schmid, R., and Wood, B. J. (1978) Thermodynamic properties of almandine–grossular garnet solid solutions. *Contrib. Mineral. Petrol.* **67**, 397–404.

Frank, F. C. (1951) Capillary equilibria of dislocated crystals. *Acta Cryst.* **4**, 497–501.

Freund, F. (1981) Mechanism of the water and carbon dioxide solubility in oxides and silicates and the role of O⁻. *Contrib. Mineral. Petrol.* **76**, 474–482.

Green, II, H. W., and Radcliffe, S. V. (1975) Fluid precipitates in rocks from the earth's mantle. *Geol. Soc. Amer. Bull.* **86**, 846–852.

Griggs, D. T. (1974) A model for hydrolytic weakening in quartz. *J. Geophys. Res.* **79**, 1653–1661.

Haasen, P. (1979) Solution hardening in f.c.c. metals, in *Dislocations in Solids*, Vol. 4, edited by F. R. N. Nabarro, pp. 155–190. North Holland, Amsterdam.

Hamaya, N., and Akimoto, S. (1982) On the mechanism of the olivine–spinel transformation. *Phys. Earth Planet. Int.* **29**, 6–11.

Hazen, R. M., and Finger, L. W. (1976) Effects of temperature and pressure on the crystal structure of forsterite. *Amer. Mineral.* **61**, 1280–1293.

Hazen, R. M., and Finger, L. W. (1978) Crystal structures and compressibilities of pyrope and grossular to 60 kbar. *Amer. Mineral.* **63**, 297–303.

Helgeson, H. C., Delaney, J. M., Nesbitt, H. W., and Bird, D. K. (1978) Summary and critique of the thermodynamic properties of rock-forming minerals. *Amer. J. Sci.* **278A**, 229 pp.

Hirth, J. P., and Lothe, J. (1968) *Theory of Dislocations*, 1st Ed. McGraw Hill, New York.

Hobbs, B. E. (1981) The influence of metamorphic environment upon the deformation of minerals. *Tectonophysics*, **78**, 335–383.

Holder, J., and Granato, A. V. (1969) Thermodynamic properties of solids containing defects. *Phys. Rev.* **182**, 729–741.

Kirby, S. H. (1983) Rheology of the lithosphere. *Rev. Geophys. Space Phys.* **21**, 1458–1487.

Kirby, S. H., and Kronenberg, A. K. (1984) Hydrolytic weakening of quartz: Uptake of molecular water and the role of microfracturing. *Trans. Amer. Geophys. Union* **65**, 271.

Knipe, R. J., and Wintsch, R. P. (1982) Feldspar behaviour in a mylonite: An example of the interaction between deformation and metamorphic processes. *Mitt. Geol. Inst. ETH, Neue Folge* **239a**, 161–163.

Lasaga, A. C. (1980) Defect calculations in silicates: Olivine. *Amer. Mineral.* **65**, 1237–1248.

Lasaga, A. C. (1981) The atomistic basis of kinetics: Defects in minerals, in *Kinetics of Geochemical Processes. Reviews in Mineralogy, 8,* edited by A. C. Lasaga and R. J. Kirkpatrick. Mineral. Soc. America, 261–319.

Lasaga, A. C., and Cygan, R. T. (1982) Electronic polarizabilities of silicate minerals. *Amer. Mineral.* **67**, 328–334.

Madon, M., and Poirier, J. P. (1983) Transmission electron microscope observation of, and $(Mg,Fe)_2SiO_4$ in shocked meteorites: Planar defects and polymorphic transitions. *Phys. Earth Planet. Int.* **33**, 31–44.

Meagher, E. P. (1980) Silicate garnets, in *Reviews in Mineralogy, 5, Orthosilicates,* edited by P. H. Ribbe. Mineral. Soc. America, 25–66.

Miyamoto, M., and Takeda, H. (1983) Atomic diffusion coefficients calculated for transition metals in olivine. *Nature* **303**, 602–603.

Poirier, J. P. (1975) On the slip systems of olivine. *J. Geophys. Res.* **80,** 4059–4061.

Poirier, J. P., and Vergobbi, B. (1978) Splitting of dislocations in olivine, cross-slip controlled creep and mantle rheology. *Phys. Earth Planet. Int.* **16,** 370–378.

Price, G. D., and Parker, S. C. (1984) Computer simulations of the structural and physical properties of the olivine and spinel polymorphs of Mg_2SiO_4. *Phys. Chem. Minerals* **10,** 209–216.

Rabier, J., Veyssiere, P., and Garem, H. (1981) Dissociation of dislocation with a/2 111 Burgers vectors in YIG single crystals deformed at high temperature. *Phil. Mag. A.* **44,** 1363–1373.

Rabier, J., Veyssiere, P., and Grilhe, J. (1976) Possibility of stacking faults and dissociation of dislocations in the garnet structure. *Phys. Stat. Sol. (a)* **35,** 259–268.

Ross, J. V., Bauer, S. J., and Carter, N. L. (1983) Effect of the quartz transition on the creep properties of quartzite and granite. *Geophys. Res. Lett.* **10,** 1129–1132.

Smith, B. K. (1982) Plastic deformation of garnet: Mechanical behavior and microstructures. Unpublished PhD thesis, University of California, Berkeley.

Spinnler, G. E., Self, P. G., Iijima, S., and Buseck, P. R. (1984) Stacking disorder in clinochlore chlorite. *Amer. Mineral.* **69,** 252–263.

Stocker, R. L., and Smyth, D. M. (1978) Effect of enstatite activity and oxygen partial pressure on the point defect chemistry of olivine. *Phys. Earth Planet. Int.* **16,** 145–156.

Suzuki, H. (1957) The yield strength of binary alloys, in *Dislocations and Mechanical Properties of Crystals,* edited by J. C. Fisher, pp. 361–390. Wiley, New York.

Turnbull, D. (1956) Phase changes. *Solid State Phys.* **3,** 226–309.

Van der Biest, O., and Thomas, G. (1974) Cation stacking faults in lithium ferrite spinel. *Phys. Stat. Sol.* **24,** 65–74.

Van der Hoek, B., Van der Eerden, J. P., and Bennema, P. (1982) Thermodynamical stability conditions for the occurrence of hollow cores caused by stress of line and planar defects. *J. Cryst. Growth* **56,** 621–632.

Vander Sande, J. B., and Kohlstedt, D. L. (1976) Observation of dissociated dislocations in deformed olivine. *Phil. Mag.* **34,** 653–658.

Vaughan, P. J., and Coe, R. S. (1981) Creep mechanisms in Mg_2GeO_4: Effects of a phase transition. *J. Geophys. Res.* **86,** 389–404.

Vaughan, P. J., and Kohlstedt, D. L. (1981) Cation stacking faults in magnesium germanate spinel. *Phys. Chem. Minerals* **7,** 241–245.

Vaughan, P. J., Green, II, H. W., and Coe, R. S. (1982) Is the olivine–spinel transformation martensitic? *Nature* **298,** 357–358.

Weidner, D. J., and Hamaya, N. (1983) Elastic properties of the olivine and spinel polymorphs of Mg_2GeO_4, and evaluation of elastic analogues. *Phys. Earth Planet. Int.* **33,** 275–283.

Yund, R. A., Smith, B. M., and Tullis, J. (1981) Dislocation-assisted diffusion of oxygen in albite. *Phys. Chem. Minerals* **7,** 185–189.

Zeuch, D. H., and Green, II, H. W. (1978) Dislocation of substructures of experimentally deformed dunite. *Trans. Amer. Geophys. Union* **59,** 375.

Chapter 5
A Natural Example of the Kinetic Controls of Compositional and Textural Equilibration

R. J. Tracy and E. L. McLellan

Introduction

Textural and compositional disequilibrium have been documented in both regional metamorphism (Griffin, 1971; Tracy, 1982) and contact metamorphism (Hollister, 1969; Loomis, 1976). However, the significance of kinetics in controlling assemblages, compositions, and textures has been treated by relatively few authors (e.g., Loomis, 1976; Foster, 1982). Among the variables that control the rates of metamorphic reactions are the rate of supply (or removal) of heat, the rate of supply of matter (diffusion control), and the rate of reactant dissolution and/or product absorption during recrystallization (interface control). In the initial high-temperature stages of contact metamorphism (i.e., until cooling begins), reaction rates are more likely to be interface than diffusion controlled because both thermal energy input and diffusion coefficients will be large at the initially higher temperatures and therefore may not be rate limiting. As temperatures fall, however, heat flux and diffusion rates may become increasingly important. The rate-limiting factor will control the nature and extent of compositional and textural equilibrium. Thus, diffusion control will lead to strict local equilibrium (Fisher, 1978) whereas interface control may lead to a partial approach to equilibrium on a larger scale (Loomis, 1976).

In order to illustrate some kinetic controls on metamorphic processes, we have chosen contact metamorphosed emeries from the Cortlandt Complex, New York, as an example. These emeries are especially interesting because they represent the ultrametamorphism of previously regionally metamorphosed, relatively coarse-grained country rocks, and therefore present the opportunity to examine very rapidly heated rocks that have achieved larger average grain size than the typical hornfels. Furthermore, the intrusion of the Cortlandt Complex at crustal depths of 20–25 km into country rocks with ambient temperatures approaching 600°C requires that cooling after the thermal peak will be slower than in other cases at shallower depths where the kinetics of contact metamorphic processes have been studied (e.g., Loomis, 1976; Joesten, 1974, 1983; Jones *et al.*, 1975). Because of their peculiar thermal history, the Cortlandt emeries contain a multitude of textural and mineral compositional features that indicate varying approaches to thermodynamic equilibrium based on scale and local environment. It is the intent of this paper, in the context of the present volume, to summarize some of our key observations on kinetically controlled textural and compositional phenomena and to suggest reasonable kinetic interpretations. The reader interested in the details of emery petrology is referred to another paper for more complete exposition (Tracy and McLellan, in preparation).

Petrology of the Emeries

Emery is a rather rare metamorphic rock type consisting of such aluminous phases as diaspore, corundum, spinel, staurolite, and aluminum silicates, along with magnetite, ilmenohematite, and minor amounts of other silicates. World-wide emery occurrences fall generally into two types: regionally metamorphosed unusually aluminous bulk compositions, such as the metamorphosed bauxites of Naxos (Jansen and Schuiling, 1976) and occurrences involving contact metamorphism and accompanying metasomatism where mafic magma has intruded pelitic sedimentary or metasedimentary rocks (e.g., Barker, 1964; Gribble and O'Hara, 1967). The emery deposits of the Cortlandt Complex are composed of a variety of mineral assemblages developed in the highly aluminous residua produced by the interaction of pelitic schist xenoliths with mafic to ultramafic magma. These emeries have been the subject of many studies since the nineteenth century, the most recent of which by Friedman (1956), Barker (1964), and Caporuscio and Morse (1978) list or summarize the earlier work.

The Cortlandt Complex is a mafic to ultramafic composite intrusive body that crops out near the Hudson River about 60 km north of New York City. It is composed of at least six discrete plutons that range in composition from diorite to olivine pyroxenite (Ratcliffe *et al.*, 1982). A precise age has not been determined for the plutonic rocks, but biotite in a contact rock has been dated at 435 m.y. (Long and Kulp, 1962). The intrusive event appears to postdate the regional metamorphism and deformation of Taconian (Ordovician) age in the country rocks (Ratcliffe *et al.*, 1982; Ratcliffe *et al.*, 1983).

Emery deposits occur at a number of places near the margin of the Complex where aluminous country rocks have been engulfed by mafic magma. Typical emery lithologies are depleted in silica and alkalis, resulting in mineral associations that commonly include corundum, spinel, and sapphirine. Previous workers (e.g., Barker, 1964) have described the processes of desilication and alkali loss as metasomatic effects produced by movement of chemical components down their chemical potential gradients. Much of the metasomatism may in fact be due to melting followed by melt extraction and hybridization with the surrounding magma (Tracy and Thompson, 1979).

The two emery localities most thoroughly studied to date are those at Emery Hill and at Salt Hill (Fig. 1). Maximum temperatures achieved in the Emery Hill body appear to be about 650–700°C, based both on mineral element distribution thermometry and on the presence of apparent primary hydrous minerals (Tracy and McLellan, in preparation). On the other hand, temperatures in the emery xenoliths at Salt Hill are estimated to have been as high as 900°C or above, although they are not as well constrained here. The evidence for the higher temperature here includes widespread melting (absent at Emery Hill) and the total absence of any primary hydrous minerals. Microscopic examination of samples collected in a 50 m traverse through one xenolith shows no evidence of any temperature gradient. This observation is consistent with a heat flow calculation (Carslaw and Jaeger, 1959, p. 230), which suggests that a roughly spherical xenolith 50 m across should come completely to magmatic temperature in less than 5 yr. (The

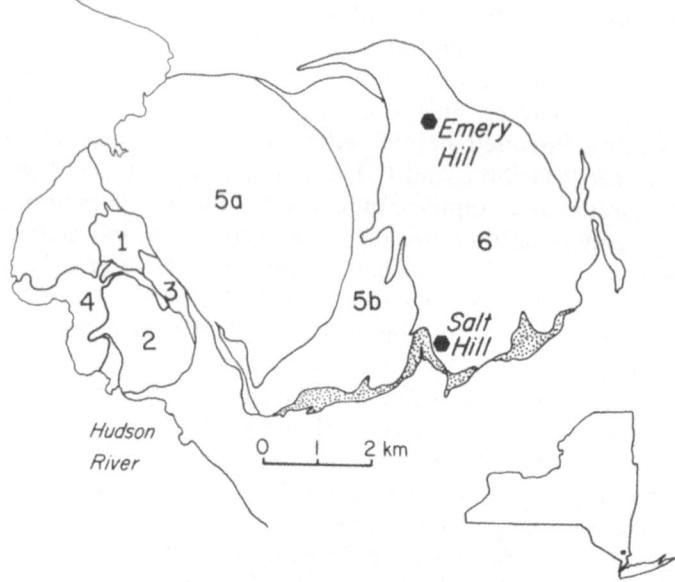

Fig. 1. Generalized geologic map of the Cortlandt Complex, New York (based on mapping by N. M. Ratcliffe, 1978–1982; used by permission of the author) showing the locations of the Salt Hill and Emery Hill emery bodies; the inset map of New York State shows the location of the Cortlandt Complex. The numbering of the plutons is that of Ratcliffe *et al.* (1983); the stippled area at the southern edge of the Complex is an area of extensive assimilation and hybrid rock development. Country rocks at the eastern end of the Complex are Manhattan Schist, units A and C.

calculation assumes an infinite reservoir of magma, but given the high solidus temperature of an olivine pyroxenite magma, it is reasonable that even with a lesser volume of magma a temperature near 900°C would rapidly have been reached within the xenolith.)

In all ensuing discussion, the term "mineral association" will be used in reference to a group of minerals that occur in relatively close proximity within small areas of a sample. The deliberate use of this term is meant to avoid the implication that these minerals are necessarily in equilibrium with each other. As will be discussed in detail below, many mineral grains within small rock volumes *may* be in equilibrium, but many others clearly are not. In some cases, grains that are physically touching are not in equilibrium, for example, the several observed instances of quartz–corundum contacts. Scales of equilibration are variable and complex for a variety of kinetic reasons.

Petrographically, individual samples from the Salt Hill xenolith consist of relatively fine-grained (100–500 micron grain size) aluminous layers of about 1–10 cm thickness separated by much coarser quartzofeldspathic lenses 1–2 cm in thickness. These lenses have been described by earlier investigators (e.g., Friedman, 1956) as quartz veins, but most contain up to 20 vol % orthoclase as a nearly continuous network surrounding 1–2 mm or larger, subrounded quartz crystals. The orthoclase is typically perthitic and has a composition of $Or_{70}Ab_{30}$. It is likely that the quartzofeldspathic lenses were melts, or at least the partially crystallized fractions of highly siliceous melts, rather than veins. Melting presumably occurred either during or immediately after the initial rapid heating.

The interspersed aluminous layers consist of somewhat variable proportions of aluminous spinel, magnetite, ilmenohematite, and sillimanite; in addition, one or more of the following may be present: garnet, cordierite, aluminous hypersthene, sapphirine, and corundum. In some samples, orthoclase occurs pervasively through the fine-grained aluminous matrix, especially adjacent to the quartzofeldspathic layers. Similarly, calcic plagioclase (about An_{70}) occurs sporadically in the aluminous layers. The distribution of silicates other than sillimanite appears to be strongly controlled by the distribution of the quartzofeldspathic lenses. Of all other silicates, only sapphirine can be found to occur at distances greater than several millimeters from the quartzose domains.

Mineral associations observed in the Salt Hill emery samples are given in Table 1; heterogeneity of most samples precludes measurement of numerical modes. Quartz and feldspars have not been included in the table because of their sporadic occurrence and localization within layers. Examination of Table 1 indicates some rather interesting correlations. Almost all of the samples that contain sapphirine also contain hypersthene; furthermore, all of the samples that contain cordierite also contain hypersthene, and all but one contain sapphirine. This strong correlation of hypersthene, sapphirine, and cordierite reflects a bulk composition effect that can also be seen in the

Table 1. Mineral associations.

Sample*		Sil	Spn	Mag	Ilm	Cor	Sapph	Gar	Crd	Opx
SH-10	(F)		x	x	x			x		x
SH-11	(F)			x	x			x		
SH-12	(F)		x	x	x			x		
SH-13	(F)			x	x			x		
SH-14	(F)	x	x	x	x			x		x
SH-15	(F)		x	x	x			x		x
SH-16	(F)	x	x	x	x	x		x		
SH-17	(F)	x	x	x	x	x		x		
SH-18	(M)	x	x	x	x			x		
SH-19	(M)	x	x	x	x	x	x	x		x
SH-20	(M)	x	x	x	x	x	x			x
SH-21	(M)	x	x	x	x	x	x			x
SH-22	(M)	x	x	x	x	x	x	x		
SH-23	(M)	x	x	x	x			x		
SH-24	(M)	x	x	x	x	x	x		x	x
SH-25	(M)	x	x	x	x		x	x	x	x
SH-26	(M)	x	x	x	x	x	x		x	x
SH-27	(M)	x	x	x	x	x	x			
SH-28	(M)	x	x	x	x	x	x		x	x
SH-29	(F)	x	x	x	x	x			x	x
SH-30	(F)	x	x	x	x	x		x		
SH-31	(F)	x	x	x	x	x				
SH-32	(M)	x	x	x	x	x		x		
SH-33	(M)	x	x	x	x			x		
77061	(F)	x	x	x	x			x		
77062	(M)	x	x	x	x			x		
77063	(M)	x	x	x	x		x	x	x	x
77064	(F)	x	x	x	x	x		x		

* (F)—Less magnesian bulk composition.
　(M)—More magnesian bulk composition.

compositions of the matrix spinels (see Fig. 3). The emeries consist of two general groups: a more magnesian compositional type containing spinels that have $Fe^{2+}/(Fe^{2+} + Mg)$ of about 0.50, and a less magnesian type in which the matrix spinels have $Fe^{2+}/(Fe^{2+} + Mg)$ of about 0.60. The former group typically contains sapphirine with or without hypersthene and cordierite. Table 1 also shows a negative correlation between sapphirine and garnet, although it may not be as strong as the above-mentioned correlations. Garnet is essentially ubiquitous in the less magnesian samples and rather rare in the more magnesian group. Close proximity of garnet and sapphirine typically involves garnet overgrowth on sapphirine indicating a reaction relation rather than stable coexistence of these phases.

The exact mineral association found in each sample apparently reflects subtle local variations in bulk composition as well as other factors such as grain size, degree of recrystallization, and local availability of silica to drive the reactions producing silicate minerals. The approximate mineral association for the magnesian group is: sillimanite − spinel − magnetite − ilmenite − sapphirine − hypersthene ± cordierite ± corundum ± garnet(rare). In the less magnesian group, the association is: sillimanite − spinel − magnetite − ilmenite − garnet ± hypersthene ± corundum.

A wide variety of textures may be observed in the emeries; a number of the more interesting ones are illustrated in Fig. 2. The aluminous layers consisting of sillimanite, spinel, magnetite, and ilmenohematite are strongly foliated, with extreme elongation of sillimanite crystals defining the foliation. Paradoxically, however, clots of oxide grains commonly show annealed textures with well-defined 120° triple junctions. In reflected light, ilmenohematite grains show complex lamellar patterns indicating multiple episodes of exsolution (see also Caporuscio and Morse, 1978, p. 1336).

The most spectacular textures have developed in the aluminous layers immediately adjacent to quartzofeldspathic lenses. Silicate minerals have grown here either as reaction rims (Fig. 2 (a),(b),(d),(e)) or as intergrowths with finely crystalline neomorphic sillimanite (Fig. 2 (c). This environment contains the overgrowths of sapphirine by garnet (noted above) as well as the much more common but similar texture of garnet overgrowths on spinel (to be discussed in detail in the next section). Growth of the secondary silicate phases is apparently caused by the breakdown of oxides (notably Fe-Ti oxide and aluminous spinel) in the environment of elevated SiO_2 activity proximal to the quartzose domains. Although SiO_2 must be introduced by diffusion in order to allow progress of the silication reactions, the textures in which the silicate minerals occur are more commonly interface controlled than diffusion controlled (see below). The variety of textures and mineral associations produced must reflect a number of important compositional and structural kinetic factors that will be elaborated upon in the following sections.

Compositional Evidence for Partial Equilibrium

There are two aspects to compositional equilibration: The first is homogeneous equilibration, that is, the relative constancy of compositions within grains of any phase, and the second is heterogeneous equilibration, the relative constancy of mineral compositions at mutual contacts within similar associations or assemblages. Loomis (1976) has presented a detailed exposition of the relationship between homogeneous and heterogeneous equilibra-

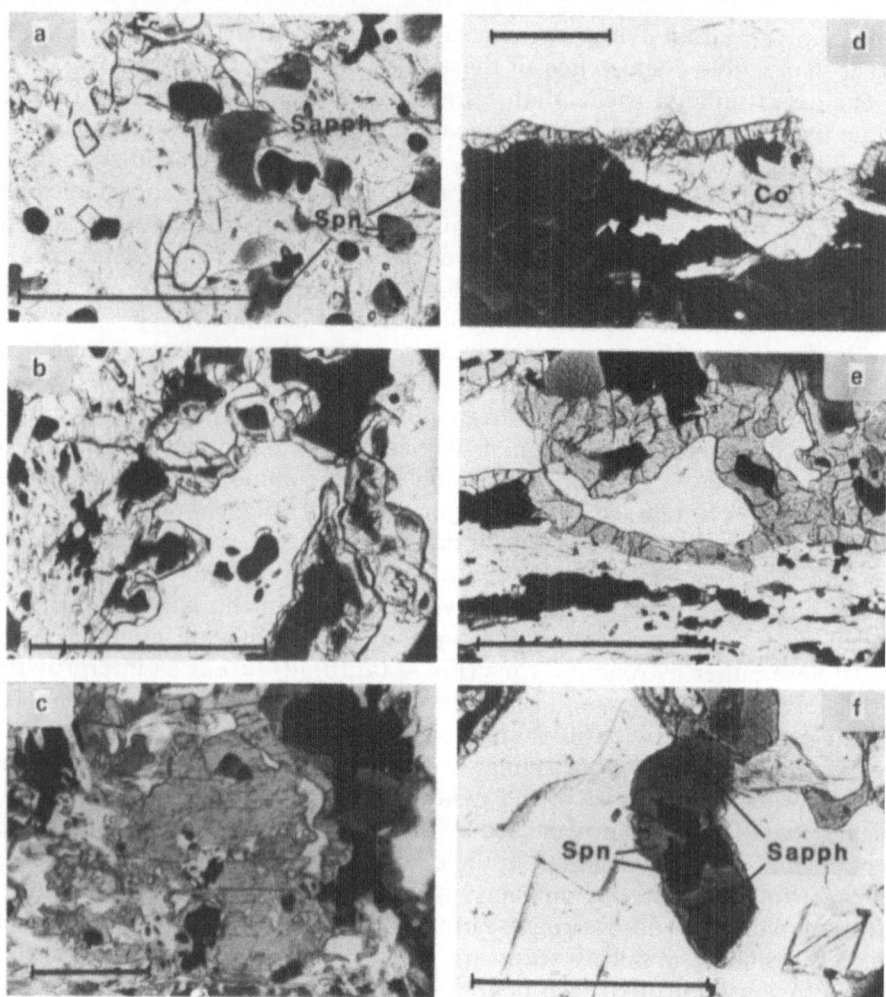

Fig. 2. Photomicrographs showing silicate reaction textures; scale bars in all photographs are 500 microns. (a) Large diffusion-controlled garnet with inclusions of both spinel and sapphirine. (b) Interface-controlled texture of garnet reaction rims overgrowing spinel; an area of quartz is in the center, and a sillimanite-rich layer on the left. (c) Aluminous hypersthene showing a composite of diffusion-controlled and interface-controlled textures; the abundant needles in the hypersthene are sillimanite. (d) Interface-controlled reaction rim of hypersthene on spinel and corundum (co); low-relief material at the top is quartz. (e) Reaction rim of hypersthene growing at the contact of aluminous layer and quartzose layer, but not localized on spinel. (f) Reaction of magnesian spinel with quartz to form sapphirine; the large sillimanite crystals were probably produced in this reaction.

tion, and the general concept of partial equilibrium for the specific case of a contact metamorphic aureole. Loomis argued that a close approach to heterogeneous equilibrium may be achieved by two or more minerals in a domain (typically reaction products) but that one or more minerals in the rock (typically reactants, but possibly products) may not participate in this equilibrium owing to kinetic reasons. The lack of equilibrium between reactants and products implies irreversibility in the reaction, which is to be expected for significant overstepping during either heating or cooling. Loomis therefore defined partial equilibrium involving certain phases in a rock as an apparent state of partition equilibrium as indicated by the compositions of these phases.

Considerable similarities exist between the mineral composition behavior observed by Loomis (1976) in the Ronda aureole and that found in the Cortlandt emeries. Specifically, with certain uncommon exceptions, the product minerals garnet, hypersthene, cordierite, and sapphirine have measured compositions that suggest a close approach to Fe-Mg partition equilibrium among coexisting pairs of these phases. There is, in fact, remarkably little variation in compositions of the above phases among all the samples, and most of this variation may be accounted for by minor bulk composition differences (it should be noted that the emery mineral associations are at least divariant in a phase-rule sense). Some domains in certain samples exhibit slightly greater variation in the compositions of the above phases, however, suggesting a greater local control of mineral compositions. In these cases, textural analysis indicates diffusion control of the reaction.

The only phases in the emeries that lack homogeneous equilibrium are garnet and hypersthene. The garnets, which are typically extraordinarily magnesium-rich for metamorphic garnets (pyrope contents commonly approach or exceed 50 mol%) and have very low calcium and manganese, are unzoned in some cases and have minor but irregular zoning in others. At present, no plausible explanation has been deduced for the garnet zoning. Zoning in hypersthene is irregular, as with the garnet zoning, but it is not in terms of Fe/(Fe + Mg) but rather in alumina content; it therefore presumably reflects variability in activity of Al_2O_3 during the growth of the hypersthene. The nominal coexistence of quartz and sillimanite with growing hypersthene should control the activity, and therefore the chemical potential, of Al_2O_3 (barring abrupt changes in P or T), so the apparent variability in Al_2O_3 activity is likely to be a kinetic effect that probably reflects difficulties in the mass transfer of Al_2O_3 through the intergranular environment. In this regard, it should be noted that the alumina-zoned hypersthene crystals are typically the larger ones that display a diffusion-controlled rather than interface-controlled texture, consistent with the above model.

The phase that shows by far the greatest variability in composition is aluminous spinel. The matrix spinel in both bulk composition types has intermediate values of $Fe^{2+}/(Fe^{2+} + Mg)$ (about 0.5 and 0.6; see above), but the range in this ratio for spinels in many samples may be from 0.25 to 0.7, depending upon the precise local environment of an individual spinel grain. Matrix spinels in both bulk composition types are never found in contact with quartz or orthoclase—overgrowths of product silicate phases always intervene. Spinel inclusions in cordierite, hypersthene, or sapphirine are generally identical with matrix spinels in composition and have a limited compositional range whereas spinel inclusions in garnet have a wide range, from matrix spinel compositions to much more magnesian values. The darker green spinels that are in contact with quartz or orthoclase have more iron-rich compositions than matrix spinels. Figure 3 summarizes the compositions of all spinels by showing the mean composition of matrix spinel in each sample as well the range in $Fe^{2+}/(Fe^{2+} + Mg)$ of matrix spinel, spinel inclusions in garnet, and spinels in contact with quartz or orthoclase.

Our interpretation of the distribution of spinel compositions is a kinetic model based upon an assumption about the mineralogy and physical state of the emeries at the peak temperatures reached, that is, *before* the reactions by which most of the silicate minerals grew. The highly aluminous residuum

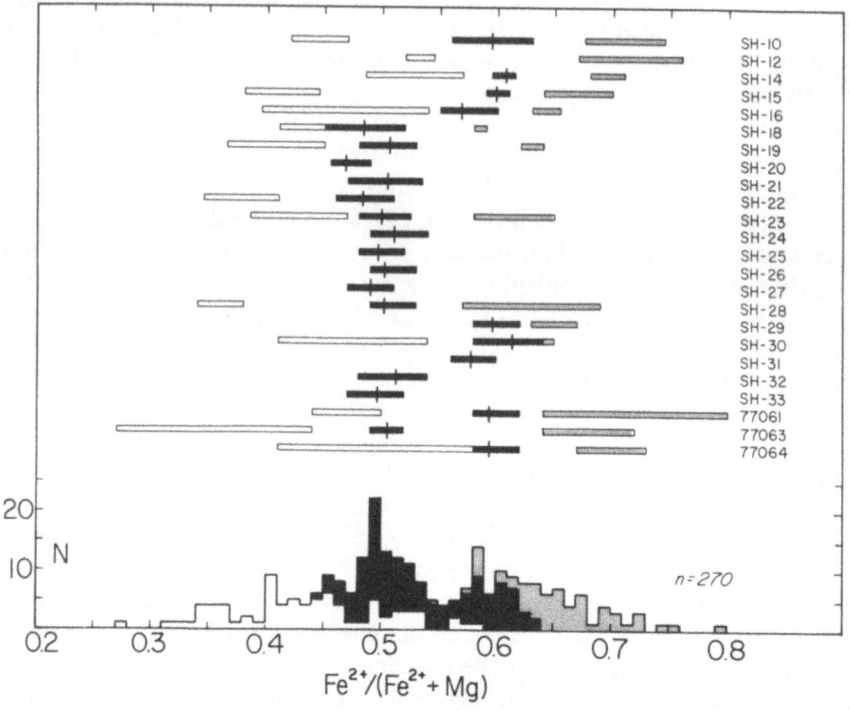

Fig. 3. Summary diagram showing the range of $Fe^{2+}/(Fe^{2+} + Mg)$ of spinels within single samples (horizontal bars) and the frequency distribution of all analyzed spinels (histogram). Matrix spinels are shown in black, spinel inclusions in garnet in white, and spinels touching quartz are shown in vertical patterning. For each sample, the mean matrix spinel composition is shown as a vertical line.

consisting of a spinel – magnetite – sillimanite – ilmenohematite hornfels (perhaps also containing mullite that later broke down to corundum + silli-manite) was created by thermal decomposition and melting of an original pelitic schist during very rapid heating. Essentially all the quartz and feld-spar were concentrated into water-undersaturated melt bodies that were redistributed as thin sills and veins during cooling. The local increase in SiO_2 activity near the sills and veins allowed the silicate-producing reactions to occur, resulting in a variety of overgrowth and inclusion textures in which spinels, or rarely sapphirines, were isolated as inclusions.

Spinels that occur as inclusions in garnets behave differently from spinels included in other silicates. Magnesium enrichment of the spinel inclusions in garnet is due to two factors, the consumption of iron by overgrowing garnet and the slowness of Fe and Mg diffusion through the newly grown garnet. This latter effect isolates the spinel from participating in any larger scale equilibrium and may, in fact, serve as a self-limit on further reaction pro-gress when the garnet rim becomes thick enough. If the change in spinel composition depends upon the extent of reaction for a particular garnet-spinel pair, then a substantial range in spinel compositions should be ex-pected, and in fact has been observed (Fig. 3). This range is actually a continuum between matrix spinel composition and some hypothetical magnesian limit.

The spinels included in sapphirine or hypersthene might similarly be ex-pected to have undergone some iron enrichment, because hypersthene and sapphirine are more magnesian than the spinels on which they nucleate. No such enrichment occurred, presumably because the diffusivities of Fe and Mg through the hypersthene and sapphirine lattices were great enough so that the spinel inclusions could remain in approximate heterogeneous equi-librium with the surroundings. Interestingly, where garnet occurs as thin, interface-controlled reaction rims (Fig. 2(b)), the spinel rarely deviates from matrix spinel compositions. In this case, garnet does not act as an effective barrier between spinel and the environment.

The origin and preservation of the iron-rich spinels in contact with quartz is problematical. Because they are typically euhedral and not obviously connected with any reaction texture, one possible explanation for their for-mation is that the siliceous melt layers were saturated with hercynite-rich spinel and that the rare small crystals crystallized from the melts. Alterna-tively, they may have formed in the initial heating event, in a manner analo-gous to formation of the matrix spinel but through decomposition of an iron-rich phase such as garnet or staurolite. In either case, their very clean contacts with quartz indicate that they were sufficiently iron-rich to be in equilibrium with silica at the conditions of emery formation, in contrast to the matrix spinels.

In order to summarize our observations on the compositional equilibra-tion of emery minerals, Fig. 4 shows compositions plotted in a quartz projec-tion within the quaternary system $MgO-FeO-Al_2O_3-SiO_2$. In the silica-under-

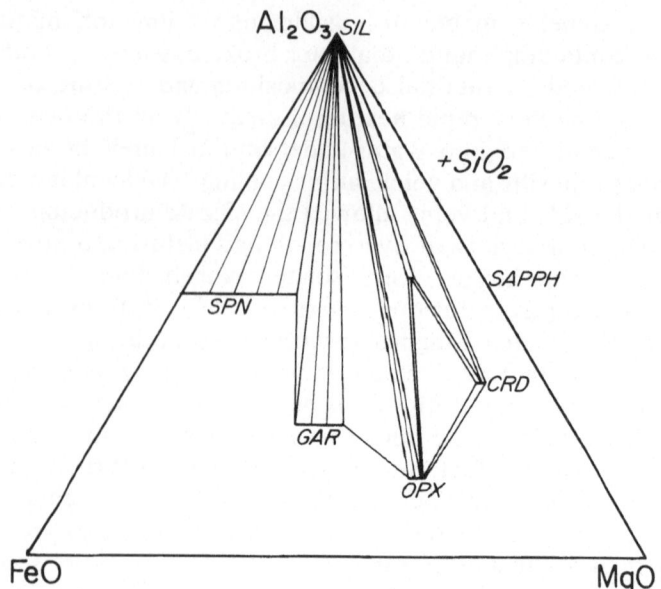

Fig. 4. SiO₂ projection within the quaternary system MgO-FeO-Al₂O₃-SiO₂, showing the generalized compositional ranges of the minerals that are found to coexist with quartz. All mineral compositions have been corrected for ferric iron calculated from stoichiometry. Note that the progressive order of magnesium enrichment is garnet-hypersthene-sapphirine-cordierite.

saturated part of the quaternary system (representative of the aluminous layers away from quartzose veins) more magnesian spinels and corundum may occur. Fe_2O_3 has been estimated stoichiometrically for all mineral analyses, and suitably corrected FeO has been used in plotting Fig. 4. All of the mineral associations shown in the diagram have been documented in the emeries. Spinel + sapphirine and garnet + sapphirine have been observed, but only in quartz-free domains; garnet and cordierite have not been found to coexist. The main point of Fig. 4 is to emphasize that mineral compositions in the emeries are typical of partial equilibrium, as defined by Loomis (1976). Depending on the exact scale or locale chosen, good examples of heterogeneous equilibrium, local equilibrium, and disequilibrium may be found. In the following sections, we will describe a kinetic model for compositional and textural equilibration that will explain this variety of behavior.

Kinetic Control of the Reaction Sequence

The data presented above indicate that magnesian spinels react with silica to produce, variously, sapphirine, orthopyroxene, garnet, or cordierite (plus sillimanite). Rimming and inclusion relations are compatible with the following reaction sequence:

1. Spinel + quartz = sapphirine *or* spinel + quartz = hypersthene (Fig. 2 (c),(d),(e),(f))
2. Sapphirine + quartz = hypersthene + sillimanite (Caporuscio and Morse, 1978, Plate 1)
3. Sapphirine + quartz = garnet + sillimanite *or* spinel + quartz = garnet + sillimanite (Fig. 2 (a),(b))
4. sapphirine + quartz = cordierite + sillimanite.

These reactions must have occurred in this order at successively lower temperatures during cooling. In view of the evidence presented elsewhere for disequilibrium, this reaction sequence presumably reflects the relative rate at which reactions occur (kinetic control) rather than a series of equilibrium states. The reaction that will occur most rapidly will be that with the lowest activation energy for nucleation (Ostwald's Step Rule). The activation energy, ΔG^*, is a balance between the release of chemical free energy upon formation on a new phase, ΔG_e, and the energy needed to accommodate that phase in the existing matrix (Putnis and McConnell, 1980):

$$\Delta G^* = \frac{16\pi}{3} \cdot \frac{\sigma^3}{(\Delta G_e + \gamma)^2} \tag{1}$$

where γ is a strain energy term reflecting the mismatch of lattice parameters between matrix and nucleus phases, and σ is a surface energy term reflecting the degree of structural similarity of matrix and nucleus phases. ΔG_e, the driving force for reaction, will increase with increasing undercooling below the equilibrium temperature (ΔT); ΔG^* is therefore proportional to $1/(\Delta T)^2$, according to the above relationship. Thus, for any given reaction ΔG^* decreases (i.e., the nucleation rate *increases*) with increasing undercooling, and reactions that were kinetically inhibited at high temperatures will become possible at lower temperatures. The restriction of the garnet-forming reaction to lower temperatures may be ascribed to this effect, as discussed below.

At a given temperature, any mechanism that decreases the surface and strain energy terms in the ΔG^* expression will make a reaction kinetically more favorable. This is most easily achieved by orienting the product nucleus such that the interface to the existing matrix is roughly coherent (i.e., elements of the structure are continuous across the interface; Putnis and McConnell, 1980). The development of such coherent interfaces may be seen in the Cortlandt emeries: Where spinel and quartz are separated by rims of sapphirine or hypersthene, the [100] planes of the product phase are oriented parallel to the [111] planes of the spinel. In this orientation the planes of cubic-close-packed oxygen atoms in both substrate spinel and overgrowing silicate are aligned. By this means, the activation energy for nucleation is reduced and the rate of reaction increased.

Thus the early breakdown of spinel plus quartz to sapphirine or hypersthene is attributed to the relative ease with which these product minerals could nucleate. Garnet, on the other hand, has less structural similarity to spinel and did not begin to form until late in the reaction sequence. It would be instructive to compare the relative degrees of undercooling below equilibrium temperature required for nucleation of garnet as opposed to hypersthene, but unfortunately there are not sufficient data to make such a calculation.

Interface-Controlled Reactions

Where reactant phases are in contact, intergranular diffusion must be relatively rapid and therefore probably not rate limiting. Failure of a product to grow in such cases must therefore be attributed to interface control, that is, to slowness of either reactant dissolution or product nucleation (Loomis, 1979). For example, the corundum–quartz associations noted in the emeries must indicate that the reaction corundum + quartz = sillimanite is interface controlled and extremely sluggish, through difficulty in corundum dissolution or sillimanite nucleation, or both.

Magnesian spinels bordering the quartzofeldspathic layers commonly develop thin rims of hypersthene ± sapphirine (Fig. 2 (d),(e)); in these cases, the reactant spinels may show limited variation in $Fe^{2+}/(Fe^{2+} + Mg)$, but there is no detectable variation in this ratio for the product hypersthene or sapphirine. Lack of chemical potential gradients for Fe and Mg does not by itself argue against diffusion control: In multicomponent diffusion, one less mobile component may limit reaction rate whereas other species diffuse readily (Shewmon, 1963). However, no other components of sapphirine (particularly Fe^{3+}) or of hypersthene (Fe^{3+} and Al) show any systematic variation with distance from, or composition of, reactant spinel. This particular reaction is therefore unlikely to be diffusion controlled. It has already been noted above that diffusion rates through the product phase were sufficiently rapid that these spinels remained in compositional equilibrium with the matrix.

A predicted growth rate was calculated for the growth of hypersthene rims on spinel (modelled as the reaction $MgAl_2O_4 + 2\ SiO_2 = MgSiO_3 + Al_2SiO_5$) assuming that development of the rims was controlled by grain boundary diffusion. The equation for growth rate that was used (Turnbull, 1956) is:

$$\dot{G} = \frac{\delta k}{h}\ T \exp\left(-\frac{Q}{RT}\right)\left[1 - \exp\left(\frac{-\Delta G_r}{RT}\right)\right], \tag{2}$$

where $\dot{G}$ is the rate of hypersthene rim growth, Q is the activation energy of grain boundary diffusion (taken as 80–90 kcal/mol), T was assumed to be 900°C (1173 K), and h and k are Planck's and Boltzmann's constants, respectively. ΔG_r was calculated from the pure phase data on H, S, and V given in Helgeson et al. (1978). δ is the displacement of the interface during one growth step; in this case it was assumed to be equal to $d(100)$ of enstatite, that is, 18.22 Å. Calculation of growth rate using the above equation and the given approximations indicates that the orthopyroxene reaction rim should grow to about 1.5 m thickness per million years.

Maximum rim thickness, however, is only about 1 mm, suggesting that some other factor is suppressing growth, even allowing for up to an order of magnitude error in the above calculation. Furthermore, the approximately constant thickness of hypersthene reaction rims, both within and between samples, implies interface control in which reaction is completed within a relatively short time. This behavior would be improbable in a diffusion-controlled reaction in which rim thickness (as a measure of extent of reaction) should be quite variable because it varies exponentially with duration of reaction at a fixed temperature (Burke, 1975). Thus, several independent lines of evidence indicate that the reaction rim texture is interface controlled. It remains to be determined whether the rate-limiting step was spinel dissolution or silicate mineral nucleation or growth at the interface.

Interface attachment may occur either by a continuous mechanism in which the crystal grows by random addition to all surfaces or by a layer mechanism whereby units are added with specific orientations to growth steps (Jackson et al., 1967). During layer attachment, atoms are added less easily to surfaces of low surface energy; higher energy surfaces grow more rapidly and thus ultimately disappear, leaving the lower energy faces as the typical forms of euhedral crystals. Continuous attachment is favored at high degrees of supersaturation and leads to high growth rates; layer attachment occurs closer to saturation and allows only slow growth rates (Jackson et al., 1967). As pointed out above, hypersthene and sapphirine reaction rims in the emery formed early in the cooling history when supercooling (and therefore supersaturation) was probably low; layer attachment is the predicted form of growth under these conditions. Indeed, many of the hypersthene crystals show well-developed [100] faces that have large d-spacing and consequent low energy, and would be expected to have formed by layer attachment. Thus, the occurrence of these faceted crystals implies that attachment of atoms to the interfaces of the product mineral was slow and may have been the rate-limiting step in this reaction. The development of garnet reaction rims around spinel is apparently analogous to the formation of sapphirine and hypersthene rims, and may likewise be interface controlled.

Diffusion-Controlled Reactions

The restriction of product assemblages to the immediate vicinity of the reactant phases is not in itself evidence that reaction is limited by the transport of necessary components. For example, the association of sapphirine and hypersthene nucleating on spinel (described above) reflects the suitability of spinel as a nucleation site rather than the limited mobility of SiO_2. Diffusion control can only be proven if it can be shown that local equilibrium prevails, i.e., that within a single thin section there are different compositional domains in which the compositions of product phases directly reflect those of the reactants (Thompson, 1959).

The hypersthene–sillimanite intergrowths surrounding sapphirine show no correlation between Fe/(Fe + Mg) of hypersthene and that of sapphirine, but there is evidence that the hypersthene closer to sapphirine has higher Al^{IV} content than that farther away. This may suggest that the formation of the intergrowths is controlled by the rate of dissolution of Al, other species being sufficiently mobile that chemical potential gradients are eliminated. Certainly the quasi-eutectoid morphology of the intergrowths is similar to that developed in metals systems where reaction occurs under diffusion control (Chadwick, 1972). The intergrowths show time-dependent growth because they are both irregular in shape and of variable size—features that are more compatible with diffusion than with interface control (as noted above). The mean size of intergrowths is greatest where nuclei are isolated; elsewhere they are relatively smaller, which may suggest that their maximum size is limited by competition for the diffusing components (Kretz, 1966). Thus, these intergrowths are controlled by the rate of diffusion of a single component, probably Al_2O_3.

In some cases the intergrowths can be seen to have developed around the hypersthene rims described in the section above (Fig. 2 (c)), suggesting that there may be a transition from interface control to diffusion control as the reaction proceeds. This is to be anticipated because development of a rim of hypersthene will partially isolate reactant spinel, with consequent lowering of diffusion gradients. Additionally, the decrease in temperature with time will lead to decreased diffusion coefficients, although this effect may be balanced in part by steepening of chemical potential gradients. In general, however, there should be a progressively reduced diffusive flux and an increased probability that diffusion of components will become the rate-limiting step in further reaction.

Sieve-textured garnets containing a large number of spinel inclusions are common in the Cortlandt emeries (Fig. 2 (a),(b)). These commonly do not have good euhedral forms and their shapes are compatible with growth along original grain boundaries, obvious loci of preferential mass transfer. The growth of these garnets is therefore likely to have been controlled by diffusion rather than by interface processes. Likewise, the Mg/(Mg + Fe) ratio in garnets directly correlates with that in the overgrown spinel; this local compositional equilibrium further supports a model of diffusion control.

Evidence Concerning Chemical Fluxes

The localization of silicate phases to the proximity of quartzose veins suggests that the ultimate control on the distribution of these phases was the availability of silica through diffusion outward from the veins. This does not require or even imply that the rate of silica diffusion is the limiting kinetic factor for reactions producing silicate phases; as noted in an above section, diffusivities of other components, notably Al_2O_3, are more likely to be limiting. However, all the reactions involved are silica-consuming, and it is clear that they could not proceed unless the local silica activity had been sufficiently elevated. Understanding this very important, though not directly kinetic, control that silica activity played in the areal distribution of emery minerals allows us to make an important inference about the diffusion rate of silica through the emery.

The distribution of silicate phases is rather consistent from sample to sample. Sapphirine has been observed to occur farthest from the veins, up to a centimeter or more away. The Si/O and Si/Al ratios of an ideal sapphirine $[(Fe,Mg)_4Al_8Si_2O_{20}]$ are 0.08 and 0.20, respectively, suggesting that the spinel to sapphirine reaction may require only a small increase in silica activity. Garnet (Si/O = 0.25; Si/Al = 1.50) and cordierite (Si/O = 0.28; Si/Al = 1.23) have the next widest distribution, up to several millimeters from the veins; note, however, that there is the above-mentioned bulk composition control on distribution of these two phases. Finally, hypersthene $[(Fe,Mg)_{1.75}Al_{.5}Si_{1.75}O_6]$ [Si/O = 0.28; Si/Al = 3.82] is restricted to the very margins of the veins. This distribution of phases makes it clear that silica activity, especially with regard to alumina activity, dropped off rapidly outward from the veins; at distances more than a centimeter or so from a vein, silica activity was so low that even a low-silica mineral such as sapphirine could not form. Figure 5 shows the deduced activity model for the formation of silicate minerals proximal to the veins and suggests why a different sequence occurs in the two bulk compositions.

There are several mechanisms through which silica diffusion outward

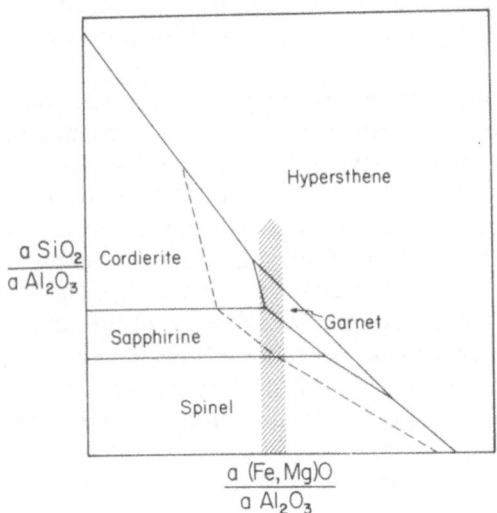

Fig. 5. Hypothetical activity–activity diagram for the conditions of emery formation, showing $aSiO_2$ versus $a(FeMg)O$, normalized to aAl_2O_3. The slopes have been accurately calculated for the model compositions given in the text. The solid field boundaries are considered to be appropriate for the more magnesian bulk compositions, while the dashed boundaries apply to the less magnesian bulk compositions. The broad arrow shows a likely path of silica activity increase adjacent to quartzose domains, and the sequence of reactions driven by this increase.

from the veins might have occurred. The most obvious of these is intergranular diffusion, whereby the transported species moved along the surface defined by grain boundaries or subgrain boundaries. Because diffusion in the emery occurred at high temperature in an apparently anhydrous material, we may assume that the grain boundary surfaces were free of a continuous intergranular film of H_2O. Alternatively, we may assume that intergranular transport was through an aqueous medium, in which case the transport rate should be much faster than in the first case (Freer, 1981). A third mechanism for silica diffusion is volume or lattice diffusion (Shewmon, 1963) which has the advantage of straighter diffusion paths, but the disadvantage (at lower temperatures) of decidedly slower rates than the first two mechanisms. However, as Freer (1981) has pointed out, the flux rate difference between intergranular and volume diffusion is smaller at high temperature, and volume diffusion may become dominant in this regime, especially if the grain boundaries are essentially dry. Given the high temperature of formation of the silicates and the apparent slow cooling of the emeries, it is remarkable that silica did not diffuse farther from the veins given the strong chemical potential gradient of SiO_2 that must have existed.

The inescapable conclusion from the distribution of silicate phases is that the silica diffusion rate must have been very low. This strongly implies that the mechanism for silica diffusion must have been either volume diffusion or intergranular diffusion along *dry* grain boundaries. It would be very difficult to distinguish between these two mechanisms, but the important point is that there was apparently no intergranular aqueous fluid. The fact that the emery is composed of anhydrous minerals does not require, but certainly is consistent with, the interpretation of dry conditions. Furthermore, the presence of H_2O-undersaturated melts will cause an at least local lowering of H_2O activity in the surroundings. It is possible that aqueous fluid became available for silica transport at a late stage when the melts were largely crystallized, but this probably occurred at sufficiently low temperature that diffusivities were greatly reduced.

There are important petrologic implications to the conclusion that silica transport distance was highly dependent upon the presence or absence of intergranular aqueous fluid. Recent studies have suggested that transport of silica may typically be a limiting factor in the kinetics of silica-consuming metamorphic reactions (Tanner et al., 1983) occurring in the presence of aqueous fluid. But in general the diffusion rate of silica has been assumed to be so rapid (e.g., Carmichael, 1969) that natural examples of a silica diffusion control of metamorphic reactions could not be found. Our observations suggest that this assumption may need to be reevaluated, particularly for rocks that have been strongly dehydrated such as hornfelses in inner contact aureoles and regional granulite facies rocks. In more common regional metamorphic rocks, the presence (at least sporadically) of an intergranular aqueous fluid probably enhances the mass transport of silica so that movement over scales of meters or more may be possible, particularly if there is bulk fluid flux.

In earlier sections, we argued for limited diffusive control of textural and compositional equilibration on a local scale. For example, it is clear that significant transport of Fe and Mg is indicated by textures such as those in Fig. 2 (d) and 2 (e) where hypersthene overgrows corundum or sillimanite. The constancy of compositions of spinels overgrown by sapphirine or hypersthene suggests that volume diffusion of Fe and Mg through these two silicates was rapid enough to allow larger scale equilibration. On the other hand, the local nature of compositional equilibration between spinel and garnet argues for very slow volume diffusion in garnet, in accord with published data (Lasaga *et al.*, 1977; Freer, 1981).

One flux that has not yet been discussed is that of heat, which may be important in determining reaction rates (Shewmon, 1963). Because the silicate-producing reactions are exothermic and occur during cooling, there may have been a kinetic control exerted by slowness of heat removal during reaction, perhaps compounded in the emery xenoliths by the large latent heat given off by the crystallizing magma. If, as has been argued, the emeries are devoid of intergranular fluid, then heat removal from reaction sites by convection was not possible, and all heat transfer was conductive, and therefore relatively slow. It may well be that the diffusion-controlled textures observed in the late reactions were at least in part controlled by heat diffusion as well as by chemical diffusion.

Acknowledgments

The authors would like to thank Richard Sack and David Walker for their help in collecting the suite of samples from Salt Hill on a very unpleasant summer day. Alan Thompson, David Rubie, and Paul Karabinos provided most helpful reviews of an earlier version of the manuscript. Financial support for the work was provided by National Science Foundation Grant EAR 81-20670 (to RJT).

References

Barker, F. (1964) Reaction between mafic magmas and pelitic schist, Cortlandt, New York. *Amer. J. Sci.* **262,** 614–634.

Bender, J. F., Hanson, G. N., and Bence, A. E. (1984) Cortlandt Complex: Differentiation and contamination in plutons of alkali basalt affinity. *Amer. J. Sci.* **284,** 1–57.

Burke, J. (1975) *The Kinetics of Phase Transformations in Metals.* Pergamon Press, London.

Caporuscio, F. A., and Morse, S. A. (1978) Occurrence of sapphirine + quartz at Peekskill, New York. *Amer. J. Sci.* **278,** 1334–1352.

Carmichael, D. M. (1969) On the mechanism of prograde reactions in quartz-bearing pelitic rocks. *Contrib. Mineral. Petrol.* **20,** 244–267.

Carslaw, H. S., and Jaeger, J. C. (1959) *Conduction of Heat in Solids,* 2nd ed. Oxford University Press, Oxford.

Chadwick, G. A. (1972) *Metallography of Phase Transformations.* Butterworths, London.

Fisher, G. W. (1978) Rate laws in metamorphism. *Geochim. Cosmochim. Acta* **42,** 1035–1050.

Foster, Jr., C. T. (1982). A thermodynamic model of mineral segregations in the lower sillimanite zone near Rangeley, Maine. *Amer. Mineral.* **66,** 260–277.

Freer, R. (1981) Diffusion in silicate minerals and glasses: A data digest and guide to the literature. *Contrib. Mineral. Petrol.* **76,** 440–454.

Friedman, G. M. (1956) The origin of spinel–emery deposits with particular reference to those of the Cortlandt Complex, New York. *New York State Museum Bulletin* **351.**

Gribble, C. D., and O'Hara, M. J. (1967) Interaction of basic magma with pelitic materials. *Nature* **214,** 1198–1201.

Griffin, W. L. (1971) Genesis of coronas in anorthosites of the upper Jotun Nappe, Indre Sogn, Norway. *J. Petrol.* **12,** 219–243.

Helgeson, H. C., Delaney, J. M., Nesbitt, H. W., and Bird, D. K. (1978) Summary and critique of the thermodynamic properties of rock-forming minerals. *Amer. J. Sci.* **278A,** 229 p.

Hollister, L. S. (1969) Contact metamorphism in the Kwoiek area of British Columbia: An endmember of the metamorphic process. *Geol. Soc. Amer. Bull.* **80,** 2465–2493.

Jackson, K. A., Uhlmann, D. R., and Hunt, J. P. (1967) On the nature of crystal growth from the melt. *J. Cryst. Growth* **1,** 1–36.

Jansen, J. B. H., and Schuiling, R. D. (1976) Metamorphism on Naxos: Petrology and geothermal gradients. *Amer. J. Sci.* **276,** 1225–1253.

Joesten, R. (1974) Local equilibrium and metasomatic growth of zoned calc–silicate nodules from a contact aureole, Christmas Mountains, Big Bend Region, Texas. *Amer. J. Sci.* **274,** 876–901.

Joesten, R. (1983) Grain growth and grain-boundary diffusion in quartz from the Christmas Mountains (Texas) contact aureole. *Amer. J. Sci.* **283A,** 233–254.

Jones, K. A., Wolfe, M. J., and Galwey, A. K. (1975) A theoretical consideration of the kinetics of calcite recrystallization produced by two basalt dykes in Co. Antrim, Northern Ireland. *Contrib. Mineral. Petrol.* **51,** 283–296.

Kirkpatrick, R. J. (1975) Crystal growth from the melt: A review. *Amer. Mineral.* **60,** 798–814.

Kretz, R. (1966) Grain-size distributions for certain metamorphic minerals in relation to nucleation and growth. *J. Geol.* **74,** 147–173.

Lasaga, A. C., Richardson, S. M., and Holland, H. D. (1977) The mathematics of cation diffusion and exchange between silicate minerals during retrograde metamorphism, in *Energetics of Geologic Processes*, edited by S. K. Saxena and S. Bhattacharji, pp. 353–388. Springer-Verlag, Berlin.

Long, L. E., and Kulp, J. L. (1962) Isotopic age study of the metamorphic history of the Manhattan and Reading Prongs. *Geol. Soc. Amer. Bull.* **73**, 969–996.

Loomis, T. P. (1976) Irreversible reactions in high grade metapelitic rocks. *J. Petrol.* **17**, 559–588.

Loomis, T. P. (1979) A natural example of metastable reactions involving garnet and sillimanite. *J. Petrol.* **20**, 271–292.

Putnis, A., and McConnell, J. D. C. (1980) *Principles of Mineral Behavior*. Blackwell, Oxford.

Ratcliffe, N. M., Bender, J. F., and Tracy, R. J. (1983) Tectonic setting, chemical petrology and petrogenesis of the Cortlandt Complex and related intrusive rocks of southeastern New York State, Field Guide, Northeastern Section. Geol. Soc. Amer. New Paltz, New York.

Ratcliffe, N. M., Armstrong, R. L., Mose, D. G., Seneschal, R., Williams, R., and Barramonte, M. J. (1982) Emplacement history and tectonic significance of the Cortlandt Complex. *Amer. J. Sci.* **282**, 358–390.

Shewmon, P. G. (1963) *Diffusion in Solids*. McGraw-Hill, New York.

Tanner, S. B., Kerrick, D. M., and Lasaga, A. C. (1983) The kinetics and mechanisms of the reaction: calcite + quartz = wollastonite + carbon dioxide (abstract). *Geol. Soc. Amer. Abstr. Progs.* **15**, 704.

Thompson, Jr., J. B. (1959) Local equilibrium in metasomatic processes, in *Researches in Geochemistry*, pp. 427–457. Wiley, New York.

Tracy, R. J. (1982) Compositional zoning and inclusions in metamorphic minerals, in *Characterization of Metamorphism through Mineral Equilibria*, edited by J. M. Ferry, *Reviews in Mineralogy, 10*. Mineral. Soc. America, 355–397.

Tracy, R. J., and Thompson, A. B. (1979) Partial fusion and magma hybridization, Cortlandt complex, New York (abstract). *Trans. Amer. Geophys. Union* **60**, 411.

Turnbull, P. (1956) Phase changes. *Solid State Physics* **3**, 225–306.

Chapter 6
On the Relationship between Deformation and Metamorphism, with Special Reference to the Behavior of Basic Rocks

K. H. Brodie and E. H. Rutter

Introduction

It has long been recognized that deformation and metamorphism are closely interlinked, but although many authors have touched upon various aspects of the subject, there has never been a completely satisfactory understanding of the interrelationships at the mechanistic level. The literature abounds with descriptions of mineralogical changes that mirror the intensity of strain in heterogeneously deformed rocks (e.g., Teall, 1885; Beach, 1973; Kerrich *et al.*, 1977) and with assessments of relative timing of deformation and metamorphism (e.g., Zwart, 1962; Spry, 1969; Vernon, 1977). A number of authors have attempted to identify mechanistic effects whereby the rate of metamorphic equilibration is affected by deformation, the resistance to deformation by particular deformation mechanisms is influenced by metamorphic transformations, or metamorphic segregation might arise during deformation (e.g., White and Knipe, 1978; Rubie, 1983; Gresens, 1966; Beach, 1982; Rutter *et al.*, in press; Heard and Rubey, 1966; Raleigh and Paterson, 1965; Paterson, 1973; Robin, 1978; Gray and Durney, 1979; review of various aspects by Vernon, 1976). The foregoing reference list does not do justice to the number of geologists who have contributed to the subject of deformation/metamorphism interrelationships, which is receiving increasing attention from structural and metamorphic geologists alike.

In this paper we present first an overview of the subject based around a small number of principal types of interactions between deformation and metamorphism. The discussion is concerned mainly with principles but is illustrated with a number of particular examples. We consider hydrous minerals and a fluid phase of pure water, recognizing that modifications are necessary to the arguments if the fluid phase is impure. Second, we examine in some detail the deformation of basic and metabasic rocks, which illustrate some of the aspects of deformation/metamorphism interrelationships. Special attention is given to the deformation of basic rocks in shear zones, where the chemistry and microstructure may be examined after varying amounts of strain.

Interaction of Deformation and Metamorphism

Rocks deform by some combination of (1) cataclasis, (2) intracrystalline plasticity, and (3) diffusive mass transfer processes. It is suggested that, at the most fundamental level, rock deformation can always be analyzed in terms of these processes no matter to what extent the effects of accompanying or following chemical and structural transformations modify or destroy microstructural indicators of deformation mechanism. In general, the microstructures of regional metamorphic rocks reflect postdeformational mineral growth and equilibration (relaxing of stress), and preexisting microstructures become obscured. Some ways in which this can occur are considered briefly below:

1. The effects of experimental cataclastic deformation of a quartzite are preserved as open intergranular microcrack arrays that are easily recognized with the optical microscope. In nature, rocks deformed in this way suffer subsequent hydrothermal cementation that infills cracks with optically continuous quartz, so that the cracks cannot be seen optically except at high magnification by virtue of micron-sized included bubble trails, or by electron-excited luminescence contrast (cathodoluminescence). Long cracks spanning several grains remain visible even after cementation but may tell only part of the story of the deformation process.
2. Deformation commonly forms part of the progress of a regional metamorphic cycle. Often the temperature peak may be inferred to have been reached after the greatest depth of burial had been attained and after the greatest deformation had been imposed. This is reflected in the common occurrence of simple metamorphic isograd patterns superimposed on complex structures. The final metamorphic reactions reflecting the temperature peak are therefore superimposed on previously deformed rock, so that a coarse texture develops, with schistosity being a mimetic feature superimposed on the genuine deformation texture.
3. Monomineralic rocks deformed by high-temperature plasticity commonly remain subject to high temperatures after the stress is relaxed and the microstructure is modified by recrystallization and perhaps grain growth.

The interaction of the deformation and the metamorphism can be considered in terms of either (1) the effect that deformation has on metamorphic reactions, or (2) the effect that metamorphic reactions may have on the deformation mechanism and path. These represent extremes of a continuous spectrum of interrelationships.

The Role of Deformation during Metamorphic Transformations

Broadly, the effect of deformation can be divided into (1) processes that will facilitate reaction kinetics towards the attainment of equilibration to the imposed $P-T-X$ conditions, and (2) the perturbation of the equilibrium conditions owing to imposed stress gradients.

Kinetic Effects

Deformation may enhance the kinetics of reactions in a variety of ways:

1. Deformation will strain the crystal structure and in doing so modify the local equilibrium point defect density. Local mean stress gradients induce gradients in the chemical potential of point defects, and in this way intracrystalline diffusion is enhanced. Plastic deformation, producing an increase in dislocation density, may enhance diffusion coefficients through the pipe effect. In addition, nucleation of new phases on defects is more energetically favorable than in an unstrained structure. Dynamic recrystallization of a strained lattice is an effective way of allowing compositional change.
2. Tectonic reduction of grain size, whether by dynamic recrystallization or cataclasis, will enhance reaction kinetics by increasing the grain surface area available for reaction.
3. If the permeability of the rock is increased either by a reduction in grain size or by dilation owing to crack propagation, fluid will have access to a previously tight rock. A rise in fluid pressure may have various effects:
   a. water is an effective catalyst (see Rubie and Thompson, this volume).
   b. hydration reactions can occur in previously anhydrous rocks, provided that the $P-T$ conditions are such that hydrous minerals are stable.
   c. rocks in which the mineral assemblage was buffering the fluid composition may become buffered by the fluid phase (e.g., Graham *et al.*, 1983). Fluid/rock ratio may become significantly altered.
   d. Chemical transport may become possible over distances substantially greater than grain size.
4. Sufficiently rapid localized deformation may produce shear heating. This has been invoked to explain the lack of retrogression along some major

presently active fault zones (Scholz *et al.*, 1979; Sibson *et al.*, 1979) but is unlikely to be significant in small-scale fault zones (e.g., Brodie, 1981).
5. The presence of a stress gradient may locally increase chemical potential gradients (e.g., Gresens, 1966; Beach, 1982; Rutter, 1983; Rutter *et al.*, in press) and hence the rates of diffusive mass transfer.

Perturbation of Equilibrium

Various mechanisms can be considered whereby it may be possible to perturb equilibrium by deforming a rock, thereby promoting the formation of a mineral or mineral assemblage outside its stability field determined under conditions of $P_{H_2O} = P_{fluid} = P_{total}$ and hydrostatic stress.

1. The effectiveness of increased stored energy of a crystal through plastic deformation in perturbing equilibrium is debatable. Wintsch (this volume) considers theoretically the effect of increasing dislocation density on the activity of quartz in solution and concludes that the increased solubility of highly strained quartz may promote mineral reactions that are metastable with respect to strain-free quartz. The dislocation densities required for this process to be viable are extremely high, several orders of magnitude greater than those normally observed in naturally deformed quartz although the latter are likely to have been reduced somewhat by recovery.
2. Hydration/dehydration equilibria have been studied only for hydrostatic conditions corresponding to $P_{H_2O} \leq P_{fluid} = \sigma_3 = \sigma_1$. Whether the more general case $P_{H_2O} \leq P_{fluid} \leq \sigma_3 \leq \sigma_1$ results in significant modification to equilibrium temperatures for local, variously oriented grain interfaces depends on the chemical properties of water films in stressed interfaces. Such properties are largely unknown. Relaxation of nonhydrostatic stress through creep will mean that any effects will be transient, yet they may be significant. This point is further discussed in the section below on syntectonic dehydration.

Bruton and Helgeson (1983) discuss the effects of departures from the condition $P_{fluid} = P_{solids}$ on mineral equilibria in rocks with pores, although they give little consideration to the permanent deformation that will tend to ensue from the resulting nonhydrostatic stress state.

The Effect of Metamorphic Transformation on Deformation

A succession of mineral and structural transformations may take place during a metamorphic cycle, each occurring over a relatively short time period that corresponds to a particular, narrow range of temperature (Fyfe, 1976). It seems very likely that many kinds of metamorphic transformations may modify rock deformability. "Deformability" is used in this context to describe the ability of a rock to deform at an increased strain rate when loaded under constant applied stress conditions, or to relax applied stress when loaded by fixed displacements applied to the boundaries. In this respect it is equivalent to mechanical softening and nothing is implied in respect of whether the deformation becomes localized into bands or is distributed throughout the rock volume. Such pulses of enhanced deformability will be superimposed on a background strain rate as a result of deformation mechanisms that would operate in the absence of metamorphic transformations. It must not be assumed, of course, that all metamorphic changes are associated with short periods of rapid strain accumulation. There are many examples where microstructural evidence indicates that transformations took place under hydrostatic conditions. Further, substantial strain rate variations may arise between pre- and posttransformation strain rates because of differences in "strength" of pre- and posttransformation mineral assemblages.

There are many possible ways in which the resistance of a rock to deformation can be altered by a syntectonic metamorphic change. These processes can produce "transformation modified deformability," noting that it cannot be assumed that they will always *enhance* the deformability:

1. Diffusion accommodated grain boundary sliding processes may be facilitated through the formation of transiently fine-grained reaction products.
2. Cataclasis may be facilitated by release of high-pressure fluid during progressive metamorphism (e.g., under conditions of rising temperature).
3. Transformation enhanced intracrystalline plasticity may occur, for example, in association with transformations involving significant solid phase volume changes.
4. Point defect chemistry may be modified either by a phase change or a change in the environmental conditions of metamorphism (e.g., P_{fluid}) and can alter deformability in various ways.
5. Variations in stress-induced chemical potential gradients can arise if a metamorphic transformation takes place at some point along that gradient, thereby enhancing creep by diffusive mass transfer processes.

These processes will now be considered in more detail.

Effects of Transiently Fine-Grained Reaction Products

Metamorphic reactions may produce transiently fine-grained products so that deformation by grain boundary sliding accommodated by diffusive mass transfer may be greatly enhanced. An excellent example of this effect is given by Rubie (1983), who describes the transformation feldspar → quartz + jadeite in a quartz diorite in the Sesia–Lanzo Zone, Western Alps. The reaction products form an ultrafine-grained intergrowth pseudomorphing feldspar grains. The pseudomorphs are seen to be preferentially deformed in small shear zones, whilst coarser, original quartz grains are carried along passively in the flowing matrix. During progressive metamorphism such products would be rapidly eliminated through grain coarsening.

It is important to point out that grain size refinement, by whatever process, will not necessarily lead to a switch to fine grain-size superplastic behavior as the dominant deformation mechanism. The combination of temperature, grain size, and grain-size stabilizing factors must be suitable for this to happen (see Rubie, 1984). For example, Schmid *et al.* (1980) found that over a 30% interval of permanent strain accumulation in the experimental deformation of Carrara Marble, the flow stress remained steady although the flow was accompanied by a profound microstructural evolution involving subgrain formation and grain refinement by dynamic recrystallization. Clearly the microstructural changes involved "passive" processes merely accompanying more fundamental intragranular processes (e.g., dislocation/dislocation interactions) that determined the flow stress. Great caution is required in the inference of deformation mechanisms and mechanical properties from microstructural studies alone.

Effects of Evolution of High-Pressure Water during Dehydration Reactions

In many cases progressive regional metamorphism involves a series of continuous and discontinuous dehydration reactions. If pore fluid pressure is initially less than the least principal stress, σ_3, the evolution of water as a reaction product will tend to raise P_{fluid}, subject to the tendency for permeation to lower it. Because dehydration reactions tend to be endothermic, the rate of production of pore fluid will be governed by the rate of heat input and the reaction kinetics (see also Ridley, this volume).

The resistance of rocks to cataclastic deformation is dramatically lowered as the effective confining pressure ($\sigma_3 - P_{fluid}$) is lowered, hence the deformability of rocks is expected to be increased dramatically if pore pressure is allowed to rise through dehydration. This effect is seen clearly in *undrained* experiments (pore pressure dissipation not permitted through the use of a totally encapsulated rock specimen) on gypsum (Heard and Rubey, 1966), serpentinite (Raleigh and Paterson, 1965), and various other hydrous minerals and rocks (Murrell and Ismail, 1976). Some of these experimental data are summarized in Fig. 1. In each case, the drop in strength associated with the dehydration reaction may be seen. The weakening is also associated with a transition from ductile flow to failure by faulting. The strength of the anhydrous product phase(s) is usually greater than that of the initially hydrous rock, provided that grain size is not so fine that diffusion-accommodated grain boundary sliding supervenes as the dominant deformation mechanism.

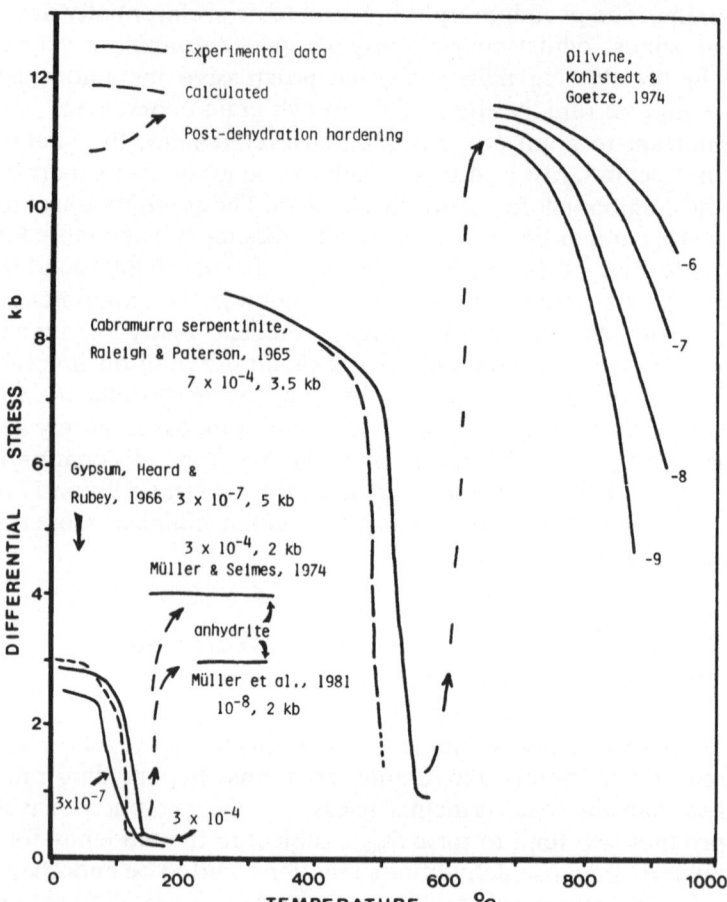

Fig. 1. Graph of differential stress at fault localization or 5% strain for serpentinite and gypsum deformed through dehydration reactions. Solid lines show experimental data of Raleigh and Paterson (1965) for serpentinite and of Heard and Rubey (1966) for gypsum. Strain rate in sec^{-1} and confining pressure in kb are indicated with each curve. Dashed lines show strength calculated assuming undrained conditions, using strength vs. confining pressure data for serpentinite of Raleigh and Paterson (1965) and for gypsum from Heard and Rubey (1966) and Handin (1966), and assuming the applicability of the effective stress principle. Dehydration data for serpentinite and gypsum were obtained from Raleigh and Paterson (1965) and Zen (1965), respectively. The gypsum dehydration produces 44% volume (relative to solids plus water) of water, and the serpentine dehydration about 35%, at the pressure/temperature conditions indicated. The strength drop accompanying each dehydration is accompanied by embrittlement of the material. The strength of rocks composed of typical dehydration reaction products are also shown for comparison. The four curves for olivine are labelled according to $\log_{10}$ strain rate (sec^{-1}).

Neglecting for the moment any effect that nonhydrostatic stress might have on the reaction equilibrium (e.g., Bruton and Helgeson, 1983), the pore water pressure will rise to progressively higher levels given by points on the univariant equilibrium curve (ignoring probable overstepping of the boundary to account for the kinetics of nucleation of new phases) as the temperature is raised above that of the onset of dehydration. Given the rate at which the strength of the rock varies with effective confining pressure, we may then predict the way in which the strength decreases because of the effective pressure effect as temperature is raised through the dehydration interval, assuming that loss of pore water does not keep pace with its production. Such estimated curves are also shown on Fig. 1.

Cataclastic deformation generally involves dilatancy, with a concomitant tendency to lower pore pressure, which leads to dilatancy hardening. Sufficient dehydration may therefore be required at any temperature to pressurize a pore volume amounting to several percent of the rock volume. However, most dehydration reactions can provide more than 10% of the rock volume as water at the appropriate temperature and pressure. Even pelitic rocks can yield $\sim 10\%$ of their volume as water. Experimental data indicate that rock permeability increases with decreasing effective confining pressure (e.g., Brace et al., 1968), and if fluid is produced sufficiently rapidly that σ_3 is exceeded by P_{fluid}, hydraulic fracture of the rock may dramatically increase rock permeability. Walther and Orville (1982) appeal to hydraulic fracture as a major mechanism for dewatering. However, despite the implied increase in permeability accompanying dehydration, in nature rate of fluid loss need not exceed its rate of production if, for example, fluid is impounded by overlying strata of cooler, stronger, impervious rock. The common occurrence of hydrothermal mineral veins implies the existence of the zero effective pressure condition from time to time. A further factor offsetting the effect of permeability increase is the possibility of intracrystalline plasticity at high temperatures (greater than 600°C in the case of rocks dominated by amphibole and pyroxene) that will facilitate compaction. A decrease in porosity will tend to increase P_{H_2O}, favouring cataclasis, which in turn will enhance dilatancy so that a dynamic balance is achieved. This effect has been demonstrated experimentally for limestone (Rutter, 1972, 1974). The effectiveness of these various competitive processes during synmetamorphic deformation can in principle be explored further by numerical modelling.

In addition to the *mechanical* effects of elevated pore pressure, we might anticipate that the new mineral products of the dehydration reaction may be transiently fine-grained, leading to enhanced deformability of the aggregate through diffusion-accommodated grain boundary sliding processes (see *Effects of Transiently Fine-Grained Reaction Products,* above). The possibility also exists that in response to the change in the metamorphic fluid environment deformability of particular solid phases by intracrystalline plasticity may be modified via the processes described below in the section entitled *Effect of Metamorphic Fluid on Plasticity through the Modification of Point Defect Chemistry.*

It is less clear whether the deformability of rock during dehydration can be modified under strictly *drained* conditions, i.e., when the pore fluid pressure is buffered by external factors, so that dehydration will not lead to the impounding of high pore pressures. If intergranular water can be trapped within a stressed interface between two grains such that its activity is less than that which would obtain if the interfacial normal stress were exerted by the water alone, then dehydration of the solid will begin at a lower temperature than predicted by the univariant curve for $P_{H_2O} = P_{total}$ conditions. Thus, dehydration under nonhydrostatic conditions might be compared with dehydration under hydrostatic conditions when water forms only one of the components of the fluid phase (e.g., Greenwood, 1961). Raleigh and Paterson (1965) interpreted aspects of their experimental results on the strength of partially dehydrated serpentinite in this way. The transient existence of overpressured water in the most highly stressed grain interfaces might then facilitate cataclastic deformation. Experimental work is currently underway by the authors on the dehydration of serpentinite in order to investigate this question. It is already clear that a dramatic weakening does result from dehydration, without any rise of pore pressure, the effect being most pronounced at low strain rates. However, it is not yet apparent what physical mechanisms are involved.

The final microstructure of a rock is likely to be imposed under low stress conditions immediately after the dehydration and strain accumulation period and may reflect the kinetic factors in mimetic crystallization and grain coarsening rather than the deformation processes operative during the dehydration reaction interval.

Enhancement of deformability during dehydration reactions probably represents the most dramatic expression of the interrelationships between deformation and metamorphism, as witnessed, for example, by the ductile deformations resulting in the formation of large amounts of schists from pelitic cover rocks in orogenic belts. It may be of considerable significance in the deformation of hydrated metabasic and ultrabasic rocks of the oceanic lithosphere. These are expected to dehydrate following subduction, leading to enhanced deformability of a subducted slab, whether through ductile deformation or by facilitating deep focus seismicity (e.g., Raleigh and Paterson, 1965). Raleigh (1977) discusses the role of dehydration reactions in facilitating crustal seismicity generally. Evolved water is also thought to play a significant role in the generation of magmas in the subduction environment (e.g., Wyllie, 1971).

Transformation-Enhanced Plasticity

Transformations can enhance the plasticity of a rock in various ways. Both displacive and reconstructive phase changes involve a volume change at the transition, and it has been suggested that these can generate stresses that will assist in overcoming the resistance to intracrystalline plastic flow (Greenwood and Johnston, 1965). Poirier (1982) reviews this phenomenon and presents an analysis aimed at estimating strain rate increase or stress drop as a material is driven through a phase change. Various authors have speculated on the importance of this process, especially with regard to changes occurring in the earth's deep crust and mantle. Gordon (1971) suggested that the olivine–spinel transformation might lead to enhanced plasticity deep in the mantle and described experiments on rubidium iodide to illustrate the effect. Gordon was careful to separate the strain resulting from the volume change from any resulting from the additional deformability accompanying the transformation. Sammis and Dein (1974) described experiments on caesium chloride that was sheared as it passed through a phase transition and presented microstructural evidence that the softer phase was plastically more deformed in the vicinity of the phase boundary. At higher levels in the mantle, the basalt–eclogite transition may be associated with similar effects.

The production of large strains by this process seems to require repeated cycling through the phase transition. Hence its significance in geological situations in which only one cycle might be possible is questionable.

Reactions between different solid phases and/or fluids can in principle produce products that are softer than the reactants, thereby enhancing the plasticity of the rock *after* the reaction (White and Knipe, 1978). However, prograde reactions tend to produce minerals that are more resistant to plastic deformation than their hydrated "parents" (see *Effects of Evolution of High-Pressure Water during Dehydration Reaction*, above) so this process will mainly be effective during retrogression.

Effect of Metamorphic Fluid on Plasticity through the Modification of Defect Chemistry

The concentration and mobility of intracrystalline point defects will affect the rate of any flow process that involves diffusion, whether impurity or self diffusion. The kinetics of creep involving dislocation glide and climb, dynamic recrystallization and grain boundary migration, and flow by diffusive mass transfer processes are diffusivity dependent. The concentration and mobility of point defects can be profoundly affected by extrinsic environmental factors. Recently, attention has been focussed on chemical equilibrium between point defects in minerals and the intergranular fluid (e.g., Hobbs, 1981; Dennis, 1981). The concentration of defects that are important in diffusion creep, e.g., oxygen and cation vacancies, can prove to be heavily dependent on the chemical potentials of the components of water and dissolved ions, through defect-forming reactions between the solid phase and the environmental fluid. Further, the mobility of point defects may be altered through the modification of the electronic structure, which determines bonding.

Because of its importance in the framework of many crustal rocks, quartz has received special attention in this respect, through the phenomenon of hydrolytic weakening, in which the strength and ductility are sensitive to the concentration of structurally bound "water" or the pressure of intergranular water (e.g., Tullis *et al.*, 1979; Kekulawala *et al.*, 1981; Dennis and Atkinson, in preparation). Thus the resistance of quartz to plastic flow may be changed in response to a metamorphic reaction that also modifies the metamorphic fluid environment (e.g., dehydration reactions). Of relevance to basic and ultrabasic rocks there is a growing body of evidence on the interrelationships between the creep properties, electrical properties, and defect chemistry of olivines. The defect state depends upon the amount of iron in solid solution (Schock, 1983). The creep behavior of pure fosterite is therefore independent of environmental P_{O_2}, whilst the creep rate of Fo_{90} is proportional to $P_{O_2}^{1/6}$ (Ricoult and Kohlstedt, 1983). It seems likely that the creep properties of many more rock-forming silicates will be similarly sensitive to metamorphic fluid environment. It should be remembered, however, that not all impurity-controlled defects soften materials. Some can produce hardening. The effect may be seen in the contrasting behavior of germanium doped respectively with gallium (p-type) and arsenic (n-type) (Patel and Chaudhuri, 1966). The former raises the upper yield stress whereas the latter lowers it, in proportion to the amount by which the extrinsic defect concentration exceeds the intrinsic concentration at a particular temperature. Westwood and Goldheim (1968) also noted the reduction of the mobility of near-surface dislocations in calcium fluoride through the influence of surface adsorbed water.

Effect of Metamorphic Transformations on Stress-Induced Diffusive Mass Transfer

When flow by water-assisted diffusive mass transfer processes leads to precipitation of overgrowths that are of different mineralogical composition to the host grains, i.e., they are the products of a metamorphic reaction, the deformation mechanism can be termed "incongruent pressure solution" (Beach, 1982). The chemical potential gradient driving the diffusive mass transfer may be steepened through the volume change terms in the expression for the free energy of the reaction, thus enhancing the strain rate through the diffusion kinetics (Beach, 1982; Rutter, 1983; Rutter *et al.*, in press). Preserved overgrowths are usually the products of a retrograde metamorphic reaction, produced through the partial destruction of a phase assemblage stable at higher temperature. The preservation of such textures indicates that complete chemical equilibrium has not been attained. Products of reactions under increasing temperature are not normally preserved as overgrowths because such reactions often go to completion, destroying evidence of earlier events in the deformation history. Further, incongruent pressure solution involving precipitation of product phases with smaller molar volumes than the reactants probably *slows* the creep rate by a small amount, relative to creep under conditions of reprecipitation of the same phases.

So far we have considered possible ways in which deformation and metamorphism may interact or influence each other, but the effectiveness of many of these processes in nature is unknown. The mineral phases that constitute the rock together with the environmental conditions will dictate which of these processes may be most significant.

Metabasic Rocks

The interplay of some of the processes outlined above can be illustrated by considering the deformation and metamorphism of basic rocks. A knowledge of the mechanical behavior of the mineral phases involved is essential in order to deduce likely deformation mechanisms from the observed microstructures, and these are briefly reviewed in Appendix I. The mineral assemblage (including mineral compositions) in the rock can be used to infer likely environmental conditions prevailing during deformation, or at least the relative changes that have occurred in these conditions. Particular attention will be given to microstructures observed in retrogressive shear zones because these often represent a situation in which the microstructure related to syntectonic metamorphism is preserved, although it too may be modified by later effects, and where the undeformed mineralogy and microstructure is generally preserved outside the shear zone.

The general features of the mineralogy and microstructures of deformed metabasic rocks are outlined below.

Mineral Assemblage and Composition as a Function of External Parameters

The mineral assemblage in metabasic rocks is generally rather constant through greenschist, blueschist, and lower amphibolite facies: amphibole + plagioclase + chlorite + epidote + Ti phase + Fe oxide + quartz, carbonate, mica (the "common assemblage" of Laird and Albee, 1981; Laird, 1982). In many metabasic rocks at amphibolite and granulite facies the assemblage amphibole + plagioclase + pyroxene + Ti phase + Fe oxide ± garnet can be considered as common. The fluid phase is considered to be pure water, but the effects of introducing CO_2 are discussed, for example, by Harte and Graham (1975). The effect is significant for small mole fractions of CO_2 when solid-phase carbonates are present.

While the mineral assemblage remains relatively constant, the mineral

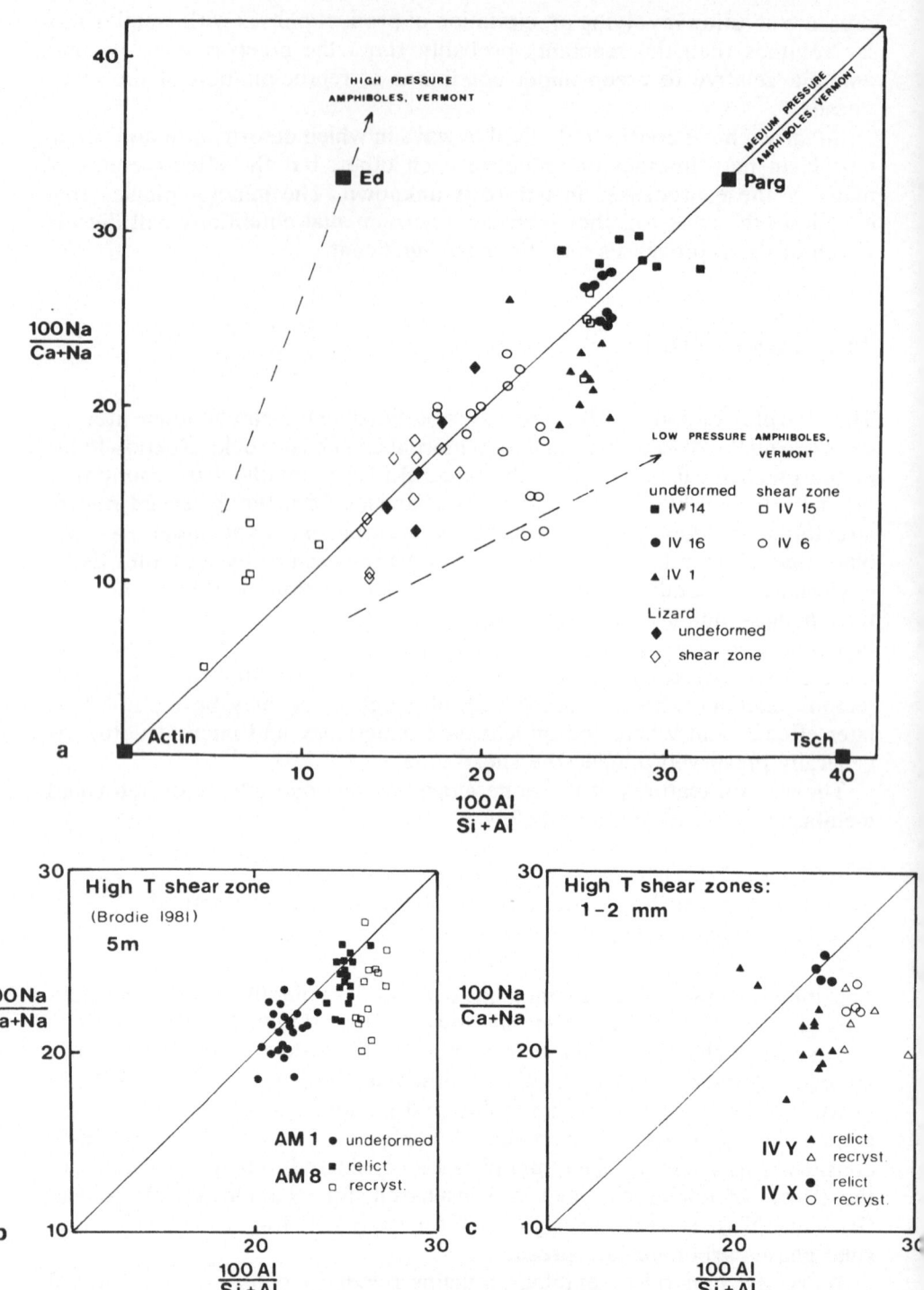

compositions and modal concentrations vary continuously. These variations have been discussed by many authors (e.g., Miyashiro, 1961, 1973; Laird, 1982) and only a few general points will be made here.

Studies of the mineral chemistry of amphibole and plagioclase in metabasic rocks from many metamorphic terrains indicate that systematic variations in composition occur as a function of P and T (e.g., Wiseman, 1934; Cooper and Lovering, 1970; Laird, 1980; Laird and Albee, 1981). These variations have been confirmed by experimental work on rocks of basaltic composition (Liou et al., 1974; Spear, 1981; Moody et al., 1983), but as yet attempts to quantify the exchange equilibria between amphibole, plagioclase, and chlorite have met with only limited success (Spear, 1980). The major problems in utilizing mineral compositional variations in basic rocks to indicate $P-T$ variations are the marked influences that bulk rock composition, and composition of the fluid phase and f_{O_2}, have on the mineral chemistry and mineral stability fields. Amphibole composition is a reliable indicator of grade only if the amphibole forms part of a "common" mineral assemblage (Laird and Albee, 1981). In the examples used in the subsequent discussion, comparison of mineral compositions between different samples is only used to infer relative $P-T$ changes when similarities of mineral assemblage and rock composition can be demonstrated. In these examples, many of the compositional changes observed in amphiboles and plagioclases during retrogression in deformed zones appear to be the reverse of those observed on a regional scale during prograde metamorphism (Fig. 2). Diagrams showing the variation in total Na/Ca with total Al/Si in amphibole are used here to illustrate this point (Fig. 2). Obviously other cation exchanges can be used but, as pointed out by Laird and Albee (1981), this diagram is indepen-

---

Fig. 2. Variation in Na/(Na + Ca) with Al/(Al + Si) content of amphibole for: (a) Representative Ivrea Zone metabasic rocks IV 1, 16 and 14 (in order of increasing metamorphic grade) and equivalent medium temperature (IV 6, 15) to low temperature (IV 15) shear zones. The retrograde metamorphic changes seen localized in shear zones produce changes in amphibole composition that are similar to, but the reverse of, the regional prograde variations. The field of amphiboles from the Lizard shear zone is also indicated, which shows no systematic variation between deformed and undeformed grains. (b) High-temperature mylonite (from Anzola, Ivrea Zone) showing variation in amphiboles from undeformed (AM 1) and deformed (AM 8) rocks. The shear zone is ~5 m wide and isochemical (described in Brodie, 1981; an intermediate sample AM 6 is omitted for clarity). (c) Two small-scale (1 to 2 mm wide) high-temperature shear zones. The variation in composition between relict and recrystallized grains indicates the deformation occurred at high temperature and possibly rather low pressure than the conditions under which the undeformed metagabbro reequilibrated. No new minerals were formed in these shear zones. The trends for amphiboles from high-, medium-, and low-pressure regional metamorphic terranes as described by Laird and Albee (1981) are indicated.

dent of the method of calculation of Fe^{2+}/Fe^{3+} and of assignment of cations to sites within the structure.

Microstructural Characteristics

In order to interpret microstructures in metabasic rocks in terms of deformation mechanism, attention must be given to the mechanical properties of the main minerals concerned: amphibole, plagioclase, and pyroxene. A brief summary of the available information concerning the mechanical behavior of these minerals under a range of physical conditions is given in Appendix I. This mainly considers the effect of temperature (and/or strain rate) on the mechanisms of deformation. The possibility of a negative pressure sensitive behavior of these minerals during plastic deformation is discussed in Appendix II.

When one or other of these minerals occurs together with carbonates, quartz, or olivine in naturally deformed rocks, they are always seen to be relatively resistant to plastic deformation. Such observations concur with the results of limited experimental studies (e.g., Rooney *et al.*, 1974; Tullis, 1983; Dollinger and Blacic, 1975; Coe and Kirby, 1975).

Even for the case of deformation by brittle fracture and faulting, basic rocks are notoriously tough. Fracture mechanics data compilations show many basic rocks have room-temperature fracture toughnesses greater than 3.0 $MPa.m^{1/2}$, in contrast to many granitic rocks whose fracture toughnesses range between 1.5 and 2.5 $MPa.m^{1/2}$ (Atkinson, 1982).

Many microstructural features of regionally metamorphosed rocks are due to recrystallization under increasing, or simply high, temperature and provide little information on the deformation mechanism, particularly in high grade examples. Most information on deformation mechanisms comes from study of retrograde metamorphic microstructures where reactions rarely go to completion, or from the preservation of intragranular deformation induced microstructure (always remembering that observable deformation features are not always indicative of the process that led to greatest strain accumulation). "Retrograde" is used here and subsequently in the sense of the production of a mineral assemblage that would be formed under conditions of decreasing temperature, all other factors being constant. It is recognized that variations in other factors, e.g., total pressure, water pressure, oxygen fugacity, etc., at constant temperature can result in similar "retrograde" assemblages. The continued presence of elevated water pressure during retrogression to catalyse deformation and mineral reaction may critically determine whether deformation microstructures are preserved.

Prograde metamorphism of basalt through greenschist to amphibolite facies must involve initial hydration of the igneous rock and progressive dehydration, whilst metamorphism of a gabbro in the granulite or amphibolite facies involves mainly microstructural adjustments or hydration reactions, respectively. Most greenschist facies metabasic rocks exhibit an intense schistosity resulting from the alignment of platy chlorite and acicular actinolite. These minerals rarely show evidence for strain and have grown mimetically along an existing schistosity (Plate 1(a)). Fine (mineralogical) compositional banding may possibly be present because of original banding, but in most cases as a result of metamorphic differentiation produced by lateral movement of components in response to normal stress gradients (e.g., Robin, 1978). This can often be demonstrated when the mineral banding transects original coarser features such as lithic fragments, lava pillows, or coarse sedimentary banding. In some cases the extreme drawing out of originally larger grains of feldspar or ferromagnesian minerals may produce a similar microstructure.

With increasing grade the proportion of amphibole increases partly at the expense of chlorite and is initially of actinolitic composition. Actinolite generally occurs as thin prisms, often of euhedral habit, and the growth of these mimetically along an existing schistosity commonly imparts a good lineation in upper greenschist to amphibolite facies rocks (Plate 1(b)). The transition from actinolitic to hornblendic amphibole with increasing temperature is reflected in a change to more equidimensional grains or stubby prisms with a poor development of prism faces. Hence higher grade metabasic rocks (amphibolites) tend to be more massive and equigranular. The breakdown of amphibole to pyroxene similarly produces a more granoblastic habit (Plate 1(c)). The higher temperatures also favor grain boundary adjustments to approach more closely an equilibrium microstructure. Rocks containing predominantly hornblendic amphibole may still show a good linear fabric. Increasing grade is generally accompanied by a substantial increase in mean grain size.

The relative proportion of amphibole to plagioclase in these rocks exerts a strong influence on the microstructure, particularly at high grades. In amphibole-rich rocks the microstructure is largely controlled by the amphibole, and plagioclase occurs as subhedral grains between a network of amphibole grains. Plagioclase-rich metabasic rocks tend to have a more gneissose texture because of the "softer" nature of the plagioclase. Associated minor minerals such as epidote, sphene, and garnet tend to have little influence on the microstructure.

Retrograde metamorphism may involve the limited introduction of fluids (mainly water) into anhydrous high-grade assemblages. Commonly the products of such reactions are fibrous (e.g., actinolitic amphibole) in contrast to the more equidimensional crystal habits of minerals of prograde assemblages. This partly reflects the nucleation of the minerals epitaxially and lack of subsequent grain growth or microstructural readjustment.

Where complete retrogression occurs, the products often appear strongly deformed, possibly reflecting the weakening of the rock. Numerous examples have been described of retrogression restricted to fractures where fluids had easy access, e.g., during ocean floor metamorphism, since the rate-

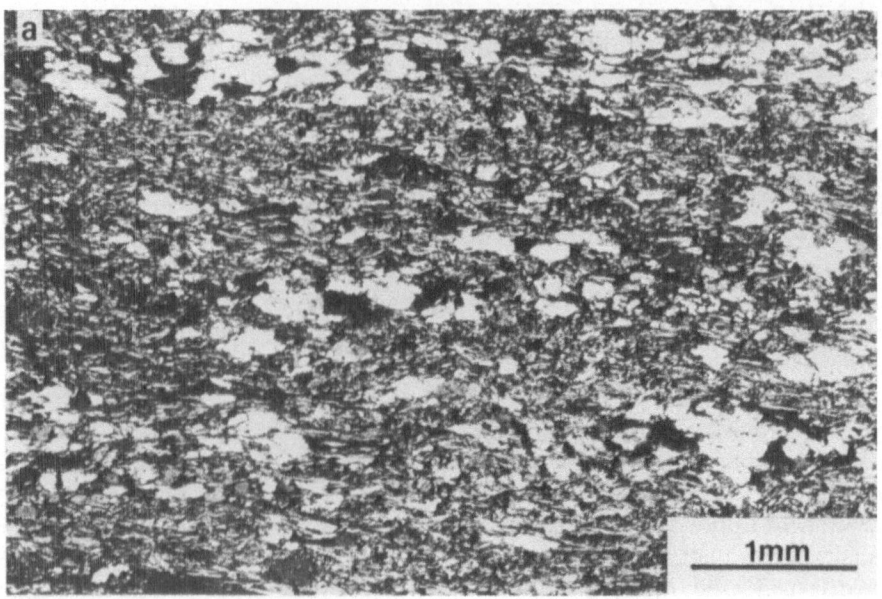

Plate 1. Typical microstructures of regionally metamorphosed metabasic rocks (crossed polars): (a) Greenschist facies (Start Point, southwest England), (actin, chl, Na plag, ep, opaque). (b) Upper greenschist to lower amphibolite facies (IV 1, Ivrea Zone), (Plag (An$_{35\pm3}$), magnesio-hbl, sphene, Fe and Fe–Ti oxides, Fe–sulphide).

Plate 1. (c) Upper amphibolite to granulite facies (IV 16, Ivrae Zone), (plag ($An_{55\pm1}$), ferroan pargasite, cpx, Fe–Ti oxide).

limiting step for the hydration is probably transport of water to the reaction site (Fyfe, 1976).

Static alteration with the formation of corona structures is common in high-grade rocks and these are only preserved where the rock has suffered no subsequent deformation. Coronas may form as a result of changes in several parameters, e.g., cooling, with or without fluid introduction, or changing pressure.

Deformation and Metamorphism in Metabasic Shear Zones

Shear zones have received particular attention because, unlike the case of progressive syntectonic metamorphism, both the deformed and the unde-formed rock is available for microstructural and chemical study (e.g., Beach, 1973; Kerrich *et al.*, 1980; Brodie, 1980). The formation of a hydrated assem-blage, possibly also involving metasomatic alteration, in highly localized shear zones that appear to provide channelways for transport of volatiles, suggest that the deformation and metamorphism in the shear zone are inti-mately related and that the deformability may be reaction enhanced.

Such high strain zones are particularly well developed in virtually anhydrous meta-igneous rocks. Many such zones are assumed to represent syn-tectonic hydration, but it must be remembered that some mechanism must be found initially to introduce the fluid into a tight rock. The possibility must be considered that in some cases the infiltration of fluid is posttectonic, prior deformation being necessary to enhance the permeability, the hydration then taking place under static conditions. Most recognizable shear zones are formed during retrograde metamorphism, often accompanied by hydration, and can also be considered to provide examples of the kinetic effect of deformation enhancing reequilibration to the prevailing P–T–X conditions.

Prograde Shear Zones

Prograde shear zones or shear zones formed at the peak of metamorphism are rare mainly because of the fact that increasing temperature tends to result in homogenization of strain, and some mechanism for localizing the strain into a narrow zone is required, e.g., propagation from lithological heterogeneities or inclusions or through strain-softening mechanisms. In addition, increased temperature following the deformation will tend to oblit-erate such zones by grain growth. It is likely that these finer grained zones will undergo rapid grain growth and may even finish coarser than the host rock, especially if chemical contrasts have been established with respect to the host rock during the deformation period. One example has been de-scribed of shear zones formed under increasing pressure and/or decreasing temperature, where dolerite sills have transformed into eclogite along their deformed margins (Griffin and Raheim, 1973).

An example of a high-grade, apparently prograde shear zone in a meta-gabbro from the Ivrea Zone of Northern Italy has been described (Brodie, 1981) where amphibole and plagioclase recrystallizing in the shear zone reequilibrate to higher temperature conditions with respect to the unde-formed gabbro (see Fig. 2(b)). TEM of amphiboles from these rocks shows evidence of dislocation climb (Plate 2(a), Appendix I). The change of mineral compositions of the amphibole and plagioclase between the shear zone and the undeformed metagabbro was used to infer an apparent increase in tem-perature (bulk chemistry and mineral assemblage remaining constant (Bro-die, 1981)). On a regional scale an increase in An content of plagioclase plus an increase in Na/Ca and Al/Si content of the amphibole is observed with increasing metamorphic grade (Fig. 2(a)). Similar compositional changes have been found in amphiboles recrystallized in this high-grade shear zone and in similar small-scale shear zones (1 to 2 mm wide) and along grain boundaries (Fig. 2(c)). A decrease in oxygen fugacity could produce a similar change in amphibole composition (Spear, 1981) but would also reduce the iron oxide phases, which is not observed. In other shear zones at the same locality, evidence has been found for reaction along the grain boundaries, both within the shear zone and in the undeformed host metagabbro (but more advanced in the former because of the greater grain boundary area), producing a fine-grained intergrowth of plagioclase and orthopyroxene (Plate 2(b)). This reaction, involving the production of orthopyroxene from amphibole, indicates either that high temperatures prevailed after the deformation ceased, presumably also modifying the microstructure of the intergrown orthopyroxene and plagioclase to some extent, or that there was a drop in P_{H_2O}. If the former is true, evidence for dynamic recrystallization observed in the shear zone (Plate 2(c)) may reflect later recovery under conditions of stress relaxation (stress annealing in the sense of Hobbs, 1968) and provide

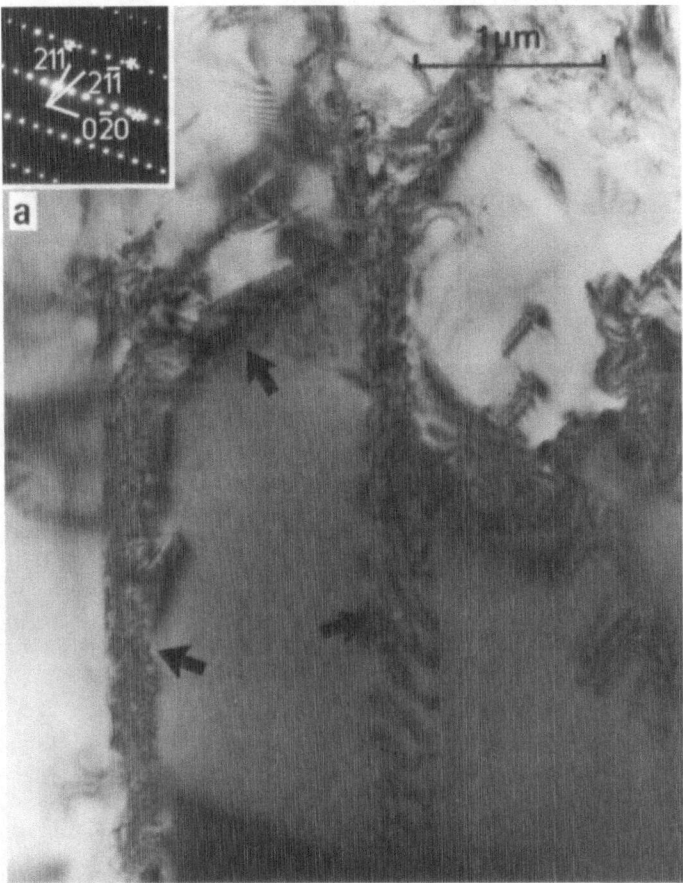

Plate 2. Higher temperature and prograde shear zone microstructures in metabasic rocks from the Ivrea Zone, Western Alps. (a) Transmission electron micrograph of dislocation walls (arrowed) in an amphibole, outlining a euhedral subgrain structure. Dislocation processes were important in the grain-size reduction of amphibole that occurred during the high-temperature deformation.

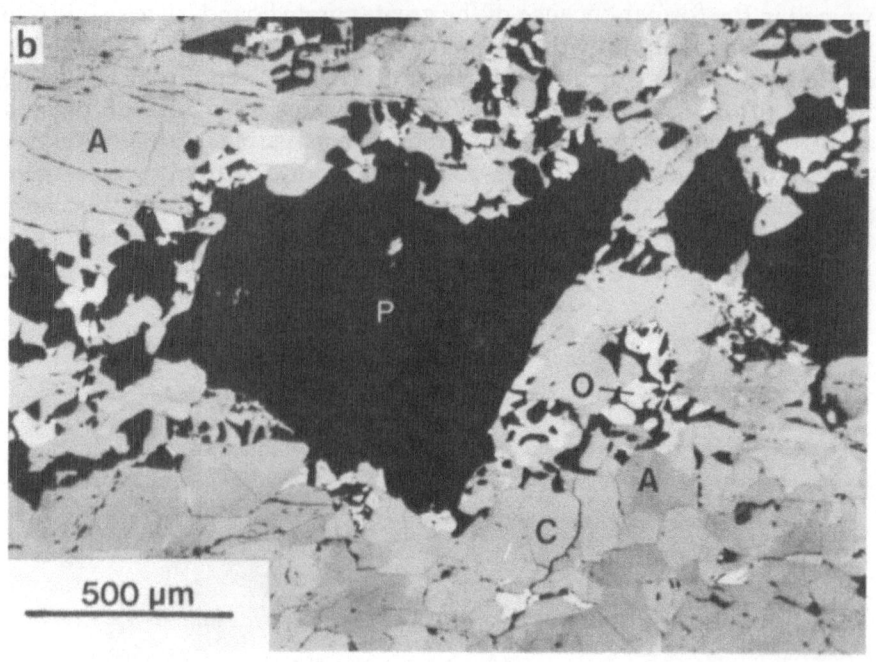

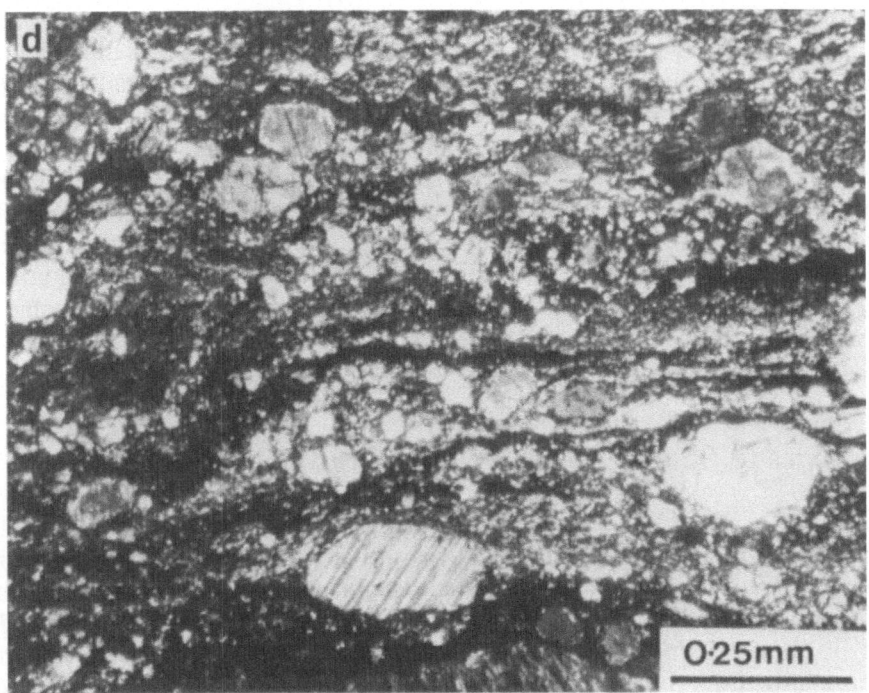

Plate 2. (continued) (b) Micrograph of a high-temperature shear zone in a metagabbro showing a fine-grained intergrowth of orthopyroxene and plagioclase formed along the grain boundaries after the deformation. Imaging by atomic number contrast from backscattered electrons on the SEM. The plagioclase appears black owing to the conditions necessary to bring amphibole and pyroxene into contrast. A: pargasitic hornblende, P: plagioclase (An$_{55}$), O: orthopyroxene, C: clinopyroxene.

(c) Optical micrograph showing microstructure of recrystallized amphibole (ferroan pargasite) in a high-temperature prograde shear zone (AM 8). The formation of a near-equilibrium microstructure with triple junctions and low index rational grain boundaries implies recrystallization but does not preclude the possibility of initial cataclastic deformation followed by high-temperature annealing.

(d) Mylonite in a plagioclase-rich metabasic rock. The light-coloured elongate areas are very fine-grained recrystallized plagioclase whilst the porphyroclasts are of clinopyroxene, amphibole, and garnet. Grain-size reduction of these latter phases appears to be dominantly by fracturing and abrasion, leaving rounded grains of identical composition to the initial grains, with the bulk of the strain in the rock being accommodated by plastic deformation and dynamic recrystallization of the plagioclase matrix.

no information on the deformation mechanisms operative during shear zone propagation, as is probably the case with most prograde regional metamorphic microstructures. The presence of this fine-grained material along the grain boundaries provides an example of the transiently fine-grained reaction products such as may form during dehydration reactions (see earlier).

The compositional change in recrystallized grains (relative to undeformed grains in the host rock) reflects deformation-enhanced kinetics of reequilibration to the prevailing conditions. The amphibole porphyroclasts show chemical zoning suggesting that the increased plastic strain in these grains led to enhancement of intracrystalline diffusion. Hence variation in composition between relict and recrystallized grains cannot be used, in this case, to infer that recrystallization occurred by a process of nucleation and growth.

Retrograde Shear Zones: Medium Grade or Low P_{H_2O}/σ_3

Shear zones are often observed in medium- to high-grade metabasic rocks that appear to involve little or no hydration and where plagioclase and amphibole recrystallize with similar or only slightly different compositions. The amphibole is commonly magnesio-hornblende and the plagioclase intermediate in composition ($An_{60\pm10}$). Microstructures often suggest recovery and recrystallization, and the grains tend to be equidimensional. In other cases (see below), a combination of cataclastic deformation with subsequent infill of voids by transported material can dominate the microstructure. Lack of hydration means that the shear zone was formed at the same temperature as that at which the original host rock equilibrated, or P_{H_2O} was low in absolute terms, or P_{H_2O}/σ_3 was low.

Some of the metagabbro shear zones from the Ivrea Zone involve grain-size reduction of magnesio-hornblende without appreciable compositional change, in contrast to the small recrystallized grains from the high-temperature shear zones. The resultant grain shapes are rather irregular (Plate 3(a)). Using backscattered electron imaging on the SEM (for discussion of the method see Hall and Lloyd, 1981) it appears that this texture is produced by overgrowths of similar (though not identical) composition amphibole on initially cataclastically deformed amphibole. In parts of the sample, narrow fine-grained zones similar to stylolytic stripes cut the rock, producing shape elongation to the amphiboles by truncation. These narrow zones contain fine-grained sphene and minor brown mica in addition to amphibole and plagioclase, suggesting deformation at possibly lower temperatures than those under which the undeformed metagabbro equilibrated, and provide possible evidence that diffusive mass transfer was occurring. The plagioclase composition is variable as a result of alteration after the microstructural development. Irregular marginal alteration to a more sodic composition (e.g., An_{50} to An_{25}) is common in many of the plagioclases investigated here, and similar behavior has been observed in other cases (e.g., Borges and

White, 1980). In the case of fractured plagioclase it is not always clear to what extent the compositional change observed along fractures is due to diffusive alteration or to cementation by new plagioclase.

In rocks rich in minerals such as hornblendic amphibole and pyroxene, which are extremely strong and resistant to creep, it seems likely that cataclasis may be widespread, even at moderate to high temperatures. In the above example, temperatures of ~600°C maybe inferred from the mechanical behavior of the amphibole, or lower temperature conditions but with low P_{H_2O} and high shear stress. An initial cataclastic microstructure may be obscured if, for example, P_{H_2O} rises sufficiently high during subsequent hydration such that dilatant voids become infilled by transported material, as in the foregoing instance. Alternatively, void infill through solid-state creep may occur if high temperatures outlast deformation.

Plate 3. Lower temperature and retrograde shear zone microstructures in metabasic rocks. (a) Optical micrograph (plane polarized light) of a medium-grade metabasic shear zone (Ivrea Zone IV 6) showing the irregular microstructure of fine-grained amphibole (A) and plagioclase (P). No variation in composition is observed between the larger relict grains and the fine-grained material. Backscattered SEM imagery shows amphibole grains overgrown by structurally continuous amphibole with slightly different composition, and the grain-size reduction is inferred to have occurred dominantly by cataclasis.

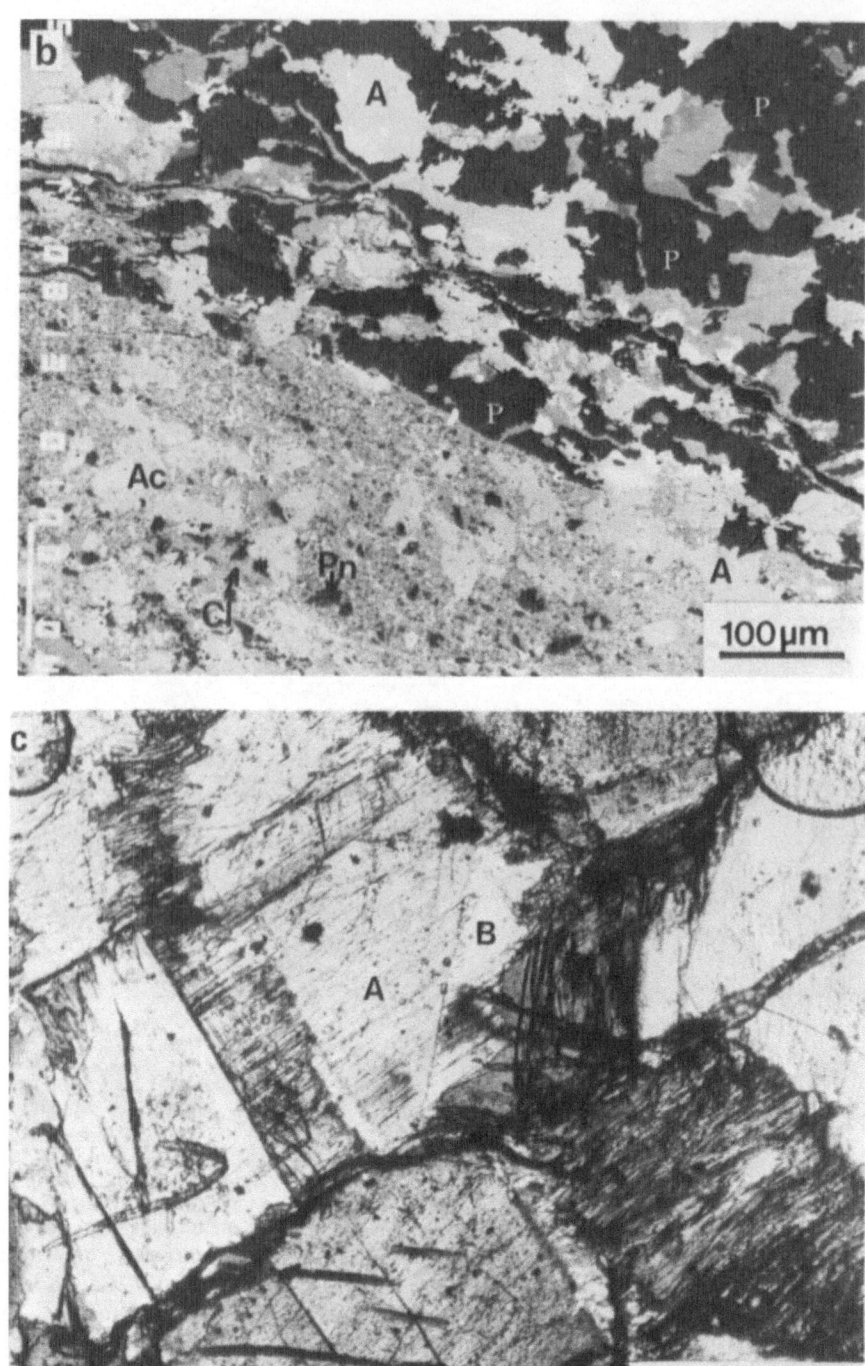

Plate 3. (*continued*) (b) Backscattered electron micrograph of a low-temperature hydrated shear zone (lower left-hand side) containing chlorite (Cl), actinolite (Ac), sodic plagioclase (P), and sphene, cutting amphibolite (Ivrea Zone, IV Y). The amphibolite is cataclastically deformed and the hydrated shear zone is located along a preexisting cataclastic zone. Note the sharp boundary to the zone and the lack of obvious shape orientation of grains within the hydrated zone. (c) Optical microstructure of Loch Alsh amphibolite (northwest Scotland), in which the initial magnesio-hornblende (A) has undergone brittle fragmentation, and the dilatant voids thereby produced have been infilled with oriented growth of chlorite and actinolite. The fragments show marginal alteration to actinolite and a magnesium silicate, probably talc. (d) Backscattered electron micrograph of a deformed amphibolite from the Ivrea Zone (IV Y), showing pervasive cataclasis of the plagioclase (P) and amphibole (A). The plagioclase is altered from An_{70} (grey areas) to approximately An_{25} (dark grey) along the filled cracks, whilst the angular fragments to amphibole (white) are recemented by chlorite (Cl, light grey). Black areas are holes in the section. This initial cataclasis localizes later, lower temperature hydration along small-scale shear zones such as that shown in Plate 3b.

Grain-size reduction either by cataclasis or by recrystallization may allow further flow by diffusion-accommodated grain boundary sliding. An example where this may have been important is an amphibolite facies metagabbro shear zone from the Lizard, Cornwall. The shear zone contains small, often tabular undeformed grains of magnesio-hornblende and plagioclase (An_{37-45}) of compositions similar to those in the undeformed metagabbro, and streaks of iron oxides. An alternative explanation of the tabular grain shapes would be that grain growth had been modified by impingement of the boundaries on the Fe-oxide streaks, although the latter are not always optically visible.

In metabasic shear zones where plagioclase is locally the dominant phase, the plagioclase may recrystallize to very small grain sizes without any change in composition, whilst the associated amphibole and pyroxene (also garnet) are either passively rotated or deform in a brittle manner even at relatively high temperatures. They deform by rupture and sliding (e.g., Allison and LaTour, 1977) or become finely granulated and strung out (Plate 2(d)). An experimentally produced example of such behavior was described by Rutter *et al.* (in press). In the latter case the mafic fragments are carried about in a cataclastically flowing matrix of feldspar fragments that is subsequently completely cemented by structurally contiguous feldspar overgrowths, resulting in a tight rock with a "dynamically recrystallized" aspect. Care is required in the interpretation of strong shape fabrics of minerals such as plagioclase; they are unlikely to be the result of dynamic recrystallization.

Pervasive cataclasis in naturally deformed basic rocks is not too difficult to explain when accompanying hydration suggests that $P_{H_2O} \rightarrow \sigma_3$, so that effective confining pressure is low. However, its pervasive development in rocks such as those described above that have not been retrogressed (to a more hydrous assemblage) suggests that effective confining pressure may have been high and P_{H_2O} low. If, as seems likely, the frictional behavior of rocks over geological time periods is similar to that over the laboratory time scale, shear stresses of several 100 MPa may be required to explain the observed textures.

Retrograde Shear Zones: Medium and Low Grade and High P_{H_2O}/σ_3

Many shear zones in medium- to high-grade metabasic rocks involve the formation of a retrograde greenschist facies assemblage, commonly containing chlorite + actinolite + epidote + plagioclase + sphene ± biotite. In these examples, actinolite is often fibrous and disequilibrium (partially transformed) assemblages are preserved. Hydrous, amphibolite facies assemblages in shear zones may also be produced by hydration of a completely anhydrous host rock, such as a pyroxene granulite. In such amphibolite facies shear zones, some degree of grain coarsening and boundary adjustment often destroys microstructural indicators of deformation mechanisms, so that only a mimetic foliated texture and the more hydrated nature of the assemblage indicate the existence of the shear zone.

Retrograde reactions involving hydration generally involve an increase in solid phase volume that must be accommodated by other processes; for example, it may enhance the role of intracrystalline plasticity if the temperature is still relatively high (see *Transformation-Enhanced Plasticity,* above). During the development of such retrograde shear zones, water is consumed and this may inhibit cataclastic processes, although the latter may have been important in the initiation of the shear zone.

It cannot always be assumed that hydration and deformability in such zones are related. In some examples, it can be observed that hydration has spread sideways, into the undeformed country rock, and it is not always clear whether such lateral spreading is the result of posttectonic hydration of the country rock around the previously hydrated shear zone, hydration precursory to the lateral spreading of the shear zone, or whether *all* the hydration is posttectonic. For hydration to occur posttectonically some process is required for preferential infiltration of the fluid along the zone. A fine-grained rock is not necessarily more permeable than a coarser one and examples have been recorded in ultrabasic rocks where undeformed peridotite becomes serpentinized whilst the fine-grained recrystallized olivine in shear zones remains unaltered (Peters, 1963). However, the frequently observed limited degrees of hydration of basic rock masses testifies to the very low levels of permeability of high-grade rocks, in the absence of special circumstances or processes leading to permeability enhancement. Most deformation zones in metabasic rocks show hydration, implying that water is available to localized volumes of enhanced permeability.

In some cases where the rocks are deformed under conditions of either syntectonic hydration or dehydration, it is possible to recognize a clear relationship between the degree of strain and a metamorphic change. This is when deformation is by "incongruent pressure solution" (Beach, 1982) and strain is accomplished by a water-assisted diffusive mass transfer process. The overgrowths that develop in potentially dilatant volumes are not of host rock minerals but are a product of a metamorphic reaction. Beach (1982) describes and interprets several examples of this process in low-grade metasediments. The same process is observed in natural syntectonic hydration of basic rocks (e.g., LaTour and Kerrich, 1982).

Experimental syntectonic hydration of basaltic rock is reported by Rutter *et al.* (in press). In these experiments overgrowths of amphibole, preferentially developed on interfaces oriented with respect to the applied forces such that they sustained maximum dilation, formed on preexisting feldspar, pyroxene, and olivine. It was concluded that although the development of overgrowths with preferred orientation pointed to direct interrelationship between the progressive deformation and mineral reactions, on the basis of the mechanical behavior there was no substantial enhancement of deformability. The data obtained were interpreted in terms of a mixture of cataclasis with incongruent pressure solution.

Such reactant/product relationships in nature are preserved either because deviatoric stresses could be relaxed at low strains (of the order of a few tens of percent) or because of kinetic factors (e.g., low temperatures or lack of water for complete hydration). The kinetics of the flow by diffusive mass transfer may be enhanced because the chemical potential gradients that drive the grain boundary diffusion may be steepened by the affinity of the chemical reaction (Beach, 1982; Etheridge and Hobbs, 1974; Rutter, 1983). In many cases this effect may be small (Rutter *et al.*, in press) but given favorable circumstances it may occur to such an extent that factors other than the kinetics of diffusion, such as rate of supply of interstitial water, may control the strain rate. Fyfe (1976) suggests that rate of "solution" or equilibration of solid phases with interstitial water at sites of consumption of the original solid phases is likely to be very rapid and hence not rate controlling.

Where overgrowths have been formed syntectonically they tend to be preferentially developed in dilatant interfaces. Retrograde minerals such as amphibole that exhibit a fibrous habit tend to grow in the local dilation direction provided crystal growth can keep pace with dilation rate. In contrast, posttectonic growth of fibrous minerals, or growth under conditions where the crystal growth rate cannot match the local dilation rate, occurs either epitaxially on the host structure, normal to the grain boundaries, or randomly in the fluid-filled voids.

An example of the syntectonic formation of fibrous overgrowth may be seen in a deformed hornblendite from Loch Alsh, Scotland. Magnesio-hornblende is overgrown on dilatant grain boundaries by oriented chlorite and actinolite, the relict grains showing marginal bleaching (optically discernable) and alteration at the contacts with the fibrous overgrowths (Plate 3(c)). Microprobe analyses show the bleaching and alteration products to be actinolite and an intergrowth of an actinolitic-amphibole with another magnesium silicate, probably talc.

Low-temperature hydrated shear zones are similarly present in the Ivrea zone metagabbros and the role of cataclasis in enhancing the influx of fluids may be inferred. Narrow zones (up to 2 cm wide) of completely hydrated ultracataclasite transect wider zones of less intense cataclastic flow (Plate 3(b)). The ultracataclasite zones cut across or form subparallel to earlier higher temperature deformation zones, brecciating the magnesio-hornblende and plagioclase (Plate 3(d)). The plagioclase is altered along fractures from An_{70} to more sodic compositions (An_{25-20}) and fractures in the amphibole are infilled with chlorite, actinolite, and sphene. Transgranular fractures cutting plagioclase are infilled with fibrous chlorite. Locally, coarse-grained epidote veins have formed parallel to the foliation.

In contrast to the example of the Loch Alsh hornblendite (above) where the accumulation of strain and mineral growth remained roughly in step (but following an initial phase of grain cracking without significant fragment displacement), the hydrated shear zones of the Ivrea zone metabasite involved large-scale, rapid cataclasis in which large intergranular voids were produced. The backscattered SEM images show that these voids were passively filled by new metamorphic minerals growing into open voids. It has been suggested by Beach (1973) that dilatancy-producing cataclasis is a necessary precursor or process accompanying shear zone development with associated hydration and other metasomatism and serves to admit the water into a rock that might otherwise remain anhydrous. Our observations of the frequent occurrence of cataclasis in hydrated basic shear zones are in accord with this inference. In the next sections we explore the associations further.

Problems of Interpretation of Deformation Mechanisms of Basic Rocks

Some generalized comments can be made on the problems involved in the interpretation of deformation mechanisms from the microstructures observed in metabasic rocks. These microstructures apply to rocks where crystallization and recrystallization have occurred syntectonically. Static hydration and alteration of metabasic rocks will produce a range of disequilibrium (partially transformed) microstructures often dominated by the growth of ragged amphibole porphyroblasts, epitaxial alteration of ferromagnesian minerals to actinolite and chlorite, alteration of plagioclase to white mica and epidote, etc. These microstructures are mainly controlled by the initial microstructure and the degree of permeation of hydrating fluids and will not be considered further here. There may be ambiguity in the interpretation of some deformation features, and these are discussed further below. It should also be remembered that the degree of development of a particular diagnostic microstructural feature cannot generally be taken to correspond to its efficiency in accumulating strain.

The distinction between grain-size reduction produced by cataclasis as opposed to recrystallization is important yet often difficult to make in aggregates of amphibole, pyroxene, and, to a lesser extent, plagioclase. Most of the criteria applied to quartz aggregates cannot be extended to metabasic rocks because the mechanical behavior of the minerals involved is often very different. Microstructures in fine-grained amphibole aggregates are often the most ambiguous, for cleavage fragments produced by cataclasis may be difficult to distinguish from recrystallized grains where the lower interfacial energy of the low index cleavage faces makes them relatively stable. Subgrain structures similarly tend to be euhedral (Plate 2(a)) so recrystallization by subgrain rotation may produce euhedral grain shapes. A polygonal network of amphibole (e.g., Plate 2(c)) or pyroxene grains is usually taken to indicate that solid-state recrystallization has occurred and that the temperature was relatively high, but often no evidence is preserved to indicate whether initial grain-size reduction was by dynamic recrystallization or by cataclasis followed by annealing or by cementation by structurally contiguous overgrowth of the same phase. Very fine-grained trails of amphibole or pyroxene in a matrix of plagioclase (Plate 2(d)) may also be produced by cataclastic grain-size reduction rather than by dynamic recrystallization. Analyses show that often such grains have the same composition as the porphyroblasts.

Grain shape elongation in plagioclase may be indicative that some form of cementation of trails of fragments by diffusive mass transfer is occurring, because elongate grains are rarely produced by crystal plastic deformation of plagioclase (e.g., Rutter *et al.*, in press, see earlier). A similar example is described above where broken amphibole is overgrown by optically continuous amphibole. It appears that often the effects of cataclasis that are frequently obscure with the optical microscope can be brought into sharp relief using backscattered electron imaging on the SEM. Examples of this from naturally deformed metabasic rocks have been described above (e.g., Plate 3(d)).

Of all the textures of syntectonically hydrated basic rocks perhaps the simplest to interpret is oriented fibrous overgrowths of amphibole developed on preexisting phases. The deformation mechanism is inferred to be some combination of interparticle sliding accommodated by diffusive mass transfer, which also provides the vehicle for the retrogressive metamorphic transformation (e.g., Plate 3(c)).

Preferred crystallographic orientation is often a significant attribute of deformed metabasic rocks. In low-grade examples and in some higher grade rocks, in which amphiboles may exert a fibrous or prismatic habit, preferred orientation may arise from mimetic crystallization or the operation of other mechanisms leading to oriented growth. Although attempts have been made to interpret such fabrics in terms of thermodynamically stable "minimum energy" configurations (e.g., DeVore, 1968), factors influencing their development are not well understood. The thermodynamic approach to prediction of preferred crystallographic orientation ignores the role of the mechanistic path whereby the orientation distribution is achieved. For example, preferred orientation produced by intracrystalline plasticity generally is different from that predicted from thermodynamic considerations alone.

Minerals that characteristically form aggregates of equiaxed grains (e.g., quartz, calcite, olivine, plagioclase, pyroxene, and amphibole at high temperatures) can develop preferred crystallographic orientation as a result of deformation by intracrystalline plasticity. Minerals of deformed metabasic rocks only rarely seem to exhibit intense crystallographic preferred orientation, except as a consequence of growth anisotropy. The infrequent occurrence of crystallographic preferred orientation is probably a reflection of the resistance of amphiboles to intracrystalline plasticity. However, the high-temperature prograde shear zones of the Ivrea Zone do contain equiaxed amphiboles which, on the basis of their pleochroism behavior, exhibit preferred orientation. This observation is consistent with other microstructural features that point to significant deformation by intracrystalline plasticity.

Summary and Conclusions

Interrelationships between deformation and coeval metamorphic transformations have been reviewed in general terms. Deformation processes can (1) facilitate reaction kinetics and enhance the rate of attainment of chemical equilibrium, and (2) perturb equilibrium conditions through the effects of stress gradients. Conversely, the occurrence of a metamorphic transformation may modify the deformability of a rock. Five main interactions were considered:

1. Facilitation of diffusion-accommodated grain boundary sliding processes through the formation of transiently fine-grained reaction products.
2. Facilitation of cataclasis by evolution of high-pressure water during progressive metamorphism.
3. Transformation-enhanced intracrystalline plasticity resulting from solid-state volume changes.
4. Modification of solid phase point defect chemistry in response to changes in pore fluid chemistry.
5. Effects of reaction-induced variations in chemical potential gradients during diffusion in stressed aggregates.

In addition to these effects the pre- and post-reaction mineral assemblages are expected to exhibit different rheological behavior and may be either stronger or weaker, according to the particular transformation and the grain sizes of the minerals concerned.

The mineralogical and microstructural characteristics of metabasic rocks and the mechanical properties of their most common mineral constituents were outlined, and some of the processes of interplay between deformation and metamorphism were illustrated with reference to the behavior of such rocks. Deformation in localized shear zones is of particular importance because the final chemistry in the deformed rock can be compared with that of the undeformed host rock. Often it can be inferred that there was gain or loss of particular elements. The behavior of shear zones in basic rocks formed during prograde and retrograde metamorphism were compared with the help of examples.

Though varying in relative importance, the effects of flow by diffusive mass transfer, intracrystalline plasticity, and, surprisingly, cataclastic deformation even at moderate to high-grade metamorphic conditions are all observed in deformed basic igneous rocks. It is often not possible from study of optical sections alone to infer with certainty whether dynamic recrystallization or cataclasis followed by cementation/annealing was the dominant grain-size reducing process. Use of the backscatter detector on an SEM to image atomic number contrast may facilitate making a distinction under favorable circumstances, but the possibility of deformation initially by cataclasis followed by annealing cannot be ruled out even at high temperatures.

Acknowledgments

We are grateful to David Rubie, Alan Thompson, Stefan Schmid, and Bruce Yardley for discussion and critical comments on an earlier version of this paper. Robert Knipe, Paul Dennis, Norman Fry, Stan White, and others who attended a meeting to discuss deformation/metamorphism interrelations at Imperial College in January 1984 provided much useful discussion that has helped us to improve the presentation of the paper.

This work was supported through U.K. Natural Environmental Research Council grants GR3/3548 and GR3/3848.

Appendix I

Mechanical Behavior of Amphibole, Feldspar, and Pyroxene

Amphibole

Little is known about the plastic properties of amphiboles, and hornblendic amphibole is the only one that has been experimentally investigated. High-pressure/high-temperature triaxial experiments indicate that it is one of the strongest silicate minerals, and in the presence of weaker layer silicates (e.g., chlorite, biotite) remains unaffected while the layer silicates deform (Rooney et al., 1974). It appears that in any rock, whether naturally or experimentally deformed, it behaves passively during deformation unless it is the phase forming the load-supporting framework.

Most experimental deformation of amphibole indicates that mechanical twinning on (101) is often the dominant mode of deformation, with brittle failure occurring in some cases (e.g., Rooney et al., 1974). Kink bands have been observed in some highly deformed grains suggesting glide on (100)[001]. Dollinger and Blacic (1975) conclude from experimental deformation studies on a hornblendite that at temperatures around 600–750°C amphibole deforms by slip on (100)[001] producing kink bands, whereas at higher temperatures (800°C) slip occurs on the (100) plane but in variable directions. They did not observe any mechanical (101) twinning and point out that some optically observed (101) lamellae that have been interpreted as mechanical twins may in fact be exsolution features.

Plastic strain features in naturally deformed amphiboles are relatively rare, but slip on (100)[001] is thought to occur. Transmission electron microscope examination of amphibolites from the Ivrea Zone of northern Italy that were deformed under upper amphibolite to granulite facies conditions show that dislocation processes played an important role in the deformation. Many of the dislocations appear to be dissociated and have related planar stacking faults, and abundant dislocation walls are observed. These often form euhedral subgrain structures (Plate 2(a)) suggesting that dynamic recrystallization occurs at least partly by subgrain rotation. Exsolution lamellae of clinopyroxene are relatively common in undeformed amphibolites, but no mechanical twinning has been observed.

As a broad generalization, metabasic rocks deformed at temperatures in excess of 600°C show evidence of plastic deformation of amphibole (providing it is the load-supporting phase), whereas at lower temperatures the amphibole deforms in a brittle manner or contributes to creep by grain boundary diffusion. LaTour and Kerrich (1982) describe two "spatially related" shear zones in basic rocks in which plastic deformation is dominant in the one formed at temperatures of 600°C, whilst fibrous overgrowths formed in the lower temperature one (540°C, temperature estimates from oxygen isotopes). Where amphibole is not load supporting it either behaves passively as a hard inclusion or deforms by fracture, "eroded" by the flow of the surrounding matrix. Allison and LaTour (1977) described hornblende within a gneiss deformed at garnet metapelite grade by the formation of oriented fractures, sliding along fractures, and consequent rotation of fragments producing a preferred orientation.

Cataclastic deformation of amphibole, even at relatively high temperatures, is probably more widespread in nature than has previously been recognized because of the difficulty in recognizing cataclastic microstructures optically. Examples of this are described in the text.

Plagioclase

Plagioclase has been observed to deform both experimentally and in nature by brittle fracture, mostly along cleavage planes, by mechanical twinning, and by slip resulting from generation and motion of dislocations. The mechanical behavior of plagioclase is partly dependent on the structural type and state of order (a recent review by Tullis (1983) provides a useful summary of the literature).

At low temperatures, plagioclase is relatively weak, and the dominant deformation mechanism is grain scale cracking or distributed microcracking. Brittle fracture in plagioclase is often obvious under the optical microscope owing to offset of twin lamellae. At higher temperatures (>300°C) it is relatively much stronger than other phases and unless it constitutes a major fraction of the rock tends to remain relatively undeformed when associated with weaker minerals such as quartz and phyllosilicates. Mechanical twinning on the albite and pericline twin laws is a common deformation mechanism in experimentally deformed plagioclase at the lower temperatures or faster strain rates. Evidence from naturally deformed rocks indicates that twinning is also important at the higher metamorphic grades where the structure is disordered.

At temperatures >550°C there is abundant evidence for slip (Marshall and McLaren, 1977) accompanied by recovery and recrystallization in both experimentally and naturally deformed plagioclase. A number of slip systems have been observed in experimentally deformed plagioclase, and cross slip appears to be common. Moving dislocations leave behind a planar fault in the structure, which defines the slip plane. The fault vector is normal to the fault plane and is due to either destruction of short-range order or imperfect reconstruction of Al/Si-O bonds after the passage of the dislocation (Marshall and McLaren, 1977). Dissociation of dislocations is predicted for certain feldspar structures, but no examples have been recorded.

Recrystallization appears to occur by a mechanism of subgrain rotation in disordered intermediate plagioclase at upper amphibolite to granulite facies conditions, but by a nucleation mechanism at lower temperatures when ordered (for review see Tullis, 1983). The very fine grain size of recrystallized plagioclase may allow diffusion-accommodated grain boundary sliding to occur and possibly lead to superplastic behavior, as for example inferred by Kerrich *et al.* (1980) for the Mieville mylonites, if the fine grain size is maintained by dynamic recrystallization. Recrystallization accompanied by small chemical changes has been observed in naturally deformed plagioclase (e.g., White, 1975; Borges and White, 1980). Generally, however, these changes in composition are in response to changing imposed physical conditions of metamorphism. Relict plagioclase grains do not become highly flattened even at high strains of the aggregate. Plagioclase appears to undergo relatively rapid alteration to more sodic compositions along fractures and grain boundaries, and this may influence its mechanical behavior during subsequent deformation.

Most studies on the mechanical behavior of plagioclase have involved rocks where a softer mineral such as quartz is the dominant phase. In contrast, in metabasic rocks at medium to high metamorphic grades, plagioclase is often the softest mineral (relative to pyroxene and amphibole) and accommodates much of the strain.

Pyroxene

For both monoclinic and orthorhombic pyroxenes the crystal structure suggests that easy slip should occur on the (100)[001] system, and this has been verified in experimentally and naturally deformed pyroxenes (e.g.,

Etheridge, 1975; Ave Lallemant, 1978; Ross and Neilsen, 1978). In experimentally deformed pyroxenes slip on (001)[100] has also been observed, and various other minor systems have been identified. Mechanical twinning producing (100) and (001) twins have been observed in experimental deformed diopside, and Ave Lallemant (1978) describes a transition from deformation dominated by (100)[001] mechanical twinning at low temperatures and high strain rates, to relatively nonselective translation gliding, polygonization and recrystallization at higher temperatures and slower strain rates. The latter appears to dominate in nature. In enstatite a shear-induced transformation can produce clinoenstatite lamellae parallel to (001) planes (e.g., Coe and Kirby, 1975).

Climb is relatively difficult in pyroxenes because of the dissociation of dislocations (Kohlstedt and Van der Sande, 1973) into partial dislocations separated by stacking faults. Hence under most crustal conditions pyroxenes deform plastically by slip of dislocations producing such features as deformation bands and grain shape elongation by slip on (100). The separation between partial dislocations is commonly variable, for example 0.1 to 100 μm in naturally deformed orthopyroxene (Kohlstedt and van der Sande, 1973). In mantle-derived kimberlite nodules, pyroxene recrystallization to a very fine grain size has been observed and interpreted in terms of deformation by grain boundary sliding with "superplastic" behavior (Boullier and Gueguen, 1975).

Appendix II

Possible Pressure Sensitivity in Resistance to Plastic Deformation

At least in the case of halite it has been determined that the resistance to plastic flow decreases with increasing hydrostatic pressure by about 30% per 1 GPa (for discussion see Nicolas and Poirer, 1976, p. 231). This is attributed to the facilitation by pressure of recovery by cross-slip of dissociated dislocations. It has been determined that stacking fault formation involves local dilation of the crystal structure by ~30% (Fontaine, 1968), hence the formation of stacking fault ribbons is suppressed by elevated pressure. Similar negative pressure sensitivity has been noted in the experimental deformation of calcite rocks where intracrystalline plasticity was dominant (Heard, 1960; Rutter, 1972) but the cause of the effect remains unknown.

Both amphiboles and pyroxenes develop dissociated dislocations, and although it is not known in these cases if dissociation involves a volume increase, the possibility that it does, leading to enhancement of recovery rates at high pressure, should be considered. It appears, for example, from data published by Ave Lallemant (1978, fig. 5) that in the experimental deformation of diopside single crystals there is a strength drop as confining pressure is raised from 500 MPa to 1.5 GPa at 800°C, when deformation is by intracrystalline plasticity. Pressure enhanced weakening of Hale Albite was described by Tullis *et al.* (1979), and although interpreted in terms of facilitation of hydrolytic weakening by pressure, alternative explanations such as the above might obtain. Observations on experimentally deformed plagioclase (Marshall and McLaren, 1977) showed that planar faults associated with dislocations have fault vectors normal to the fault plane and hence might represent dilation across the fault plane (Clareborough, 1973). Poirier and Vergobbi (1978) discuss the possible role of dislocation dissociation in olivine and postulate a negative contribution to the hydrostatic pressure sensitivity of the resistance to flow controlled by cross-slip of screw dislocations.

Experimental deformation of dolerite (Caristan, 1982) at various pressures and temperatures up to 1000°C displayed negative pressure sensitivity of strength for part of the high temperature regime of behavior. A pressure increase of 300 MPa was associated with about 20% drop in strength. It is possible that such effects are due to some dilatancy-producing feature associated with intracrystalline slip in plagioclase or pyroxene, and it remains to be seen how real and significant negative pressure effects are in the fully plastic behavior of metabasic rocks.

References

Allison, I. S., and LaTour, T. E. (1977) Brittle deformation of hornblende in a mylonite: A direct geometrical analogue of ductile deformation by translation gliding. *Can. J. Earth Sci.* **14**, 1953–1959.

Atkinson, B. K. (1982) Subcritical crack propagation in rocks; theory, experimental results and applications. *J. Struct. Geol.* **4**, 41–56.

Ave Lallemant, H. G. (1978) Experimental deformation of diopside and websterite. *Tectonophysics* **48**, 1–27.

Beach, A. (1973) The mineralogy of high temperature shear zones at Scourie, N.W. Scotland. *J. Petrol.* **14**, 231–248.

Beach, A. (1982) Deformation mechanisms in some cover thrust sheets from the external French Alps. *J. Struct. Geol.* **4**, 137–150.

Borges, F. S., and White, S. H. (1980) Microstructural and chemical studies of sheared anorthosites, Roneval, South Harris. *J. Struct. Geol.* **2**, 273–280.

Boullier, A. M., and Gueguen, Y. (1975) S-P mylonites: Origin of some mylonites by superplastic flow. *Contrib. Mineral. Petrol.* **50**, 93–104.

Brace, W. F., Walsh, J. B., and Frangos, W. T. (1968) Permeability of granite under high pressure. *J. Geophys. Res.* **73**, 2225–2236.

Brodie, K. H. (1980) Variations in mineral chemistry across a phlogopite shear zone. *J. Struct. Geol.* **2**, 265–272.

Brodie, K. H. (1981) Variation in amphibole and plagioclase composition with deformation. *Tectonophysics* **78**, 389–402.

Bruton, C. J., and Helgeson, H. (1983) Calculation of the chemical and thermodynamic consequences of differences between fluid and geostatic pressure in hydrothermal systems. *Amer. J. Sci.* **283A**, 540–588.

Caristan, Y. (1982) The transition from high temperature creep to fracture in Maryland diabase. *J. Geophys. Res.* **87**, 6781–6790.

Clareborough, L. M. (1973) Slip plane dilation and disorder in a copper–aluminum alloy. *Phys. Stat. Sol.* **18**, 427–438.

Coe, R. S., and Kirby, S. H. (1975) The orthoenstatite to clinoenstatite transformation by shearing and reversion by annealling: Mechanism and potential applications. *Contrib. Mineral. Petrol.* **52**, 29–55.

Cooper, A. F., and Lovering, J. F. (1970) Greenschist amphiboles from Haast River, New Zealand. *Contrib. Mineral. Petrol.* **27**, 11–24.

Dennis, P. F. (1981) Defect chemistry and diffusion in quartz and olivine. *Progress in experimental petrology, NERC. Pub: Series D, 18.*

DeVore, G. W. (1968) Elastic strain and preferred orientation in monoclinic crystals. *Lithos* **2**, 9–24.

Dollinger, G. S. and Blacic, J. D. (1975) Deformation mechanisms in experimentally and naturally deformed amphiboles. *Earth Planet. Sci. Letts.* **26**, 409–416.

Etheridge, M. A. (1975) Deformation and recrystallization of orthopyroxene from the Giles Complex, Central Australia. *Tectonophysics* **25**, 87–114.

Etheridge, M. A., and Hobbs, B. E. (1974) Chemical and deformational controls on recrystallization of mica. *Contrib. Mineral. Petrol.* **43**, 111–124.

Fontaine, G. (1968) Dissociation des dislocations sur les plans (110) dans les cristaux ioniques de type NaCl. *J. Phys. Chem. Sol.* **29**, 209–214.

Fyfe, W. S. (1976) Chemical aspects of rock deformation. *Phil. Trans. Roy. Soc. London* **283A**, 221–228.

Gordon, R. B. (1971) Observation of crystal plasticity under high pressures with applications to the earth's mantle. *J. Geophys. Res.* **76**, 1248–1254.

Graham, C. M., Greig, K. M., Sheppard, S. M. F., and Turi, B. (1983) Genesis and mobility of the H_2O-CO_2 fluid phase during regional greenschist and epidote amphibolite facies metamorphism: A petrological and stable isotope study in the Scottish Dalradian. *J. Geol. Soc.* **140**, 577–600.

Gray, D. R., and Durney, D. W. (1979) Crenulation cleavage differentiation; Implications of the solution-redeposition process. *J. Struct. Geol.* **1**, 73–80.

Greenwood, G. W., and Johnston, R. H. (1965) The deformation of metals under small stresses during phase transformations. *Proc. Roy. Soc. London* **283A**, 403–422.

Greenwood, H. J. (1961) The system $NaAlSi_2O_6$-H_2O-Argon: Total pressure and water pressure in metamorphism. *J. Geophys. Res.* **66**, 3923–3946.

Gresens, R. I. (1966) The effect of structurally produced pressure gradients on diffusion in rocks. *J. Geol.* **74**, 307–321.

Griffin, W. L., and Raheim, A. (1973) Convergent metamorphism of eclogites and dolerites, Kristiansund area, Norway. *Lithos* **6**, 21–40.

Hall, M. G., and Lloyd, G. E. (1981) The SEM examination of geological samples with a semiconductor back-scattered electron detector. *Amer. Mineral.* **66,** 362–368.

Handin, J. (1966) Strength and ductility, in *Handbook of Physical Constants* edited by S. P. Clark. Geol. Soc. Amer. Mem. **97,** 223–289.

Harte, B., and Graham, C. M. (1975) The graphical analysis of greenschist to amphibole facies mineral assemblages in metabasites. *J. Petrol.* **16,** 347–370.

Heard, H. C. (1960) Transition from brittle fracture to ductile flow in Solnhofen limestone as a function of temperature, confining pressure, and interstitial fluid pressure, in *Rock Deformation,* edited by D. T. Griggs and J. Handin. *Geol. Soc. Amer. Mem.* **79,** 193–226.

Heard, H. C., and Rubey, W. W. (1966) Tectonic implications of gypsum dehydration. *Geol. Soc. Amer. Bull.* **77,** 741–760.

Hobbs, B. E. (1968) Recrystallization of single crystals of quartz. *Tectonophysics* **6,** 353–401.

Hobbs, B. E. (1981) The influence of metamorphic environment on the deformation of minerals. *Tectonophysics* **78,** 335–383.

Kekulawala K. R. S. S., Paterson, M. S., and Boland, J. N. (1981) An experimental study of the role of water in quartz deformation. *Geophys. Monograph* **24.** American Geophysical Union, 49–60

Kerrich, R., Fyfe, W. S., Gorman, B. E., and Allison, I. (1977) Local modification of rock chemistry by deformation. *Contrib. Mineral. Petrol.* **65,** 183–190.

Kerrich, R., Allison, I., Barnett, R. L., Moss, S., and Starkey, J. (1980) Microstructural and chemical transformations accompanying deformation of granite in a shear zone at Mieville, Switzerland: With implications for stress corrosion cracking and superplastic flow. *Contrib. Mineral. Petrol.* **73,** 221–242.

Kohlstedt, D. L., and Goetze, C. (1974) Low-stress, high temperature creep in olivine single crystals. *J. Geophys. Res.* **79,** 2045–2051.

Kohlstedt, D. L., and Van der Sande, J. B. (1973) Transmission electron microscopy investigation of the defect structure of four natural orthopyroxenes. *Contrib. Mineral. Petrol.* **42,** 169–180.

Laird, J. (1980) Phase equilibria in mafic schists from Vermont. *J. Petrol.* **21,** 1–37.

Laird, J. (1982) Amphiboles in metamorphosed basaltic rocks: Greenschist to amphibolite facies, in Veblen, D. R., and Ribbe, P. H., *Amphiboles, Petrology and Experimental Phase Relations. Reviews of Mineralogy, 9B,* Mineral. Soc. America, 113–137.

Laird, J., and Albee, A. L. (1981) Pressure, temperature and time indicators in mafic schist: Their application to reconstructing the polymetamorphic history of Vermont. *Amer. J. Sci.* **281,** 127–175.

LaTour, E., and Kerrich, R. (1982) Microstructures, mineral chemistry and oxygen isotopes of two adjacent mylonite zones: A comparative study, in *Issues in Rock Mechanics,* edited by F. E. Heuze and R. E. Goodman. Proc. 23rd U.S. Symposium on Rock Mechanics, 389–396.

Liou, J. G., Kuniyoshi, S., and Ito, K. (1974) Experimental studies of the phase relations between greenschist and amphibolite in a basaltic system. *Amer. J. Sci.* **274,** 613–632.

Marshall, D. B., and McLaren, A. C. (1977) The direct observation and analysis of dislocations in experimentally deformed plagioclase feldspars. *J. Mater. Sci.* **12**, 893–903.

Miyashiro, A. (1961) Evolution of metamorphic belts. *J. Petrol.* **2**, 277–311.

Miyashiro, A. (1973) *Metamorphism and Metamorphic Belts*. Wiley, New York.

Moody, J. B., Meyer, D., and Jenkins, J. E. (1983) Experimental characterization of the greenschist/amphibole boundary in mafic systems. *Amer. J. Sci.* **283**, 48–92.

Muller, P., and Siemes, H. (1974) Festigkeit, verformbarkeit und gefugeregelung von anhydrite-experimentelle stauchverforming unter manteldrucken bis 5 kbar bei temperaturen bis 300°C. *Tectonophysics* **23**, 105–127.

Muller, W. H., Schmid, S. M., and Briegel, U. (1981) Deformation experiments on anhydrite rocks of different grain size: Rheology and microfabric. *Tectonophysics* **78**, 527–543.

Murrell, S. A. F., and Ismail, I. A. H. (1976) The effect of decomposition of hydrous minerals on the mechanical properties of rocks. *Tectonophysics* **31**, 207–258.

Nicolas, A., and Poirier, J. P. (1976) *Crystalline Plasticity and Solid-State Flow in Metamorphic Rocks*. Wiley, New York.

Patel, J. R., and Chaudhuri, A. R. (1966) Charged impurity effects on the deformation of dislocation free germanium. *Phys. Review* **143**, 601–608.

Paterson, M. S. (1973) Non-hydrostatic thermodynamics and its geologic applications. *Rev. Geophys. and Space Phys.* **11**, 355–389.

Peters, T. (1963) Mineralogie und Petrologie des Totalp Serpentins, bei Davos. *Schweiz. Min. Pet. Mitt.* **43**, 529–685.

Poirier, J. P. (1982) On transformation plasticity. *J. Geophys. Res.* **87**, 6791–6797.

Poirier, J. P., and Vergobbi, B. (1978) Splitting of dislocations in olivine, cross-slip controlled creep and mantle rheology. *Phys. Earth Plan. Interiors* **16**, 370–378.

Raleigh, C. B. (1977) Frictional heating, dehydration and earthquake stress drops, in Proceedings of Conferences. II. Experimental studies of rock friction with application to earthquake prediction, edited by J. F. Evernden. U.S. Geological Survey, Menlo Park, CA, 291–304.

Raleigh, C. B., and Paterson, M. S. (1965) Experimental deformation of serpentinite and its tectonic implications. *J. Geophys. Res.* **70**, 3965–3985.

Ricoult, D. L., and Kohlstedt, D. L. (1983) Experimental evidence for an effect of chemical environment on the creep rate of olivine. *Trans. Amer. Geophys. Union* **64**, 494.

Robin, P-Y. F. (1978) Pressure solution at grain to grain contacts. *Geochim. Cosmochim. Acta* **42**, 1383–1389.

Rooney, T. P., Gavasci, A. T., and Riecker, R. E. (1974) Mechanical twinning in experimentally and naturally deformed hornblende. *Environ. Res. Papers, No.* **484** Air Force Cambridge Res. Labs. 1–21.

Ross, J. V., and Neilsen, K. C. (1978) High temperature flow of wet polycrystalline enstatite. *Tectonophysics* **44**, 233–261.

Rubie, D. C. (1983) Reaction enhanced ductility; the role of solid–solid univariant reactions in the deformation of the crust and mantle. *Tectonophysics* **96**, 331–352.

Rubie, D. C. (1984) The olivine–spinel transformation and the rheology of subducting lithosphere. *Nature* **308**, 505–508.

Rutter, E. H. (1972) The influence of interstitial water on the rheological behaviour of calcite rocks. *Tectonophysics* **14**, 13–33.

Rutter, E. H. (1974) The influence of temperature, strain rate and interstitial water in the experimental deformation of calcite rocks. *Tectonophysics* **22**, 311–334.

Rutter, E. H. (1976) The kinetics of rock deformation by pressure solution. *Phil. Trans. Roy. Soc. London,* **283A**, 203–219.

Rutter, E. H. (1983) Pressure solution in nature, theory and experiment. *J. Geol. Soc. London* **140**, 725–740.

Rutter, E. H., Peach, C. J., White, S. H., & Johnston, D. (in press) Experimental 'syntectonic' hydration of basalt. *J. Struct. Geol.*

Sammis, C. G., and Dein, J. L. (1974) On the possibility of transformational super-plasticity in the earth's mantle. *J. Geophys. Res.* **79**, 2961–2965.

Schmid, S. M., Paterson, M. S., and Boland, J. N. (1980) High temperature flow and dynamic recrystallization in Carrara Marble. *Tectonophysics* **65**, 245–280.

Schock, R. N. (1983) Olivine electrical conductivity as a function of oxygen fugacity. *Trans. Amer. Geophys. Union* **64**, 494.

Scholz, C. H., Beavan, J., and Hanks, T. C. (1979) Frictional metamorphism, argon depletion and tectonic stress on the Alpine fault, New Zealand. *J. Geophys. Res.* **84**, 6770–6782.

Sibson, R. H., White, S. H., and Atkinson, B. K. (1979) Fault rock distribution and structure within the Alpine fault zone, New Zealand: A preliminary account, in The origin of the Southern Alps, edited by R. I. Walcott and M. M. Cresswell. *Bull. Roy. Soc. New Zealand* **18**, 55–65.

Spear, F. S. (1980) NaSi–CaAl exchange equilibrium between plagioclase and amphibole. *Contrib. Mineral Petrol.* **72**, 33–41.

Spear, F. S. (1981) An experimental study of hornblende stability and compositional variation in amphibolite. *Amer. J. Sci.* **281**, 697–734.

Spry, A. (1969) *Metamorphic Textures.* Pergamon Press, Oxford.

Teall, J. J. H. (1885) The metamorphism of dolerite into hornblende schist. *Quart. J. Geol. Soc. London* **41**, 133–145.

Tullis, J. (1983) Deformation of feldspars, in *Reviews in Mineralogy, 2* (2nd Ed.), edited by P. H. Ribbe. Mineral. Soc. America, 297–323.

Tullis, J., Shelton, G. L., and Yund, R. A. (1979) Pressure dependence of rock strength: Implications for hydrolytic weakening. *Bull. Minéral.* **102**, 110–114.

Van der Molen, I., and Paterson, M. S. (1979) Experimental deformation of partially melted granite. *Contrib. Mineral. Petrol.* **70**, 299–318.

Vernon, R. H. (1976) *Metamorphic Processes.* George Allen and Unwin, London.

Vernon, R. H. (1977) Relationships between microstructures and metamorphic assemblages. *Tectonophysics* **39**, 439–452.

Walther, J. V., and Orville, P. M. (1982) Volatile production and transport in regional metamorphism. *Contrib. Mineral. Petrol.* **79**, 133–145.

Westwood, A. R. C., and Goldheim, D. L. (1968) Occurrence and mechanism of Rehbinder effects in calcium fluoride. *J. Appl. Phys.* **39**, 3401–3406.

White, S. H. (1975) Tectonic deformation and recrystallization of oligoclase. *Contrib. Mineral. Petrol.* **50**, 287–304.

White, S. H., and Knipe, R. J. (1978) Transformation and reaction enhanced ductility in rocks. *J. Geol. Soc. London* **135,** 513–516.

Wiseman, J. D. H. (1934) The central and south-west highland epidiorites; a study in progressive metamorphism. *Quart. J. Geol. Soc. London* **90,** 354–417.

Wyllie, P. J. (1971) *The Dynamic Earth*. Wiley, New York.

Zen, E-an (1965) Solubility measurements in the system $CaSO_4$–$NaCl$–H_2O at 35°, 50° and 70°C and one atmosphere pressure. *J. Petrol.* **6,** 124–164.

Zwart, H. J. (1962) On the determination of polymetamorphic mineral associations, and its application to the Bosost area, Central Pyrenees. *Geol. Rundschau.* **52,** 38–65.

Chapter 7
Heterogeneous Deformation, Foliation Development, and Metamorphic Processes in a Polyphase Mylonite

R. J. Knipe and R. P. Wintsch

Introduction

Study of the interactions between deformation and metamorphic processes is fundamental to the understanding of the origin of fabrics and foliations developed in tectonites. Mineral changes that accompany deformation may exert an important influence on the rheological behavior of rocks, and deformation processes play a significant role in the mechanisms of mineral reactions that take place in a deforming aggregate (White and Knipe, 1978; Brodie and Rutter, this volume). Recent studies that have combined micro-structural and microchemical analyses have been able to outline the important facets of these interactions and highlight areas requiring further attention (e.g., Bell and Vernon, 1979; Beach, 1980; Kerrich et al., 1980; Knipe, 1981; Rubie, 1983; Rutter, 1983). Most studies of mylonite foliations developed along ductile fault zones have concentrated on establishing the deformation mechanisms and strain paths from crystallographic fabrics, together with estimating stress levels from microstructural features (White et al., 1980; Weathers et al., 1979; Garcia Celma, 1982; Lister and Price, 1978; Law et al., 1984). Few studies have attempted to correlate the deformation and metamorphic processes that occur synchronously during mylonite evolution (see, however, Vernon, 1974; Rubie, 1983; Bell and Rubenach, 1983; White et al., 1982; Dixon and Williams 1983). This is despite the numerous examples of mylonites forming along zones of concentrated deformation that bring deeper level rocks to higher crustal levels where the original mineral assemblage is no longer stable. This paper attempts a detailed analysis of the interaction between deformation and metamorphic processes that contribute to the development of a mylonitic foliation. Specifically the aims of the study were:

1. To establish the patterns of heterogeneous flow and the distribution of deformation mechanisms and strain histories on a granular scale during mylonite generation from a polymineralic parent rock.
2. To determine which minerals were stable during the evolution of the tectonite and which were involved in reactions.
3. To investigate how deformation affects the reaction and growth mechanisms in the deforming aggregate and to evaluate the specific role of the grain boundary fluid in these processes.
4. To attempt to identify which microstructural features preserved in the final tectonite developed at particular stages in the mylonite evolution.

The paper is divided into three sections. The first presents a description of the microstructures in the tectonites selected for study. The second section presents an interpretation of the microstructural and microchemical features observed and discusses this information in the context of the aims outlined above. The third section attempts to synthesize the data obtained into a model for the foliation development.

Foliation Description

The tectonites selected for study form part of a mylonite sequence developed in the pre-Cambrian Lewisian gneiss during the evolution of the Moine Thrust zone. A review of the tectonic evolution of this Caledonian thrust zone and its deep structure is given by McClay and Coward (1981) and Butler and Coward (in press). A review of the structure and evolution of mylonites in this area is given by White et al. (1982), Evans and White (in press), and Dixon and Williams (1983). The specimens studied here are from the northern part of the thrust belt at Loch Eriboll (Fig. 1). This area was selected because of the large variation in foliation types and intensities exposed over a small area. Lewisian rocks least affected by the Caledonian deformation are coarse-grained (~3 mm) gneisses composed of K-feldspar (~40%), plagioclase (~20%), quartz (~30%), muscovite (~10%) together with chlorite, epidote, and Fe-oxides. These rocks have a weak foliation defined by irregular patchy domains containing a few grains. Near the major thrust planes the gneisses are converted to a fine-grained mylonite (see Dixon and Williams, 1983, and White et al., 1982), with compositional layering on a millimeter scale and numerous internal folds. The specimens studied in detail here are L-S tectonites that show an intermediate amount of foliation development between the mylonites at the base of the thrust sheet and the undeformed Lewisian. These tectonites exhibit compositional banding on a centimeter scale. They are blastomylonites, made up of quartz/feldspar-rich bands that alternate with phyllosilicate-rich bands. The rock contains fine-grained quartz/feldspar mylonitic bands as well as coarse-grained

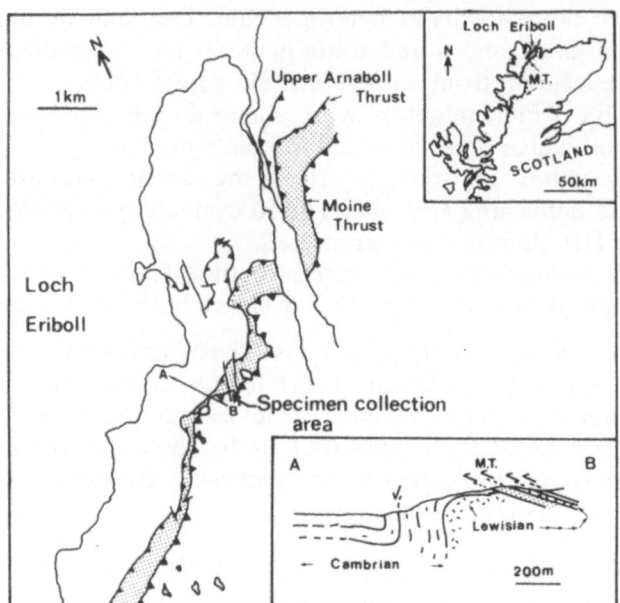

Fig. 1. Map of the northern Moine Thrust zone, showing the location of specimens studied. The inset shows a cross section through the area sampled. The specimens collected are located on a sheet of sheared Lewisian basement located at the base of the Moine Thrust sheet.

quartz/feldspar bands and thin chlorite-rich layers; all are parallel to the main foliation. Internal folds (wavelength 10 cm) are occasionally present that, like the mylonitic bands, increase in frequency towards the base of the thrust sheet.

The 22 oriented specimens collected have been subjected to combined microstructural and microchemical analyses including optical microscopy, transmission and scanning electron microscopy, together with microprobe analysis and chemical analysis in the scanning-transmission electron microscope.

The selected tectonites are composed of quartz, microcline, muscovite, plagioclase, chlorite with accessory magnetite, apatite, epidote, and zircon. Four microstructural domains may be defined on the basis of mineralogy, grain size, and internal structure (see Fig. 2). The domains are elongate lenses approximating to bands parallel to the main foliation and are:

1. *Q.F.P. Domains*—containing quartz, feldspar (K-feldspar and plagioclase), and phyllosilicates (mica and chlorite). These domains contain large feldspars (up to 5 mm) set in a fine-grained matrix.
2. *P. Domains*—fine grained phyllosilicate-rich ($\geq 60\%$) domains.

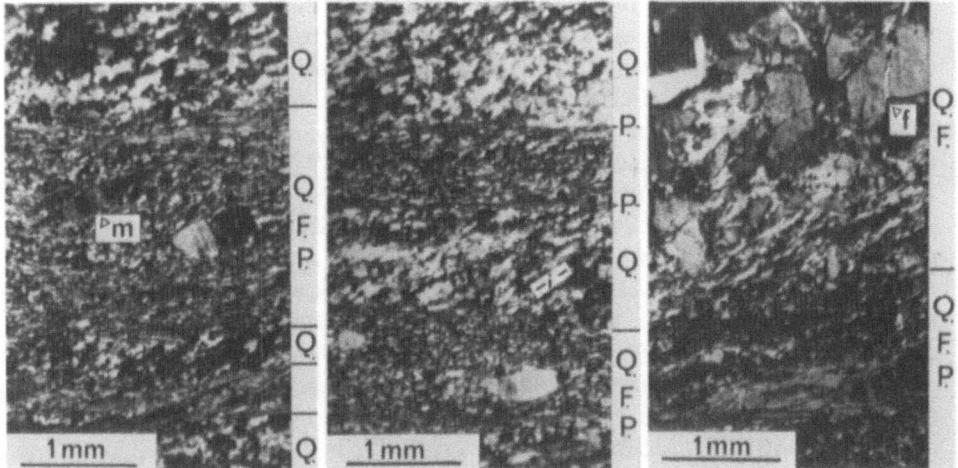

Fig. 2. Microstructural domains parallel to the main banding present in the tectonites. Q.F.P. (quartz, feldspar, phyllosilicate) domains, Q. (quartz) domains, P. (phyllosilicate) domains. Note the internal fabrics (i) within the Q. domains, the fractures (f) across feldspar grains within Q.F., and the elongate microclines (m) present within the Q.F.P. domains.

3. *Q. Domains*—thin mylonitic domains usually of pure quartz but also occasionally composed of quartz and feldspar (usually plagioclase).
4. *Q.F. Domains*—coarse-grained (~1 mm) quartz/microcline domains.

These domains should be considered only as end members that aid the description of the microfabric, for although type examples are easy to identify, transitional or intermediate types are common, especially between the Q.F.P. and P. domains. In addition P. domains often occur inside Q.F.P. domains. Individual domain types exhibit a large range in the structure, orientation, shape, and distribution of phases, and a detailed description of each domain is given below before the interrelationships between the domains and their origin are discussed.

Q.F.P. Domains

Lenticular domains composed primarily of quartz (20–30%), K-feldspar (20–30%), plagioclase (~10%), and muscovite (20–30%) dominate the fabric of the rocks studied. These domains are up to 1 cm thick and tens of centimeters long and have a bimodal grain-size distribution. Feldspars (primarily microcline) occur as large grains (up to 5 mm) set in a matrix of quartz, plagioclase, microcline, muscovite, and chlorite where the grain size is less than 0.3 mm. A foliation is present in the matrix and is defined by the orientation of muscovite grains. This foliation is slightly oblique to the domain boundaries (which coincide with the main foliation in hand specimen) and has the geometry of a foliation developed by shearing parallel to the domain boundary (see also Simpson and Schmid, 1984). The indicated direction of shearing verges towards the west-northwest.

The detailed microstructural features of the individual phases present in these Q.F.P. domains are described below:

Quartz grains within the Q.F.P. domain matrix are up to 0.1 mm long, slightly elongate (<3 : 1), parallel to the internal, (oblique) foliation and occur in elongate clusters up to 1 mm in length separated by phyllosilicate grains. Undulatory extinction and subgrains are present within the grains, and grain boundaries are dentate. Where quartz occurs in pressure shadows, associated with the large feldspars, the grain size is larger (up to 0.2 mm) and the grains are more equant. Occasionally coarse-grained irregular-shaped patches of pure quartz up to 1.5 mm in diameter also occur in these Q.F.P. domains. Grains within these patches are either fibrous or extremely irregular in shape with dentate boundaries and usually exhibit marked undulatory extinction and/or deformation bands.

K-feldspar occurs as both large grains (up to 5 mm) and as small grains within the matrix. The large microclines are elongate (up to 7 : 1) and have an elongate/rhomboid shape in sections cut perpendicular to the foliation and parallel to the lineation. The orientation of the rhomboid grains is best described by the orientation of the long sides of the grain that lie parallel to the local foliation defined by phyllosilicates. Large microcline grains (>1 mm), spanning a Q.F.P. domain, thus tend to have this dimension parallel to the orientation of the main foliation, while smaller grains have this dimension approximately parallel to the local foliation within the domain. Equant-shaped microclines are occasionally present that show evidence of rotation in the curved shape of their pressure shadows. The inferred sense of rotation is consistent with the shear sense indicated by the other features. Occasional undeformed, elongate grains, oriented at high angles to the foliation, are also present. These grains usually lack any marked pressure shadows and tend to be the smaller members of the grain-size population. The larger K-feldspars tend to have (1) more internal deformation features, (2) better developed pressure shadows, and (3) phyllosilicate-rich concentrations on both sides of the grain (Fig. 3).

Internal features within the K-feldspars visible with the optical microscope include microcline twins and undulatory extinction, all of which are more common in the large grains. Albite and mica inclusions are present in some grains. Analysis of the microclines with the TEM revealed a core and mantle dislocation substructure. The mantle is usually less than 30 microns wide (Fig. 4) and characteristically marked by subgrains that replace the microcline twins (>0.5 micron wide) common in the core of the grains.

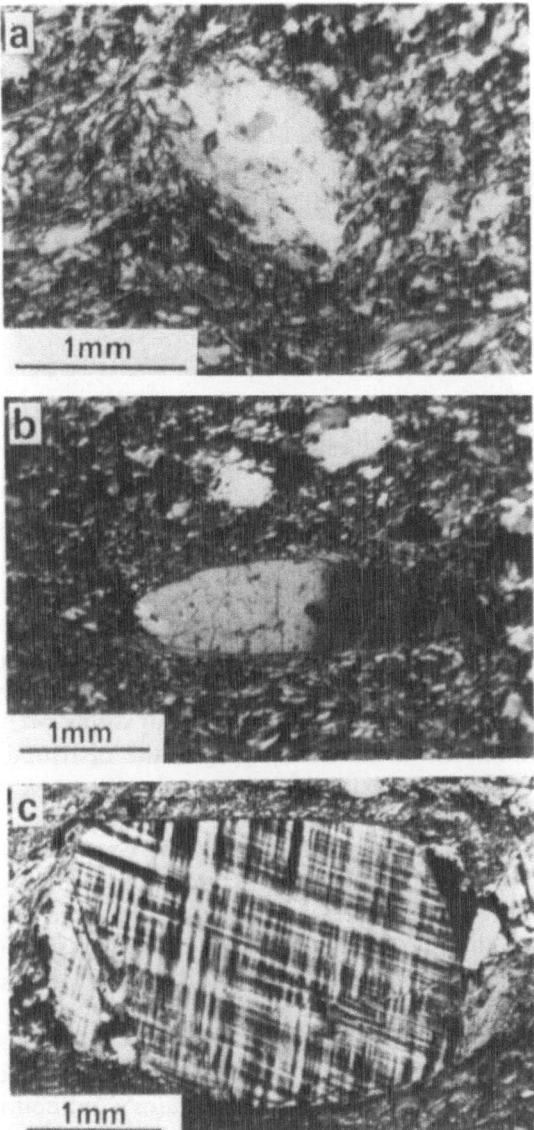

Fig. 3. Microstructures present in microcline within Q.F.P. domains. (a) Illustrates a small, undeformed microcline with no well-developed pressure shadow, which may represent a newly crystallised grain (see text for details). (b) Shows a larger microcline with undulatory extinction. (c) Illustrates a large microcline with well-developed twins and pressure shadow. Note also the phyllosilicate concentration at the grain boundary parallel to the foliation.

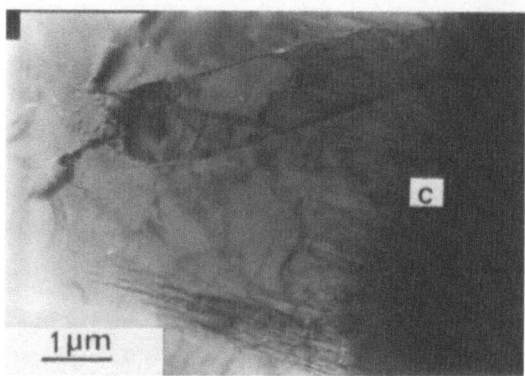

Fig. 4. Electron micrograph of the edge of a microcline grain. The core region (c) is composed of twins, while the grain boundary region contains subgrains.

Only a small percentage (~15%) of the K-feldspar grains in these domains show evidence of fracturing, but where present the majority of fractures are at high angles to the foliation and many record shear displacements consistent with other west-northwest shear indicators. A few fractures at high angles to the foliation record extension, and these are quartz filled.

The microclines present in these Q.F.P. domains are almost pure $KAlSi_3O_8$. However, they do contain small concentrations (<2 wt%) of Na_2O, CaO, and iron oxide, and detailed mapping of the distribution with the microprobe reveals that these oxides form irregular zones within the crystals (see Fig. 5).

Plagioclase grains occur primarily in the matrix of the Q.F.P. domains as equant to slightly elongate grains (<3 : 1). Occasional large grains up to 300 μm occur, and these cause a deviation of the adjacent foliation and have associated quartz-rich pressure shadows. The plagioclase grains contain discontinuities in the concentrations of CaO, K_2O, and iron oxide identified from detailed microprobe mapping. An example is shown in Fig. 6 (see also Wintsch and Knipe, 1983).

These plagioclase grains contain very few internal features visible in the optical microscope. Twins are rare, and undulatory extinction and fractures are only pronounced in a few of the larger grains. Electron microscopy supports this observation but reveals that dislocation substructures are concentrated in a narrow zone (1 μm wide) adjacent to the grain boundaries. The dislocation substructures usually take the form of high dislocation densities, generally in slip bands traceable to the grain boundary (Fig. 7). Occasional subgrains or cells are also present. These observations like those in the microcline suggest that a small amount of plastic deformation is concentrated near the grain boundary of the plagioclase.

Phyllosilicate grains that occur in the Q.F.P. domains are primarily phengitic with a grain size of 0.1 mm. Small (<50 micron) chlorite grains are

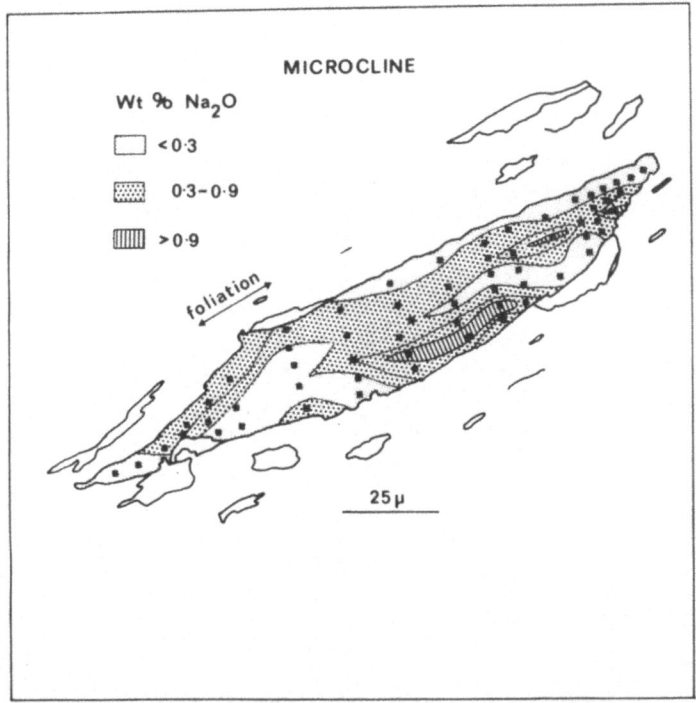

Fig. 5. Microchemical map of the Na$_2$O content of a microcline grain. (The irregular zoning was mapped using an EDS system on a microprobe and selected regions checked using a WDS system.) The points analysed are indicated.

also present. The phyllosilicate grains define the irregular and oblique foliation present in these Q.F.P. domains. Where the phyllosilicate concentration is low (i.e., <20%), phyllosilicate grains are usually isolated and oriented at angles of (43–30°) to the domain boundary. In areas where the phyllosilicates account for more than 30% of the domain, the phyllosilicates form anastomosing zones, up to 0.1 mm wide oriented at low angles (<20°) to the domain boundary. In the latter type of domain, K-feldspar is absent or rare. This general pattern of phyllosilicate orientation and concentration is modified near the large feldspar clasts where the muscovites concentrate along the grain boundaries parallel to the local foliation and help enclose quartz-rich pressure shadows. Small-scale folding of the foliation also occurs away from large feldspar clasts. Such folds (wavelength ≪ 200 μm) are asymmetric, occasionally overturned in the same west-northwest transport direction indicated by the pressure shadows. Concentrations of muscovite are again associated with the limbs of these folds. An important feature of some of these folds is that they appear to have developed by movement along micro-detachment planes marked by P. domains (Fig. 8).

There is a general decrease in the feldspar content of the Q.F.P. domains associated with the increase in the phyllosilicate content and the intensity of

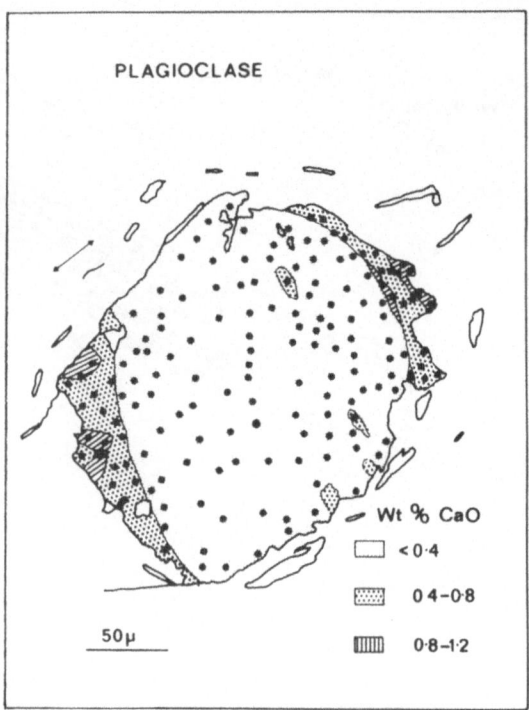

Fig. 6. Microchemical map of CaO distribution in a plagioclase grain. Note the CaO-rich overgrowths.

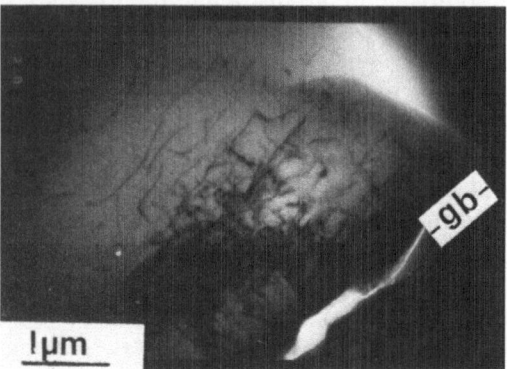

Fig. 7. Electron micrograph illustrating the concentration of dislocations near a grain boundary (gb) of a plagioclase grain.

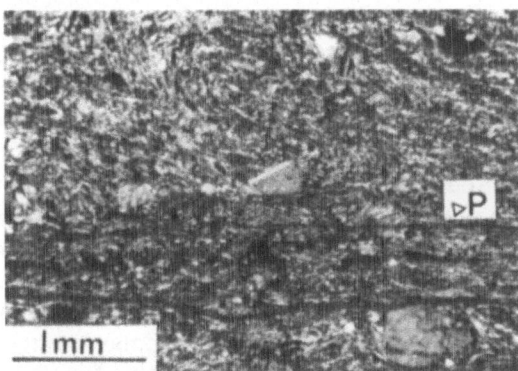

Fig. 8. Asymmetrical microfold (top center, with outline dashed) developed by detachment along an adjacent P. domain (P). Note also the undeformed microcline
(porphyroblast?) in the core of the syncline.

the internal foliation. That is, there is a complete gradation between domains
containing quartz, large microclines together with distributed phyllosilicates
and plagioclase grains to domains with anastomosing phyllosilicates concentrated into zones composed almost entirely of phyllosilicates. The phyllosilicate-rich (~80%) end member zones are considered as separate microstructural domains below.

P. Domains

Phyllosilicate-rich (~60%) domains are an important component in the blastomylonites studied and are of two types. The first is muscovite-rich and
helps define the main foliation in the rock. The second type are more chlorite-rich and represent very late stage brecciated movement zones that often
truncate the main foliation at low angles.

The muscovite-rich P. domains exhibit a range in internal structure related
to the thickness of the domain. The thin domains (~0.3 mm wide) are composed of grains oriented almost exactly parallel to the domain boundaries
and range in length from short (0.3 mm) discontinuous domains within the
Q.F.P. domains to longer (<3 cm) domains. Wider domains (~3 mm) contain slightly less phyllosilicate material, more quartz and microcline, and
have grains oriented slightly oblique (≲15°) to the domain boundaries. These
latter domains are transitional in type with the Q.F.P. domains described
above.

Where one P. domain terminates another adjacent P. domain commonly
starts, and there is usually a short zone of overlap. This is especially true of
thin, long P. domains. The sense of overlap between the domains may be

either in the form of a step up (i.e., the adjacent domain is above the termination of the other P. domain) or as a step down (where the next P. domain in the movement direction is below the P. domain that terminates). Large feldspar grains, or quartz or phyllosilicate aggregates are commonly located at this cross-over zone (Fig. 9). Asymmetric contraction folds in the Q.F.P. domains are associated with step-up geometry and asymmetrical extensional crenulations associated with the step downs (Fig. 10). In both cases the associated concentration of phyllosilicates at the cross-over appears to be a result of the removal of the quartz-rich matrix, and the P. domains either side of the concentration appear to have acted as detachment surfaces for the production of the folds. The phyllosilicate concentration connecting the P. domains is effectively a micro ramp (Fig. 10). Small folds (wavelength ~100 μm) located above these P. domains are also found away from P. domain terminations.

Electron microscopy of the thin P. domains reveals that the domains contain well-aligned phyllosilicates together with occasional extremely fine-grained (micro-breccia) zones (Fig. 11). Occasional brick-shaped quartz grains are located between the phyllosilicates. Such quartz commonly has bubble trails perpendicular to the foliation trace representing either healed fractures or growth discontinuities; this together with the shape of these grains indicates that the quartz crystallized in extension sites between phyllosilicate grains during sliding.

A number of the thin P. domains have undulating stylolitic geometries indicating some late pressure solution across these zones.

Q. Domains

The third microstructural domain recognizable in the tectonites are mylonitic bands usually composed of quartz with minor amounts of feldspars (0–15%), primarily plagioclase, phyllosilicates (0–3%), and magnetite (0–10%). These fine-grained (<0.1 mm) domains are up to 2 mm wide and can sometimes be traced back to a pressure shadow while others can be linked to the edge of a Q.F. domain (described below). Quartz grains in these domains are slightly elongate with axial ratios generally below 2:1, although occasional grains have ratios reaching 6:1. Most grains have dentate boundaries, contain subgrains, and are typical products of dynamic recrystallization. The elongation and end-to-end arrangement of these grains defines a foliation within domains that is oblique to the main foliation (see Fig. 2).

The c-axes fabrics of these domains are girdles, asymmetrical with respect to the domain boundary, and can be interpreted as indicative of a noncoaxial strain path approximating to simple shear parallel to the domain walls. A detailed analysis of the fabrics in these domains is the subject of a separate paper (Knipe and Law, in preparation). The sense of shear is the

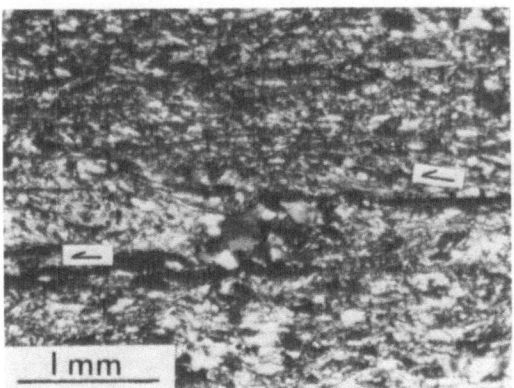

Fig. 9. Quartz concentration (center) located at a "step down" between two offset P. domains. A microfold is also present in front of this offset indicating the strain history of this area involves both extension and contraction parallel to the foliation.

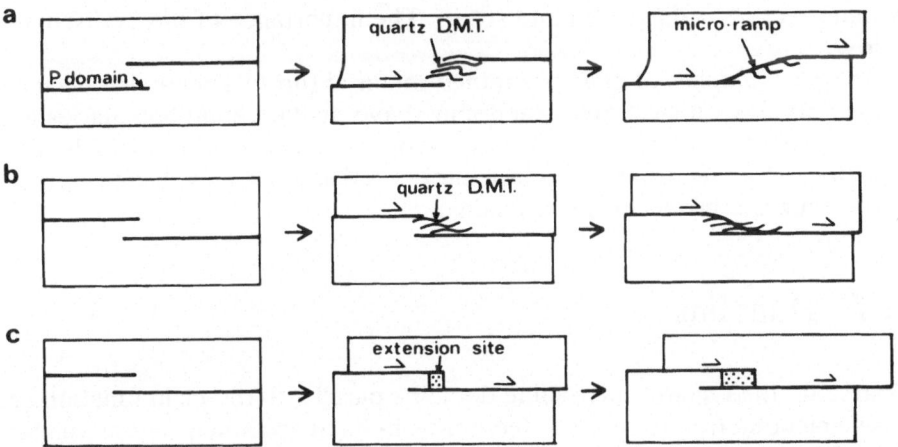

Fig. 10. Diagram illustrating the deformation features developed at the terminations of P. domains. (a) Illustrates the evolution of a microfold at a "step up" between P. domains. Removal of quartz from the limb of the fold by DMT can induce the concentration of phyllosilicates and the production of a micro ramp. (b) Shows the evolution of an extensional crenulation to link a "step-down" overlap of P. domains. (c) Illustrates the development of an extension site, ideal for the crystallization of quartz and/or feldspar. This process can produce both Q. and Q.F. domains present in the tectonite (see text for details). The processes shown in (a) and (b) are favored by slow strain rates while the extension site developed in (c) is favored by fast strain rates and lower temperatures.

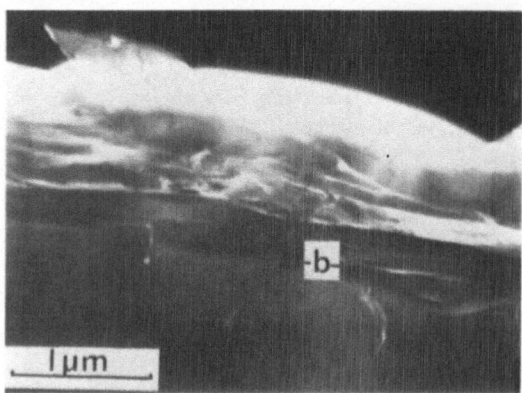

Fig. 11. Electron micrograph of a thin P. domain. Note the almost perfect alignment of grains parallel to the domain boundary, (b) and the fractured and rotated grains in the center of the zone.

same as that indicated from the other microstructural features. The angle between the internal elongation fabric and the domain walls varies from one Q. domain to another. The range in angles recorded varies from 42° to 20°. This variation in the orientation of the internal fabric indicates that each domain records a different finite strain. The importance of this is discussed below.

Irregular patches of coarser grained material (up to 1 mm) also occur in these domains. In such areas the grains have dentate grain boundaries and show undulatory extinction and large subgrains. These zones, which occasionally contain feldspar grains, are interpreted as grain growth areas formed late during the history of the domains.

Q.F. Domains

Coarse-grained quartz/microcline domains parallel to the main foliation are a conspicuous feature of these tectonites in hand specimen. These veinlike domains are up to 5 cm wide, tens of centimeters long, and contain angular feldspar grains up to 1 cm in diameter floating in a matrix of quartz. The domains are often bounded by well-developed P. domains on one or both sides. The feldspar within these domains is an almost pure K-feldspar with small amounts (~ 0.5 wt%) of CaO, Na_2O, and iron oxide. These grains show less variation in internal chemistry than do the smaller microclines present in the Q.F.P. domain matrix. The microcline grains of this domain show occasional twins near the domain walls and a small amount of undulatory extinction. However, the most conspicuous feature of the veins are the fractures, now quartz filled, that have disrupted and separated the feldspars (Fig. 2).

Individual feldspars that have been pulled apart are often in optical continuity and can be reconstructed. Many of these grains show an increased separation, rotation, and grain-size reduction towards the domain boundary, where the resultant fabric may merge into a fine-grained mylonitic Q. domain. Not all the fractured feldspars can be reassembled; some show complex dentate forms that do not fit back together, which indicates either corrosion of the fracture interface after initial fracturing (see Hammer, 1982) or synchronous crystallization of quartz and feldspar.

The quartz that fills the areas between the feldspars is generally coarse grained and shows large amounts of undulatory extinction and a mantle of subgrains. Electron microscopy reveals that the cores of these grains have extremely high dislocation densities ($\sim 10^{10}$ cm/cm^3) (Fig. 12) but no well-developed subgrain structure. These large grains of quartz may represent either original vein grains crystallized from solution with slight subsequent deformation or may represent areas of grain boundary migration that have then undergone deformation.

Discussion

The above description of the microstructures present in the Lewisian tectonites studied demonstrates the preservation of microstructures indicative of the operation of a range of deformation processes all of which have

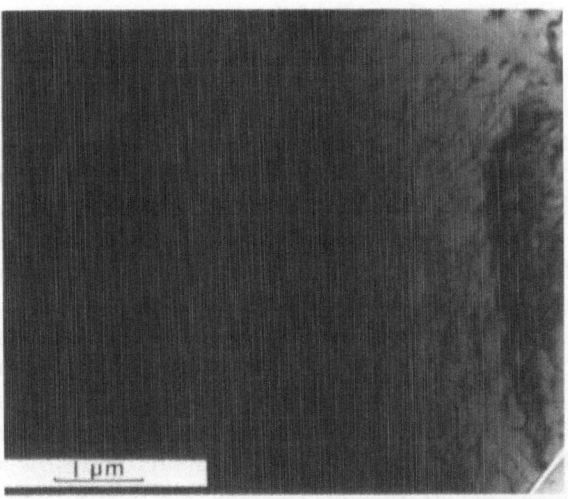

Fig. 12. Electron micrograph of quartz from within a Q.F. domain. This high dislocation density is associated with marked undulatory extinction in the optical microscope.

contributed to the final fabric. However, these processes need not have operated synchronously. The view that the final microstructure can contain elements representative of the different parts of the evolution of the tectonite is emphasized here.

The first section of the discussion attempts to interpret the microstructures preserved in the tectonites studied in terms of identifying the processes that microstructural domains/elements reflect and the sequence of the microstructural development. An attempt is then made to correlate the deformation and metamorphic processes in the second part of the discussion. A synthesis of these data into a model for the evolution of the final microstructures preserved in the tectonites is then presented in the last section of the paper.

Distribution of Deformation Mechanisms and Strain in the Microstructure

The microstructures present are indicative of the operation of a wide range of deformation mechanisms and straining processes that operated at some time during the evolution of the final preserved structure. This range in deformation behavior reflects both the different response of the different minerals in the polymineralic starting rock and a continual variation of environmental conditions of deformation. The microstructural variation observed in the final preserved microstructures is thus considered to arise from both the spatial and the temporal variations in: (1) the finite strain, (2) the strain rate, (3) the strain path, and (4) the deformation mechanisms. Although interrelated, each of these factors is discussed separately below.

Finite strain variation within individual domain types and between adjacent domains is indicated by their internal structure. For example, the Q.F. domains exhibit a range of microstructures from only slight fracturing in K-feldspar-rich domains to extensive microfracturing and grain-size reduction in other Q.F. domains. The Q. domains, composed primarily of quartz, also exhibit a range in the finite strain recorded, indicated by the variation in the angle that the internal fabric makes with the main foliation.

The variation in the finite strain preserved can be explained in terms of (1) the "life time" of the deformed elements preserved within domains and/or (2) a variation in strain rate. The feldspar-rich Q.F. domains that show only minor fracturing and the Q. domains with internal fabrics at high angles to the main foliation are interpreted as being "young" compared with the highly fractured Q.F. domains or the Q. domains with low-angle internal fabrics, which have existed for longer.

Strain rate variations between domains at one time and an overall variation in strain rate during the production of the final microstructure are indicated by a number of features. For example, the complete range of microstructures in the Q.F. domains from fracturing to mylonitic textures indicates the preservation of domains that have deformed at different rates and have existed for different time periods. The Q.F. domains that show slight fracturing usually indicate a lower finite strain than the mylonitic ones, indicating that the fractured domains are young and hard regions that represent a response of the Q.F. domains to later and/or faster deformation events. The chlorite-rich domains that truncate the foliation can also be considered to represent late, fast-strain rate events preserved in the tectonites. This range of microstructures may be interpreted as arising from either (1) an increase in the strain rate with time, which may reflect an increase in the displacement rate across the mylonite zone, or (2) may reflect the changing response of the material to deformation at lower temperatures. In this latter case, the change in deformation in the Q.F. domains may be due to a change from a higher temperature, more uniform strain rate deformation to a lower temperature, discontinuous deformation involving higher strain rate events. That is, this change in behavior may occur at a constant average displacement rate on the main fault zone but reflect the change from higher temperatures, steady-state (constant strain rate) deformation to lower temperature deformation accommodated by less frequent but higher strain rate events.

Evidence for synchronous variations in strain rate arising from the compositional differences within the tectonite can be seen in the Q.F.P. domains where large resistant feldspar grains generate local, high strain rates in the adjacent matrix. This heterogeneous deformation associated with the feldspars has important consequences for both the microstructural evolution and the rheology of the tectonite and is discussed in more detail below. In contrast to the Q.F.P. domains the almost pure quartz Q. domains appear to record a more homogeneous internal strain rate history approximating to simple shear.

Variation in the strain paths of the different domains and within individual domains over a time interval are indicated by some of the preserved microstructures. The overall strain path for the tectonite is inferred to approximate to the simple shear to the west-northwest and near parallel to the mean foliation. However, important deviations away from this ideal are recorded in the tectonite. For example: Shortening parallel to the foliation is indicated by the folding of the main foliation and is often associated with detachment along the phyllosilicate-rich P. domains (especially their terminations). The presence of P. domains folded by movement on other P. domains emphasizes the heterogeneous distribution of deformation in the tectonite in both time and space, with early localized high deformation zones becoming inactive and folded by movements on later P. domains. This change in the pattern of localization of deformation can also produce the superposition of different strain histories on small areas, for example, microfolds disrupted by extension and quartz infill (see Fig. 9).

Syntectonic changes in volume within the tectonite are also indicated.

Local volume increases are associated with the generation of Q.F. domains; the production of quartz "bricks" in the P. domains and the pressure shadows in the Q.F.P. domains. Local volume reductions are indicated by the removal of quartz and the concentration of phyllosilicates near large feldspars and microfolds. In addition, the growth of new phases (muscovite and feldspar) by replacement reactions (see below for details) in the Q.F.P. domains will also involve a change in volume. It is difficult to assess if there is an overall change in volume within the tectonites during its evolution because the relative importance of increases and decreases can not be established accurately.

The dominant deformation mechanisms vary between different domains. Many of the variations in finite strain, strain rate, and strain paths reflect the various deformation mechanisms that have contributed to the tectonite evolution. This variation in deformation mechanisms reflects primarily the differing mineralogies of the domains recognized. For example, the feldspars deform primarily by fracturing, although the subgrain structure, twinning, and core and mantle structures observed reveal that minor ductile deformation at slower strain rates has also occurred. Deformation by dislocation movement is the main deformation mechanism within the quartz grains present in the tectonite. The variation in the dislocation substructures preserved in the different domains indicates that even the type of dislocation mechanisms vary. For example, the subgrain microstructures preserved in the Q. domains are typical of dislocation creep processes, while the extremely high dislocation densities present in quartz in some of the Q.F. domains are more indicative of rapid work hardening (i.e., cold work). This distinction again supports the suggestion that the Q.F. domains with such quartz are "young" features developed at higher levels involving higher strain rates at lower temperatures late in the tectonite's history.

Microstructures indicative of diffusion mass transfer (DMT) are also present in the tectonite. There are two different DMT processes indicated, and both are associated with the removal of quartz within the Q.F.P. domains and the concentration of phyllosilicates. Both processes contribute to the production of the phyllosilicate-rich P. domains. The first of these processes involves the removal of quartz from the matrix adjacent to large feldspars and in microfold limbs. The second process involves the removal of silica from a reaction site where feldspar is being converted to mica. These two DMT processes are distinguished here because the kinetics of the first will be dependent upon the dissolution of quartz while the second will be dependent on the rate of feldspar reaction. Evidence for late, minor amounts of DMT at the edges of the P. domains are also indicated.

The operation of grain boundary sliding is also indicated in localized sections of the tectonite; i.e., within the phyllosilicate-rich P. domains where the "brick"-shaped quartz shows evidence of growth in extension sites created by sliding between phyllosilicates.

Deformation and the Growth of New Grains during the Tectonite Evolution

The microstructures preserved indicate that the growth of feldspars, quartz, and micas has occurred during the foliation evolution and that deformation influences the growth of each.

Evidence for the syntectonic growth of feldspars comes first from the Q.F. domains (essentially veins) and the range of deformation features present inside them, and second from the zoning patterns and the variation in the internal deformation of both plagioclase and microcline. The details of the zoning in plagioclase have been presented in Wintsch and Knipe (1983). The zoning present in the microclines and plagioclase indicates that the fluid composition changed during growth and the possible interactions between deformation and fluid chemistry has been discussed by Wintsch (1975) and Wintsch and Knipe (1983, and in preparation). A third piece of evidence indicating the growth of feldspars in the rocks studied are the small ($\leqslant 0.1$ mm), elongate grains with few intracrystalline deformation features, oriented across the foliation and with dentate grain boundaries (e.g., Fig. 3(a)). These grains appear to represent new crystals growing by replacement reactions in the matrix of the Q.F.P. domains.

Syntectonic growth of quartz from solution is indicated by its presence in (1) fractures between feldspars in Q.F. domains, (2) pressure shadows in the Q.F.P. domains, (3) local extension sites in the P. domains, and (4) occasionally veins cross-cutting the foliation. The large, irregular quartz grains occasionally seen in the Q. and Q.F.P. domains are not considered to represent new quartz growth from solution in extension sites as above but to develop during grain boundary migration, i.e., secondary recrystallization late during the microstructural evolution.

The growth of new phyllosilicates in the tectonite is indicated by the presence of occasional large undeformed grains and the concentration of micas in sites where feldspar appears to be breaking down.

The growth of quartz, mica, and microcline described above can be linked to the following reaction:

$$\text{Muscovite} + 6\text{SiO}_2(\text{aq}) + 2\text{K}^+ \leftrightharpoons 3 \text{ Microcline} + 2\text{H}^+$$

Evidence for the operation of this reaction in both directions is found in the tectonites, in that the breakdown and growth of feldspar is indicated. The amount of feldspar reaction at individual grains varies from zero to almost complete.

Deformation influences the operation and direction of this reaction by (1) influencing the processes operating at the reaction site, (2) by providing diffusion paths for the migration of material to and from the reaction sites, and (3) by providing locations where new growth can occur, i.e., sinks. Each of these influences is considered separately below.

The influence of deformation on the processes in the source areas of reaction components is primarily associated with the effect deformation has on the rate at which material enters the diffusion path to and from the reaction site. Where a grain boundary fluid is present then the effect of plastic deformation on the solubility is important (Bosworth, 1981; Wintsch and Dunning, 1983). There are two potential processes in the rocks studied where deformation may influence the solubility of material. The first is the increased solubility associated with a high internal strain energy, which arises from plastic or elastic deformation. The high dislocation densities recorded in the quartz and the grain boundaries in the feldspars will increase the solubility of these phases (see Wintsch and Dunning, 1983) and hence the reaction processes. The second process that can influence the solubility and thus the fluid chemistry within the tectonites is ion-exchange reactions promoted by the generation of new surfaces during fracturing (Wintsch, 1975). The grain boundary sliding and fracturing recorded in the phyllosilicate-rich P. domains are a likely location for this process. Ion-exchange in these zones would allow the exchange of H^+ for cations (e.g., K^+) on the new surface and thus increase the activity of K^+ in the fluid. Such a change in the fluid chemistry may promote feldspar crystallization (see Wintsch, 1975; Wintsch and Knipe, 1983). However, it should be noted that the phyllosilicate-rich fracture zone observed is probably representative of the last movements in the phyllosilicate-rich P. domains and may only be associated with the last feldspars to form, i.e., those that are now essentially undeformed. The growth of earlier formed feldspars (e.g., those now fractured and separated in the Q.F.P. domains) may have been driven by the operation of a similar process, but the associated fracture zones that will have healed by the precipitation of new phases (e.g., quartz) in extensional sites and by the replacement of fine-grained micas to coarser grained phyllosilicates during grain boundary migration. These earlier movement zones may now be represented by the folded P. domains.

The production of chemical potential gradients and the generation of diffusion paths during deformation has an important influence on reactions in that it enhances the movement of material between sources and sinks. In the case of the tectonites studied, deformation has increased the number of diffusion paths by reducing the grain size from the original gneiss. The generation of grain boundaries and surfaces is not considered to have been one of a progressive increase during the evolution of the tectonites; rather, the local coarse-grained areas of the Q.F.M. and Q. domains indicate there are grain growth events (of quartz, feldspar, and mica) that produce a temporary minor reversal in this trend. The effect of these events on the deformation is discussed in the synthesis below.

The generation of sites for the growth of material in a deforming aggregate is one of the most important influences deformation has on metamorphic processes. Two such deformation-induced growth sites may be recognized in the tectonites studied here. The first involves crystallization from solutions in extension sites. These occur, for example, between the fractured feldspars; in the vicinity of overlapping P. domains where displacement is transferred to a lower detachment zone; and within P. domains themselves because of grain boundary sliding. The extension created by linking sliding P. domains provides a mechanism for the generation of Q.F. domains as well as for some of the large feldspars present in the Q.F.P. domains (Fig. 10). The second site where deformation can aid the growth of a new phase is where a single undeformed grain grows to replace a finer grained deformed aggregate. Examples of large grains of quartz and feldspar apparently replacing a deformed matrix can be recognized in the microstructure and are considered to represent a late growth by grain boundary migration/reaction process (see also Knipe, 1981, p. 268). A second example of growth by replacement in the tectonite is the production of mica by the alteration of deformed feldspar. Such sites are concentrated on the sides of the feldspar grains parallel to the local foliation. The breakdown of feldspar to quartz and muscovite at these sites, accompanied by the removal of quartz and the concentration of matrix phyllosilicates by DMT, provides one mechanism for generation of the P. domains that can be traced back to the edge of a large feldspar grain.

Synthesis

Before developing a model for the microstructural evolution of the tectonites it is worthwhile first considering an overview of the conditions experienced by rocks located within thrust-related mylonite zones. A general model for the distribution of deformation processes within large thrust fault zones has been presented by Sibson (1977, 1982, 1983).

This model comprises a deep level zone where continuous ductile (quasi-plastic) deformation takes place and a higher level zone where deformation is dominated by pressure-sensitive frictional sliding and fracturing. The microstructures and fabrics developed in fault rocks that have presented evidence of an early ductile deformation and a late cataclastic deformation have been described by a number of workers (e.g., Grocott, 1977; Sibson, 1982; White et al., 1980; White et al., 1982; Watts and Williams, 1983). It is the evidence for the changes in the mode of deformation accompanying the uplift of tectonites to different levels that is considered in detail here. Attention is also given to the preservation and modification of early formed textural features and the problem of recognizing such early features in the final microstructure.

There are four changes that may occur in a deep level shear zone as deformation and erosion bring rocks to higher crustal levels. These changes, which occur before the onset of widespread cataclasis, are induced by the increased resistance to plastic deformation associated with the lower temperatures and the inability of the rock to continue deforming at the same rate by the same mechanisms. They are:

1. Localized ductile deformation may take place, within the original fault zone, in domains of favorable mineralogy and microstructure that can remain ductile to higher crustal levels.
2. Deformation may spread so that the edge of the shear zone takes up more deformation, thus allowing a constant displacement rate across the zone to be maintained but at a lower internal strain rate within the original shear zone.
3. Reactions may be promoted by the changes in P, T and chemical conditions rendering minerals unstable and producing reaction products that may be capable of deforming at the required rate. If the reaction products are particularly weak, the active deformation volume may decrease and sections of the original shear zone become inactive.
4. A new shear zone may develop and the once active zone experiences uplift without further deformation, perhaps preserving the original textures.

A combination of all these changes is considered to have occurred in the tectonites studied.

Process (1) is likely in the quartz- and mica-rich domains, because these minerals are ductile to lower temperatures than the feldspars. Process (2) is harder to assess here because the edge of mylonite zones has not been studied in detail. However, the fracturing recorded by White *et al.* (1982) and Dixon and Williams (1983) in an adjacent area on the edge of mylonite zones in the Lewisian can also be interpreted as arising from the spreading of the deformation zone at higher levels. This alternative interpretation of the low strain zones bordering large mylonite zones questions the view that the microstructures and mineralogy of such sites represent those characteristics of the initiation of the mylonite zone (Dixon and Williams, 1983). The operation of the third process listed above is evidenced in the tectonites by the breakdown of feldspar. This produces P. domains that appear more favorable to strain accommodation later in the tectonite's evolution, i.e., at high levels. In addition, the development of folding of the mylonitic foliations can also be considered to arise in part from the work hardening or reduced deformability associated with uplift to conditions where homogeneous deformation becomes more difficult, i.e., the folding is not only due to the increased anisotropy associated with the intensity of foliation development. Process (4) occurred in the area studied when thrusts developed below the movement zone studied. These have carried the tectonites studied in a "piggy back" fashion towards the foreland (see McClay and Coward, 1981, for a discussion of the thrusting sequence in the Eriboll area). It is during this stage that deformation features associated with displacement on lower thrusts may be generated in the higher level mylonite zones.

A model for the production of the final tectonite fabric is presented below.

In general this microstructural evolution involves the progressive change from "distributed" to "localized" deformation in the mylonite zone during decreasing temperature and provides an example of the development of flow partitioning (Lister and Williams, 1983). The fabric development can be divided into three stages, each of which is discussed separately below. As an aid to the discussion, Fig. 13 reviews the microstructural evolution and Fig. 14 presents an attempt to identify the strain rate patterns that accompany this evolution.

Stage A

The earliest, deepest level deformation of the tectonite studied is difficult to reconstruct because of the modification of early formed textural features by deformation at higher levels. However, it is possible to speculate briefly on the initial deformation.

During the early stages of the fabric evolution the rock is considered to

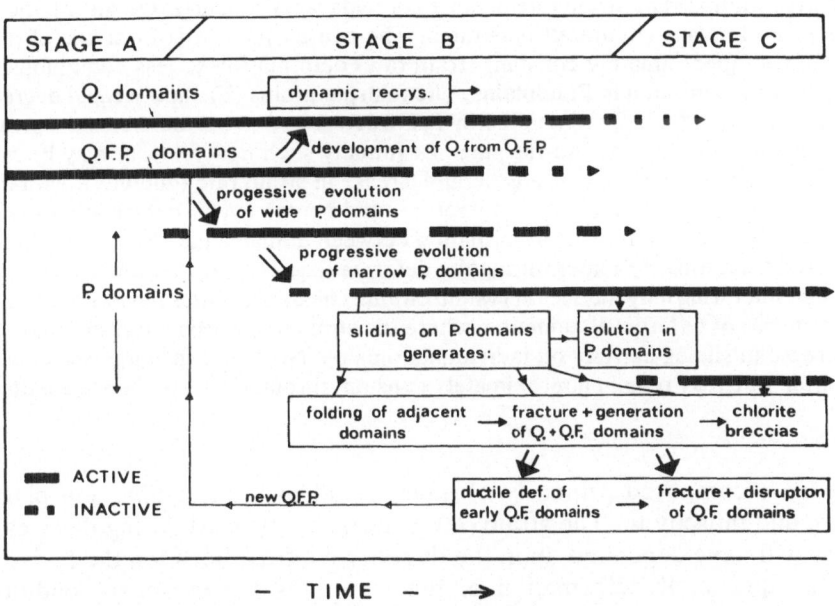

Fig. 13. Summary of evolution and interaction of microstructural domains. The general evolution involves the decrease in strain accommodation (activity) in Q.F.P. and Q. domains and the concentration of later deformation into P. domains. The diagram also emphasizes the development of some domains from others; e.g., (a) of Q. and P. domains from Q.F.P. domains, (b) of Q.F. domains from deformation of P. domains, and (c) of Q.F.P. domains from the deformation, disruption, and reaction of Q.F. domains.

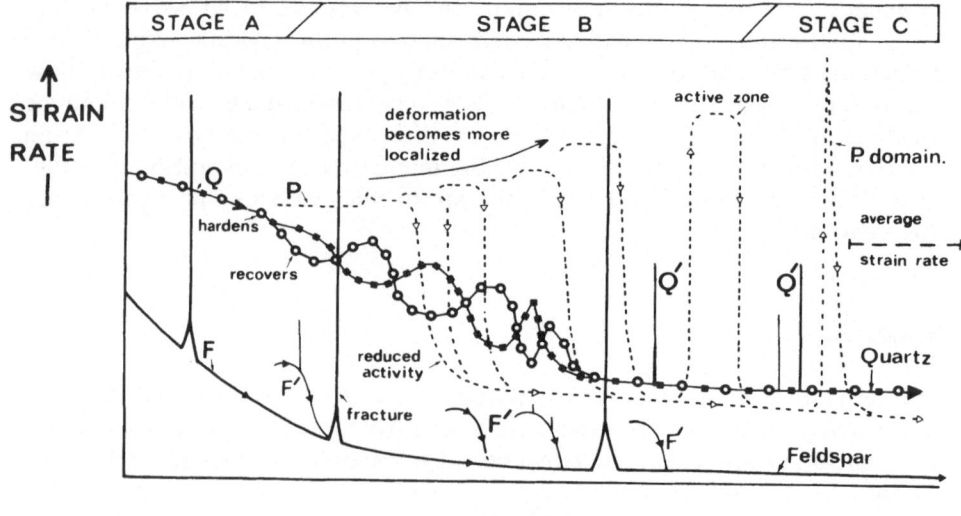

Fig. 14. Suggested strain rate–time patterns developed during the evolution of the tectonites studied. The strain rate histories of feldspar (F), quartz (Q), and phyllosilicates in P. domains (P) are all illustrated. The overall pattern is from one of more continuous (approximately constant strain rate) deformation to less continuous deformation concentrated in P. domains. The feldspar grains (F) show a rapid decrease in strain rate (hardening) interrupted by fast fracture events. F illustrates the growth and breakdown of feldspar during stage B. Initially such new feldspar may be more ductile. Quartz (Q) also shows a general decrease in strain rate (ductility), with time associated with the decreasing temperature, and shows cyclic deformation associated with the redistribution of deformation between domains (see Fig. 15) as well as increased fracturing (Q') later during the deformation. The lower strain rate periods between fracturing may induce recrystallization. The dashed lines illustrate the strain rate histories of different P. domains, where the strain rate drops are associated with a decrease in sliding activity on individual domains. Note that the large-scale, average strain rate may remain approximately constant throughout this changing pattern.

have been composed primarily of quartz, feldspar (microcline and plagioclase), and muscovite. The strain rate patterns developed during these early stages will have depended upon the ductility contrast between the feldspars and the quartz. If deformation of the tectonites began under conditions where both the quartz and feldspar were ductile the strain accommodation pattern would have been a relatively homogeneous one on the scale of a few centimeters. On a smaller scale the strain rate pattern would have been heterogeneous with the more ductile quartz accommodating more deformation, i.e., the quartz may have been able to protect the feldspar, allowing it to deform in a ductile fashion at a slower rate. Continuation of this process to large strains would have led to a mylonite composed of recrystallized quartz and feldspar. This end member microstructure is not observed in the tectonites studied, indicating that no significant strain under such conditions has occurred. A more likely situation for the tectonite studied is that deformation began under greenschist facies conditions and strain rates where feldspar was not ductile and thus the strain rate patterns will have been more complex. Given the feldspar content of the original rock, the deformation is likely to have been made up of work-hardening-fracture cycles at the feldspar grains leading to the separation of feldspar fragments and the flow or growth of quartz in the fracture zones. Under constant deformation rate conditions, this process would continue until the feldspars were sufficiently small and dispersed to avoid causing stress concentrations leading to fracture (this "fibre-loading" process is discussed in detail by Lloyd et al., 1982). At this stage both the quartz and the feldspar may deform in a ductile fashion, although the feldspar at lower strain rates.

The progression of the tectonite through stage A will have been associated with a decrease in the temperature and accompanied, by first the increasing resistance of feldspar to any plastic deformation and thus a concentration of deformation into quartz and Q. domains, second by the decrease in the deformability of quartz and therefore an increase in the flow accommodation problems around feldspars, and third a decrease in feldspar stability, promoting reaction to quartz and mica (Figs. 13 and 14).

Stage B

The deformation and microstructural evoluton during stage B is characterized by the progressive development of P. domains that act as zones of concentrated deformation. These zones vary from early wide diffuse P. domains to narrow, sharply bounded P. domains generated later during stage B.

The production of the P. domains may have been a direct consequence of the hardening centers developed around the feldspar grains arising from the reduced deformability of quartz at progressively lower temperatures. This hardening may be relieved by (1) removal of quartz by local DMT resulting in the concentration of phyllosilicates, (2) breakdown of the feldspar to quartz and mica followed by the DMT of quartz, (3) fracture of the feldspar, and (4) by the relocation of the strain accommodation to an adjacent area capable of more concentrated deformation.

Although there is evidence that all of these processes occurred during some part of stage B, it is processes (1) and (2) that dominate the creation of the P. domains and are favored by lower strain rates.

The initial P. domains developed during stage B will have been short, discontinuous features, and their use as important slip zones during later stages will have depended upon the linking of the individual small, isolated P. domains to remove the hardening on restricted flow at their terminations.

The microstructures present indicate that the "softening" processes associated with the linking of P. domains takes place under various conditions. The presence of folds (either contractional or extensional crenulations) in these sites is suggestive of deformation at high temperatures and/or slower strain rates, i.e., more likely early during stage B. At higher strain rates and/or lower temperatures (i.e., primarily later during stage B), fracturing and grain boundary sliding are interpreted as taking place to generate extension sites (Fig. 10) where crystallization of feldspar and quartz leads to the creation of Q.F. and Q. domains.

The folding of P. domains by movement along other P. domains emphasizes that the displacement on the P. domains was not synchronous but heterogeneous where different domains operated at different times. This indicates that at any one time some P. domains in the tectonite were active (accommodating deformation) while others were less active or inactive.

The progressive development of the P. domains and the concentration of strain accommodation into these zones results in the production of a flow pattern where strain rates in the Q.F.M. and early Q.F. and Q. domains progressively decreases (Figs. 13 and 14) while the strain rate in the P. domains increases. This progressive partitioning of strain between localized P. domains and into Q. domains has important implications for both the rheology and microstructural evolution. Early during this partitioning history, when there are only small differences in the deformability of domains, the deformation accommodation pattern may be able to redistribute itself continually in a complex fashion between transient (and alternating) hardening and softening centers. For example, the hardening in one domain may lead to a relocation of strain accommodation into an adjacent softer zone (Fig. 15(a)), thereby allowing a drop in the strain rate and increasing recovery in the initially hardening zone. The initially hardening zone may then recover, alter its microstructure, and become more susceptible to deformation, i.e., softer, while the adjacent initially soft zone hardens. Eventually the strain accommodation pattern may return to its original distribution (Fig. 15(a)). This cyclic redistribution of strain rate which has been observed in experimental shear zones developed in paradichlorobenzene, is potentially an important process in the localization of deformation in shear zones, and is the subject of a separate paper (Knipe, in preparation). Three examples of the possible effects of localizing deformation into selected domains during different parts of stage B are shown in Figs. 15(b), 15(c), and 15(d). The examples are presented in the order in which they are most likely to occur in the mylonites during deformation at progressively lower temperatures. The example shown in Fig. 15(b) is considered more likely to be the dominant process early during stage B, while the example shown in Fig. 15(d) is more common during the later parts of stage B.

Figure 15(b) illustrates the possible effect of a reduced strain rate in a Q. domain induced by work hardening. The reduction in the strain rate within the domain associated with a transfer of strain allows recovery and grain

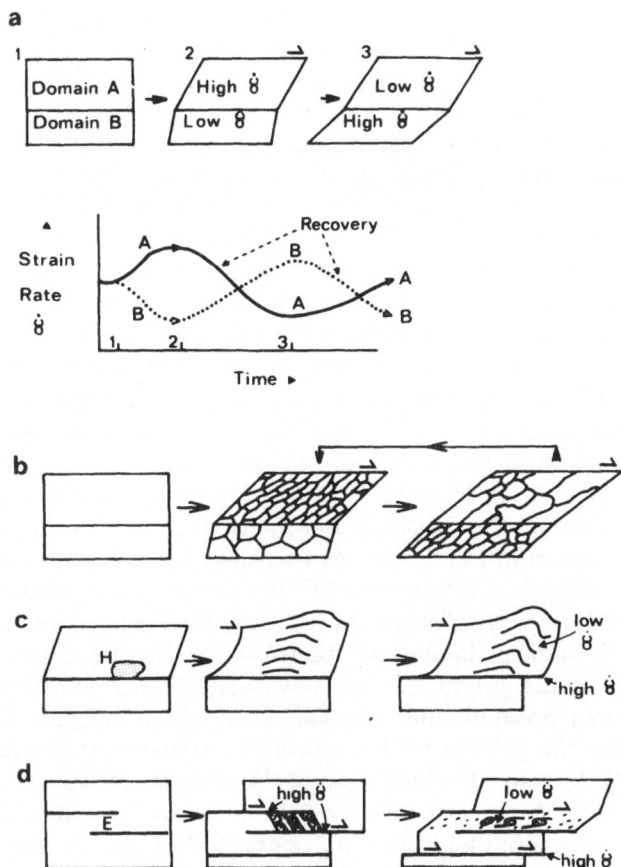

Fig. 15. Review diagram of some of the consequences of strain rate redistribution in the tectonites studied. (a) Illustrates the general pattern of strain rate arising from the cyclic relocation of deformation. Domains that initially accommodate more deformation may harden and induce transfer of deformation to a new site. Recovery processes (dislocation rearrangement and grain growth) may then allow the strain to return to its original location. (b) Illustrates the grain-size reduction–increase cycles associated by the strain rate patterns. (c) Shows the change on the strain path and location as hardening centers (H) develop and induce the development of folds and a transfer of strain to adjacent P. domains. The reduced deformability of quartz or growth of feldspar are possible hardening events (see text). (d) Successive relocation of deformation in P. domains. The extension site (E) developed allows the growth of Q.F. domains. The relocation of deformation into the lower P. domain can induce a strain rate decrease in the Q.F. domain. This reduced strain rate may eventually convert the Q.F. domain into a Q.F.P. domain.

boundary migration that replaces the initial deformed aggregate with undeformed quartz. This enables the domain to deform again and generate a new
internal fabric. The grain boundary migration event can be considered to
reset the finite strain indicator, i.e., the internal fabric (see also Means,
1981), and this process operating at different times in different Q. domains is
one way of accounting for the various internal fabric angles recorded in the
final tectonite. The occasional presence of undeformed feldspars (Fig. 3(a))
in the Q. and Q.F.P. domains also indicates that transient low strain rate
periods promote feldspar growth by a replacement reaction. In addition, this
syntectonic feldspar growth will disrupt further deformation in the domain,
which in turn may promote its subsequent breakdown. This raises the possibility of cyclic feldspar growth and breakdown during deformation.

Figure 15(c) illustrates one possible consequence of the evolution of P.
domains and the reduced ductility of quartz during the decrease in temperature that accompanies stage B; i.e., the generation of flow perturbations
leading to the generation of folds. A third example of the effect of heterogeneous deformation on the rheology and microstructural evolution later during stage B is shown in Fig. 15(d). At this time the continuing reduction in
quartz deformability is likely to amplify the strain rate differences between
the domains of localized deformation (i.e., the P. domains) and the adjacent
zones. Grain boundary sliding and fracturing in the already established P.
domains leads to the production of new surfaces and ion-exchange reactions, which can promote the crystallization of K-feldspar. Deformation
creates sites for the growth of this phase in extension zones generated by
sliding along P. domains. Continuation of this process provides a mechanism
for creating new Q.F. domains. Early Q.F. domains created in this way will
be extensively fractured and possibly converted to Q.F.P. domains by the
subsequent breakdown of feldspar. The later Q.F. domains developed will
exhibit only slight deformation.

Both the feldspar growth processes described above, associated with the
localization of deformation into the P. domains, promote a period of feldspar
growth in low strain rate and/or extension sites. This is in contrast to the
general trend of feldspar removal from some tectonites (see also Dixon and
Williams, 1983) and makes the separation of original fractured microcline
from syntectonic feldspar that has been subsequently fractured extremely
difficult.

Stage C

The latest deformation events recorded in the tectonites are the generation
of local chlorite-rich fracture zones. These are usually approximately parallel to the main foliation and carry a lineation trending west-northwest. Occasional chlorite-rich breccia zones at a high angle to the foliation are also
present. Both these zones are considered to represent movement zones

developed at high crustal levels where the influx of water along the fault zone has introduced Fe- and Mg-rich fluids needed for the growth of chlorite. This fluid differs from those inferred to be present earlier in the tectonite evolution, which appear to have been alkali-rich and may be considered to be derived from the Lewisian gneiss. The small chlorite grains ($\geqslant 5$ μm) found throughout the tectonite matrix may also represent the influx of this fluid along slightly dilatant grain boundaries at high crustal levels. This inferred increased permeability of the mylonite zone may have allowed a larger variety of fluid compositions into the fault zone, including Fe, Mg-rich fluids derived perhaps from more basic pods in the original gneiss. The model presented envisages these transient increases in permeability to increase in frequency as the mylonite preserved reached higher levels.

The occasional calcite veins present are also late features that may be associated with emplacement over the Durness limestone sequence.

Finally, the other microstructural changes likely during stage C and late during stage B are the adjustment of early formed deformation features. Such changes appear to include (1) local grain boundary migration, (2) pressure solution along P. domains, and (3) crystallization and sealing of local breccia zones developed along selected P. domains (see Figs. 13 and 14).

Conclusions

The polyphase mylonite studied contains microstructural domains of varying mineralogy and internal structures, which preserve microstructures indicative of the operation of a wide range of deformation and metamorphic processes. This range of microstructures is considered to arise from the preservation, in different domains, of different stages in the tectonite's evolution and reflects the heterogeneous flow experienced by the rock. Different flow domains show evidence of different finite strains, strain rates, and strain paths, and have been active for different time intervals.

The microstructures indicate a change in flow pattern from a more continuous and distributed ductile deformation to a more localized discontinuous deformation involving fracturing and sliding. The change to a more localized deformation is associated with deformation at progressively lower temperatures and high strain rates together with the development of phyllosilicate-rich domains from the syntectonic breakdown of feldspar. However, the localization of deformation also generates sites for the syntectonic growth of feldspar, which on subsequent straining reacts back to muscovite and quartz. In addition to this cyclic growth and deformation of feldspar there is also evidence for the cyclic deformation of quartz.

The model presented for the evolution of the final mylonitic fabric involves the cyclic redistribution of deformation between domains. The types of domain being created and the operating deformation mechanisms vary as the mylonite experiences different conditions during its uplift and deformation along the thrust fault. The changing strain rate–time pattern is crucial to the evolution, modification, and preservation of microstructures in the final foliation as well as the localization of deformation. The changing patterns of heterogeneous deformation, from a more distributed pattern of strain accommodation to a more localized pattern has allowed the preservation of microstructures developed earlier in the history of the tectonite. That is, the microstructures preserved reflect the changing deformation processes and conditions operating during its history of uplift and record attempts to produce different "end-member" microstructures. The extent to which production of an end-member microstructure is achieved depends upon the rate of microstructural production at different conditions, strain rates, etc., and the time spent at the conditions that lead to that particular end-member microstructure. The preservation of elements (e.g., microstructural domains/features), reflecting different stages in the development of a series of end-members, depends upon the later deformation, first, not completely destroying the early microstructures and/or, second, developing in localized zones thus preserving domains containing earlier deformation features.

Acknowledgments

Drs. E. H. Rutter, C. Ferguson, and S. White are thanked for comments and discussion. The authors would also like to acknowledge the support of N.E.R.C. grants GR3/4612 and GR3/4100, NSF grant EAR-8313807, as well as the Leverhulm Trust and Research Corporation. The work reported here was originally presented at the Conference on Planar and Linear Fabrics of Deformed Rocks held at Zurich in 1982.

References

Beach, A. (1980) Retrogressive metamorphic processes in shear zones with special reference to the Lewisian Complex. *J. Struct. Geol.* **2**, 257–265.

Bell, T. H., and Rubenach, M. J. (1983) Syntectonic vein and fibre growth associated with multiple slaty cleavage development in the Lake Moondarra area, Mt. Isa, Australia. *Tectonophysics* **92**, 195–211.

Bell, T. H., and Vernon, R. (editors). (1979) Microstructural process during deformation and metamorphism. *Tectonophysics* **58**, 1–220.

Bosworth, W. (1981) Strain induced preferential dissolution of halite. *Tectonophysics* **78**, 509.

Butler, R. W. H., and Coward, M. P. (in press) Geological constrains, deep geology and structural evolution of the N.W. Scottish Caledonides. *Tectonics*.

Coward, M. P. (1983) The thrust and shear zones of the Moine thrust zone and the N.W. Scottish Caledonides. *J. Geol. Soc. London,* **140,** 795–811.

Dixon, J., and Williams, G. D. (1983) Reaction softening in mylonites from the Arnaboll thrust, Sutherland. *Scott. J. Geol.* **19,** 157–168.

Evans, D., and White, S. (in press) Microstructural and fabric studies of Moine Nappe rocks, Eriboll, N.W. Scotland. *J. Struct. Geol.*

Garcia Celma, A. (1982) Domainal and fabric heterogeneities in the Cap de Creus quartz mylonites. *J. Struct. Geol.* **4,** 443–455.

Grocott, J. (1977) The relationship between Precambrian shear belts and modern fault systems. *J. Geol. Soc. London* **133,** 257–262.

Hamner, S. K. (1982) Microstructure and geochemistry of plagioclase and micro-cline in naturally deformed granite. *J. Struct. Geol.* **4,** 197–215.

Law, R. D., Knipe, R. J., and Dayan, H. (1984) Strain path partitioning within thrust sheets: Microstructural and petrofabric evidence from the Moine Thrust zone at Loch Eriboll, N.W. Scotland. *J. Struct. Geol.* **6,** 477–499.

Lister, G. S., and Price, N. J. (1978) Fabric development in a quartz-feldspar my-lonite. *Tectonophysics* **49,** 37–78.

Lister, G. S., and Williams, P. F. (1983) The partitioning of deformation in flow rock masses. *Tectonophysics* **92,** 1–33.

Lloyd, G. E., Ferguson, C. C., and Reading, K. (1982) A stress-transfer model for the development of extension fracture boudinage. *J. Struct. Geol.* **4,** 355–372.

Kerrich, R., Allison, I., Barnett, R. L., Moss, S., and Starkey, J. (1980) Microstruc-tural and chemical transformation accompanying deformation in a shear zone at Mieville, Switzerland. *Contrib. Mineral. Petrol.* **73,** 221–242.

Knipe, R. J. (1981) The interaction of deformation and metamorphism in slates. *Tectonophysics* **78,** 249–272.

McClay, K. R., and Coward, M. P. (1981) The Moine Thrust Zone: An overview. In *Thrust and Nappe Tectonics* edited by K. R. McClay and N. J. Price. Spec. Pub. Geol. Soc. London **9,** 241–260.

Means, W. D. (1981) The concept of steady state foliation. *Tectonophysics* **78,** 179–199.

Rubie, D. C. (1983) Reaction enhanced ductility: The role of solid–solid univariant reactions in deformation of the crust and mantle. *Tectonophysics* **96,** 331–352.

Rutter, E. H. (1983) Pressure solution in nature, theory and experiment. *J. Geol. Soc. London* **140,** 725–740.

Sibson, R. H. (1977) Fault rocks and fault mechanisms. *J. Geol. Soc. London* **133,** 191–213.

Sibson, R. H. (1982) Transient discontinuities in ductile shear zones. *J. Struct. Geol.* **2,** 165–175.

Sibson, R. H. (1983) Continental fault structure and shallow earthquake source. *J. Geol. Soc. London* **140,** 741–769.

Simpson, C., and Schmid, S. M. (1984) An evaluation of the criteria to deduce the sense of movement on sheared rocks. *Geol. Soc. Amer. Bull.* **94,** 1281–1288.

Vernon, R. H. (1974) Controls of mylonitic compositional layering during non-cata-clastic ductile deformation. *Geol. Mag.* **111,** 121–123.

Watts, M. J., and Williams, G. D. (1983) Strain, geometry, microstructure and mineral chemistry in metagabbro shear zones: A study of softening mechanisms during progressive mylonitization. *J. Struct. Geol.* **5**, 519–539.

Weathers, M. S., Bird, J. M., Cooper, R. F., and Kohlstedt, D. C. (1979) Differential stress determined from deformation induced microstructures of the Moine Thrust Zone. *J. Geophys. Res.* **84**, 7495–7509.

White, S., and Knipe, R. J. (1978) Transformation and reaction enhanced ductility in rocks. *J. Geol. Soc. London* **135**, 513–516.

White, S. H., Evans, D. J., and Zhong, D-L. (1982) Fault rocks of the Moine Thrust Zone: Microstructures and textures of selected mylonites. *Textures and Microstructures* **5**, 33–63.

White, S. H., Burrows, S. E., Carreras, J., Shaw, N. D., and Humphreys, F. J. (1980) On mylonites in ductile shear zones. *J. Struct. Geol.* **2**, 175–187.

Wintsch, R. P. (1975) Feldspathization as a result of deformation. *Geol. Soc. Amer. Bull.* **85**, 35–38.

Wintsch, R. P., and Knipe, R. J. (1983) Growth of a zoned plagioclase porphyroblast in a mylonite. *Geology* **11**, 360–363.

Wintsch, R. P., and Dunning, J. D. (1983) The role of defect density on "strain solution." (Abstract). *Trans. Am. Geophys. Union (E.O.S.)* **64**, 319.

Chapter 8
Aspects of Relationships between Deformation and Prograde Metamorphism that Causes Evolution of Water

S. A. F. Murrell

Introduction

During the past 25 years it has become recognized that pore-fluid pressures must play a fundamental role in crustal deformation. Although the role of pore pressures was already well known in soil mechanics, their role in rock mechanics and structural geology only came into focus in the classical papers by Hubbert and Rubey (1959, and Rubey and Hubbert, 1959) on the mechanics of overthrust faulting. These papers provided strong evidence that the formation of nappes and overthrusts could in many cases be best explained by the presence of unusually high pore-fluid pressures trapped in restricted crustal horizons (see Murrell, 1981, for a later review). At about the same time the first experimental studies took place on the effects of pore pressures in rock deformation, and Murrell (1963, 1965a, 1964b) extended his application of Griffith's microcrack theory to the brittle fracture of rock (Murrell, 1958) to show that the effective compressive stresses in the rock were reduced (and the effective tensile stresses increased) by an amount equal to the pore pressure, and that this was in good agreement with the available experimental data (Murrell, 1963, 1965). This was later extended to a fully three-dimensional microcrack model by Murrell and Digby (1970). The hydrodynamics of the maintenance of high pore pressures was dealt with by Bredehoeft and Hanshaw (1968) and Hanshaw and Bredehoeft (1968).

Rubey and Hubbert (1959) pointed out (see also Yoder, 1955) that at deep levels high pore-fluid pressures might arise by decomposition of hydrous minerals during prograde metamorphism and lead to weakening of the rock. This was confirmed in experiments by Raleigh and Paterson (1965) on serpentinite and by Heard and Rubey (1966) on gypsum. Hanshaw and Brede-hoeft (1968) developed a full treatment of the physics of the development and maintenance of high pore pressures by this mechanism.

Griggs and Handin (1960), Raleigh (1967), and Raleigh and Lee (1969) also suggested that metamorphic dehydration reactions might be responsible for some earthquakes. The tectonic effects and deformation styles associated with metamorphic reactions were discussed by Heard and Rubey (1966) and by Holland and Lambert (1969).

The experimental studies clearly showed the marked weakening that could occur during prograde metamorphism, and, in the case of serpentinite, embrittlement was also observed (Raleigh and Paterson, 1965). Thus meta-morphism certainly affects the deformation stress and the style of deforma-tion. Conversely, it has long been recognized (e.g., see Fyfe *et al.*, 1958) that the defects produced by shear or tensile deformation may accelerate the kinetics of metamorphism. It has also been recognized that when shear (deviatoric, differential, or nonhydrostatic) stresses are present the thermo-dynamics are different (Paterson, 1973), so that when an interstitial pore fluid is present "pressure solution" may occur (first recognized in rocks by Sorby; see Rutter, 1976, 1983) with a tendency to restore a condition of hydrostatic equilibrium.

In our laboratory an extensive series of experiments has been carried out on rock deformation under conditions of prograde metamorphism, some of the results of which have already been published (Murrell and Ismail, 1976a, 1976b). Experiments were carried out at pressures up to 662 MPa together with temperatures up to 780°C on five rocks containing hydrous minerals that decomposed to release water within this range of conditions, character-istic of the Earth's continental crust to a depth of ~20 km.

In an altered granodiorite the decomposition of a chlorite mineral, releas-ing water, caused partial melting at 670°C in an originally dry sealed rock specimen at a confining pressure of 448 MPa, the strength fell to a very low value, and stable flow occurred (Murrell and Ismail, 1976b). At 520°C and 448 MPa pressure the rock was weak, brittle, and unstable (differential ther-mal analysis showed that the decomposition reaction was initiated at a tem-perature <504°C).

The other four rocks studied in our laboratory were gypsum, a serpen-tinite, a partially sepentinized peridotite containing brucite, and a chloritite. In sealed specimens the commencement of decomposition of the hydrous minerals was accompanied by loss of strength, a reduction in sliding friction, and enhanced embrittlement and mechanical instability (Murrell and Ismail, 1976a). In this paper we present some further information concerning the phase changes occurring in the rocks, obtained by means of differential thermal analysis (DTA), X-ray observations on quenched samples, and opti-cal microscopy, and relate these observations to the results of our deforma-tion experiments and to published dehydration curves of the hydrous min-erals.

The authors of the papers reporting the above experiments on deformation under conditions of prograde metamorphism have all accepted that high pore pressures generated by the metamorphism are responsible for the mechanical phenomena observed. However, there has remained some doubt about the deformation mechanisms and about the mechanical role of the new phases, and at present there is no published quantitative microscopic model of the formation of water-filled pores (or cracks) by the metamorphic process. There has also remained some doubt about the magnitude of the pore pressure generated and the relationship of this to pressure and temperature conditions of metamorphism. We discuss these matters below.

Clearly, during the deformation experiments under discussion the rocks were, unavoidably, not in equilibrium either mechanically or chemically. This presents problems for petrologists in interpreting the significance of the experimental observations, and it continues to be the subject of lively controversy. A major problem is that in the experiments (which are generally of short duration) the new mineral phase assemblages are metastable and are difficult to identify. Another problem is that any natural metamorphic mineral assemblage will have developed over such a long period of time that metastable assemblages such as those produced experimentally may not have natural counterparts in outcrop (but see Trommsdorff and Evans, 1972). The experiments therefore represent snapshots of part of a metamorphic process.

Nevertheless, we believe that the experiments show that prograde metamorphism during which a fluid phase evolves can cause sharp and early mechanical effects, which may strongly influence styles of deformation (including seismicity), and this will affect metamorphic textures and chemical and mineral transport and accumulation in the crust. In the discussion below we present a model, based on the experimental data and theoretical studies, for the development of the texture of a metamorphic rock during burial metamorphism.

Concurrently with our work Fyfe et al. (1978) describe how they "were attempting to formulate realistic experiments linking metamorphic processes and rock mechanics." Their book will be a valuable aid in the design of future experiments.

Experimental Techniques

The techniques used in our high-pressure/high-temperature deformation experiments are described elsewhere in detail (Murrell and Ismail, 1976a, 1976b). The oven-dried rock samples (diameter 10 mm, length 30 mm) were jacketed in annealed copper. Stress and pressure were measured with an accuracy of ± 2 MPa and ± 3.5 MPa, respectively. Temperature differences within the specimen were less than 10 K axially and 5 K radially. All specimens were maintained at the test pressure and temperature for one hour before testing, and at the end of the experiment the temperature was reduced rapidly to atmospheric.

Differential thermal analysis (DTA) measurements were made with a Stanton Redcroft "Standata" instrument using calcined alumina as the inert standard and Pt/Pt13% Rh thermocouples. Samples were held in matched pairs of platinum crucibles in ceramic holders. The tests were carried out in air at atmospheric pressure, and the heating rate was 10 K per minute.

X-ray diffraction measurements were made under ambient conditions with a Philips powder diffractometer using CuKα radiation, and peaks were indexed using the tables and curves given in Parrish and Mack (1963) and the J.C.P.D.S. data book (1974).

Effects of Decomposition of Hydrous Minerals on Mechanical Properties

There are three principal effects in sealed specimens: weakening, embrittlement, and dilatancy hardening, and these result from the pore pressure created by water released by the decomposition reaction. These effects are discussed in detail in two earlier papers (Murrell and Ismail, 1976a, 1976b), and are illustrated in Fig. 1.

The effect on strength is shown by a reduction in the strength at fracture (indicated by the first stress-drop) or at 2% strain (when there is no stress-drop because deformation is by plastic or cataclastic flow or because dilatancy hardening has occurred), and also by a reduction in apparent friction. The fracture strength falls to values as low as the strength of unconfined rock (Fig. 1(a)) and also becomes independent of confining pressure, and the apparent coefficient of friction falls to a value close to zero (Fig. 1(b)). Such low values of strength and friction show that the pressure of the water released by decomposition into the specimen is very close to the confining pressure (Murrell and Ismail, 1976a). If the sample is drained to atmosphere, however, the strength remains high (Fig. 1(c)).

Embrittlement is indicated, not only by the reduction in strength, but also by a transition from ductile to brittle behavior (the latter being indicated by faulting at low strains, accompanied by a stress-drop) and by the occurrence of a characteristic minimum (which may be zero) in the stress-drop as a function of temperature (Fig. 1(d)). If the sample is drained to the atmosphere, embrittlement does not occur (Fig. 1(c)).

Under "undrained" conditions brittle behavior is accompanied by dilatancy hardening, as cracks grow in volume, pore pressure falls, and effective confining pressure increases (Ismail and Murrell, 1976; Murrell and Ismail, 1976a, 1976b) (Fig. 1(e)).

All of the above features are diagnostic of the release of water by decomposition of hydrous minerals in originally dry rocks.

In all of the rocks studied here there is more than one stage in the decomposition reaction and in the release of water. In cases in which the reaction or water release that occurs at the lowest temperature releases only a small quantity of water (as for the serpentinized peridotite and the chloritite) it is possible to distinguish two stages of weakening of the rock.

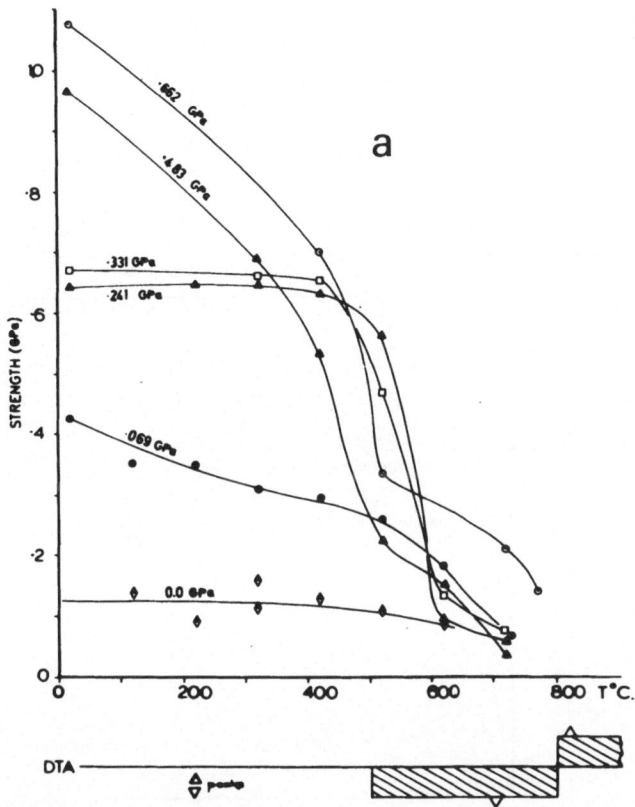

Fig. 1. Examples of weakening, embrittlement, and dilatancy hardening in rock owing to decomposition of hydrous minerals. (a) Strength, at fracture or 2% strain, of undrained specimens of Oulx serpentinite at different confining pressures and temperatures. Schematic D.T.A. data below. (b) Sliding friction of undrained specimens of Oulx serpentinite at different confining pressures and temperatures. Schematic D.T.A. data below. (c) Stress-strain curves for drained (D) and undrained (UD) specimens of Caprie serpentinite at 0.310 GPa confining pressure and 670°C. (d) Stress-drops owing to brittle faulting in undrained specimens of Oulx chloritite at different confining pressures and temperatures. Schematic D.T.A. data below. (e) Stress-strain curves for undrained specimens of Caprie serpentinite at 0.552 GPa confining pressure and various temperatures. Note brittle instability followed by dilatancy hardening at 520°C. (Based on figures from Murrell and Ismail, 1976a, reprinted with permission of Editor, Tectonophysics)

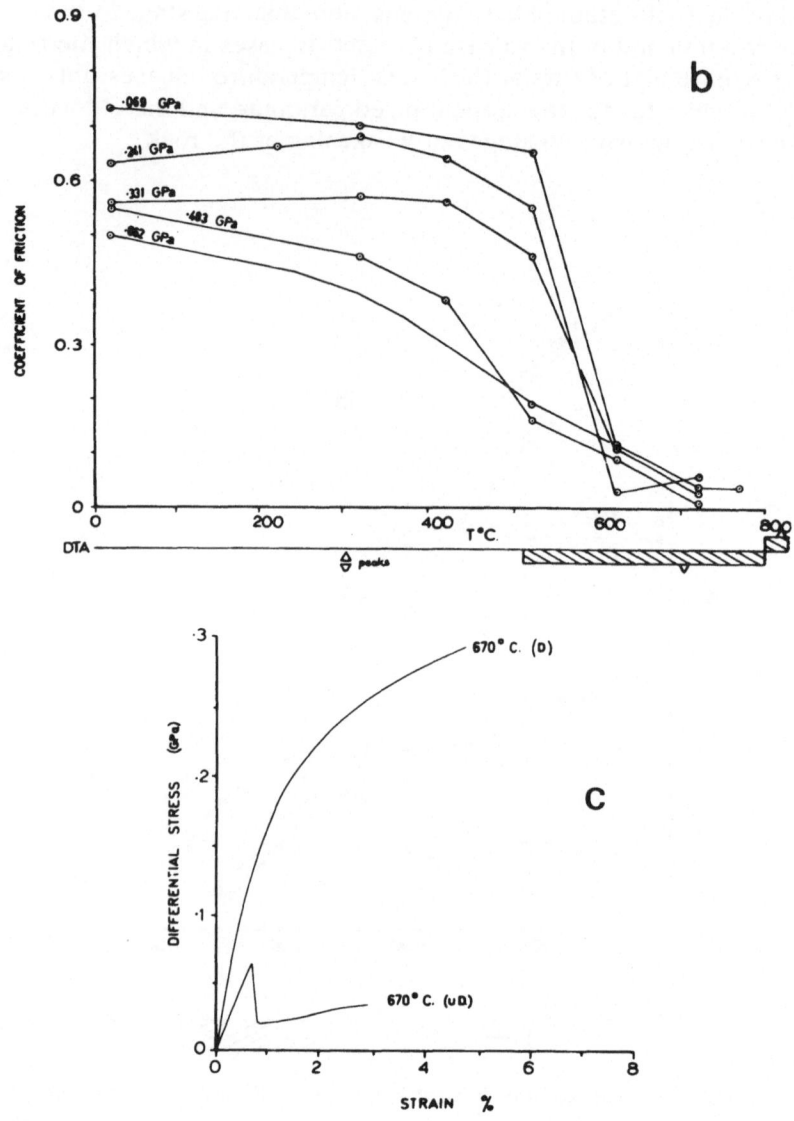

Fig. 1. (continued)

Results

Gypsum

The gypsum came from the Glebe Mine, Gotham, Nottinghamshire (Great Britain) and was donated by British Gypsum, Ltd.

The texture is fine to medium grained, and the gypsum crystals are subhe-

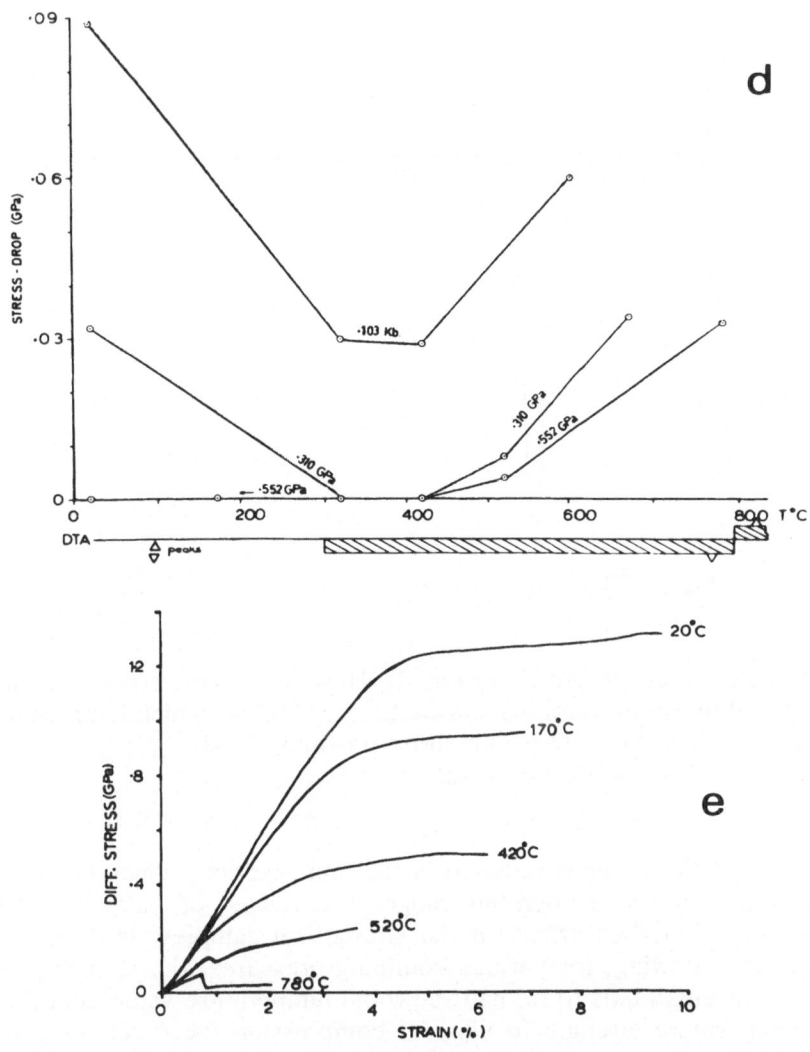

Fig. 1. (continued)

dral to anhedral in shape and colorless in thin section. The rock is locally darkened by impurities such as clays or iron oxides. X-ray diffraction confirmed that the rock consists of nearly pure gypsum.

DTA studies in air (Figure 2) show that rapid decomposition of the gypsum begins at $\sim 110°C$, and there are endothermic peaks at $\sim 161°C$ (corresponding to the breakdown to the metastable hemihydrate, bassanite, $CaSO_4 \frac{1}{2} H_2O$), and at $\sim 196°C$ (corresponding to the breakdown to anhydrite). The phase boundary for gypsum in the presence of anhydrite and pure water at atmospheric pressure has been determined as 42°C by Posnjak (1938; see also MacDonald, 1953), 40° to 46° ± 25°C by Zen (1965), and 58°C by Hardie

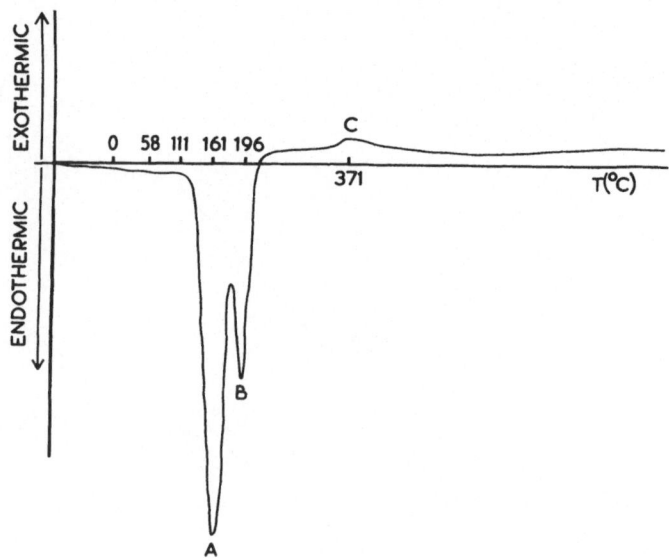

Fig. 2. DTA diagram for gypsum sample heated in air.

(1967). (See curves C1 and C2 in Fig. 3.) However, in the absence of anhydrite crystal nuclei gypsum is metastable to 97.5°C, at which temperature it converts to the metastable hemihydrate (Posnjak, 1938).

The gypsum decomposition reaction is:

$$CaSO_4 \cdot H_2O \rightleftharpoons CaSO_4 \cdot \tfrac{1}{2} \cdot H_2O + 3/2.H_2O \rightleftharpoons CaSO_4 + 2H_2O$$

Thus most of the water is released in the first reaction, which takes place over a rather narrow temperature range. This release of water into sealed samples has a marked effect on the strength of samples. In Table 1 we present data showing, for various confining pressures, the temperature at which the strength falls to (a) half the room temperature value, and (b) the room temperature strength in uniaxial compression (zero confining pressure), and also the temperatures at which were observed (c) a sharp break in the change of strength, (d) a transition from ductile to brittle behavior, and (e) the occurrence of dilatancy hardening. More extensive data are given in Murrell and Ismail (1976a). These data indicate temperatures that agree well with the first endothermic range in the DTA curve for gypsum.

X-ray diffractometry showed that samples tested at 170°C consisted mainly of the hemihydrate (bassanite), though with about 20% gypsum still present. At 270°C only anhydrite was present. A thin section from a specimen tested at 470°C showed that complete recrystallization had occurred, with the formation of a tessellated structure of small crystals, whose long axes are aligned in two approximately perpendicular directions. The photomicrograph obtained was very similar to one illustrated in Fig. 3 of Heard

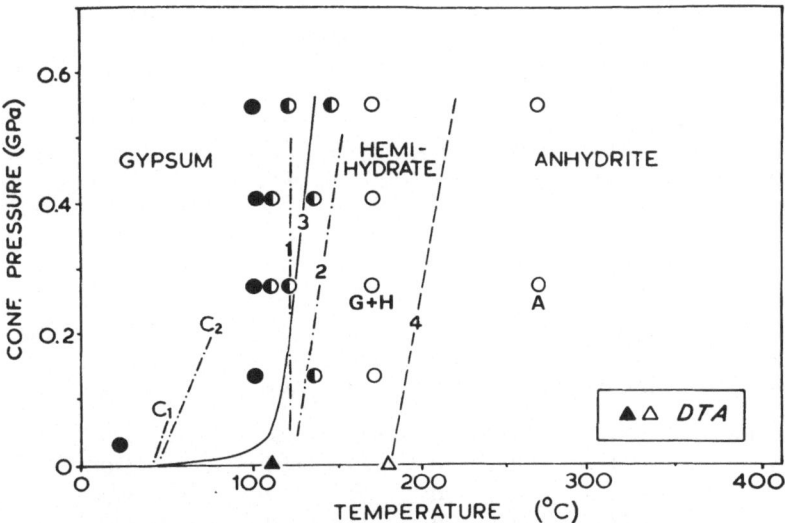

Fig. 3. Pressure–temperature stability fields for gypsum, hemihydrate, and anhydrite. C1 and C2: calculated gypsum–anhydrite boundary after MacDonald (1953) and Zen (1965), respectively; 1 and 2 experimental boundary after Heard and Rubey (1966); 3 and 4 from this work. ● ductile, ◑ transitional, ○ brittle. G + H ≡ gypsum + bassanite present, A ≡ only anhydrite present. D.T.A. points refer to rapid evolution of water from gypsum and bassanite, respectively.

and Rubey (1966), but their specimen (tested at 152°C) was thought to consist of gypsum and hemihydrate. We believe the exothermic peak at ~371°C in the DTA curve (Fig. 2) probably corresponds to the growth of the anhydrite crystals (with release of surface and strain energy).

The data are compared with published phase diagrams in Fig. 3. There is good agreement with the earlier and similar work of Heard and Rubey (1966), and the data are consistent with the phase boundary calculated by Zen (1965).

Serpentinite

This came from Sauze d'Oulx (in the Piedmont Alps of Italy), between Briançon and Susa in the Val di Susa and close to Montgenevre (see Carta Geologica d'Italia 1 : 100,000, Oulx sheet (no. 54), by Zaccagna *et al.,* 1914).

About 90% of the rock consists of serpentine, which in thin section is colorless to pale green and has low relief. There is approximately 8% of magnetite in rims around the serpentine crystals. The crystals are of random orientation, and pseudomorphous phenocrysts of olivine and pyroxene can be distinguished. The rock contains two sets of talc veins, inclined at about

Table 1. Summary of data indicating the onset of decomposition of hydrous minerals (see Murrell and Ismail, 1976a, for more detail).

Rock	Conf. Press. (GPa)	Dehydration Temperature (°C)							
		a	*b*	*c*	*d*	*e*	*f*	*g*	*h*
Gypsum	0.138	150	135	—	170	170	x	x	x
	0.276	110	120	110	145	—	x	x	x
	0.414	110	135	—	135	170	x	x	x
	0.552	100	130	100	132	145	x	x	x
Serpentinized	0.103	375	510	320	—	320	510	600	510
Peridotite	0.310	370	470	320	595	320	420	480	470
(Caprie)	0.552	370	595	—	595	520	320	490	520
Serpentinite	0.069	440	620	520	—	—	570	x	x
(Oulx)	0.241	400	620	420	695	—	570	x	x
	0.331	410	670	420	570	620	570	x	x
	0.483	380	670	420	570	520	470	x	x
	0.662	470	770	420	620	—	520	x	x
Chloritite	0.138	320	645	580	500	320	500	620	650
(Oulx)	0.345	420	720	580	645	—	580	620	630
	0.552	320	695	580	695	—	620	590	650

[a] Temperature at which strength (at fracture or 2% strain) falls to half its value at 20°C.
[b] Temperature at which strength falls to 20°C value of unconfined compressive strength (conf. press. zero).
[c] Temperature at which a sharp reduction in strength occurs.
[d] Temperature at which a transition from ductile to brittle behaviors occurs.
[e] Temperature at which dilatancy hardening is observed.
[f] Temperature at which the apparent coefficient of friction falls to half its 20°C value.
[g] Temperature at which the strength falls to half its value at 420°C.
[h] Temperature at which the apparent coefficient of friction falls to half its value at 420°C.

45° to each other, and containing occasional inclusions of serpentine and magnetite.

X-ray diffraction showed that the serpentine was mostly lizardite. Examination of material from the veins indicated the presence in these of talc and enstatite in addition to lizardite and magnetite.

The DTA curve obtained at atmospheric pressure is shown in Fig. 4. The principal endothermic peaks at ~103°C and ~702°C and the exothermic peak at ~818°C agree with the values given by Faust and Fahey (1962) for lizardite, though chrysotile has DTA peaks in the same temperature ranges. The wide and shallow exothermic peak between ~100°C and ~500°C has also been noted by these authors (but in our case may be due to the presence of magnetite) (see Faust and Fahey, 1962). The minor peak at 103°C is probably associated with adsorbed water. The main endothermic reaction,

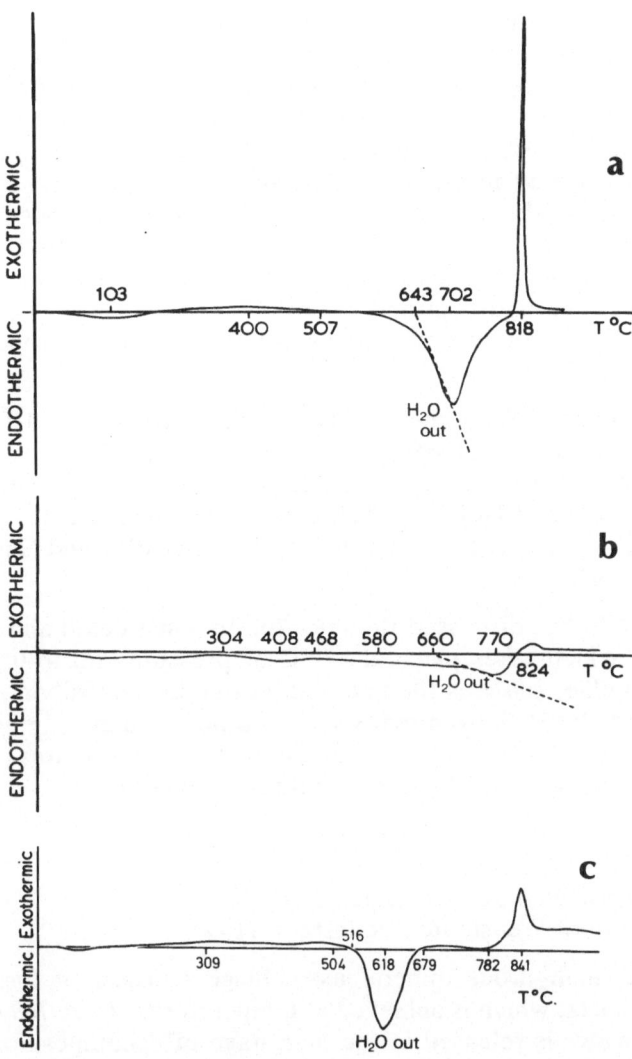

Fig. 4. D.T.A. diagrams for samples heated in air. (a) Serpentinite (Oulx). Serpentine mineral is lizardite (+ minor chrysotile). (b) Serpentinized peridotite (Caprie). Serpentine mineral is antigorite. (c) Chloritite. Chlorite mineral is ripidolite.

resulting from decomposition of the serpentine minerals, begins at ~500°C, reaching its peak at ~700°C, and the main exothermic peak at ~818°C is associated with recrystallization into a new phase assemblage. The predominance of lizardite is also indicated by the estimated characteristic dehydration temperature as defined by Weber and Greer (1965) (by drawing a tangent to the endothermic peak at the point where it has its maximum gradient and taking the temperature where the tangent cuts the DTA baseline). The value found is ~640°C, which is close to the value of 636°C, the mean for 10 samples of lizardite (or lizardite + minor chrysotile) determined by these authors with the samples in water vapor at a pressure of one atmosphere. (The mean endothermic peak temperature was 694°C, compared with our value of ~702°C.) Brindley and Zussman (1957) studied the decomposition of serpentine, heated in air, by X-ray diffraction. The first indication of decomposition was the weakening of h01 reflections, which occurred at 500–550°C in lizardite and chrysotile. Ball and Taylor (1963) showed that in runs of 1–7 days there was nearly complete decomposition of chrysotile at 520°C under both dry and hydrothermal conditions.

When brucite is not present, three reactions have been proposed for the decomposition of serpentine:

1. 5 serpentine $\rightleftharpoons$ 12 forsterite + 2 talc + 18H$_2$O (Bowen and Tuttle, 1949; Raleigh and Paterson, 1965; Scarfe and Wyllie, 1967);
2. 2 serpentine $\rightleftharpoons$ 3 forsterite + quartz + 4H$_2$O (Brindley and Zussman, 1957; Brindley and Hayami, 1965; Deer *et al.*, 1962); and
3. serpentine $\rightleftharpoons$ forsterite + enstatite + 2H$_2$O (Brindley and Hayami, 1965; Coleman, 1971).

Turner (1968) has discussed the question in some detail and shows that reaction (1), which takes place at 450°C at pressures up to 0.3 GPa (see Scarfe and Wyllie, 1967), is the first step in the decomposition.

Greenwood (1963) showed that talc decomposed to give anthophyllite at temperatures of 600–700°C and that at still higher temperatures (700–800°C) anthophyllite decomposed to give enstatite and quartz:

4. 9 talc + 4 forsterite $\rightleftharpoons$ 5 anthophyllite + 4H$_2$O;
5. anthophyllite + forsterite $\rightleftharpoons$ 9 enstatite + H$_2$O;
6. talc $\rightleftharpoons$ 3 anthophyllite + 4 quartz + 4H$_2$O;
7. anthophyllite $\rightleftharpoons$ 7 enstatite + quartz + H$_2$O.

The ultimate nonhydrous mineral assemblage consists of forsterite, enstatite, and quartz, which is achieved at temperatures above 700–800°C, but most of the water is released in the first stage of decomposition (reaction (1)). As with gypsum the release of water by decomposition of the hydrous minerals has a marked effect on the strength of sealed samples of serpentinite. Data on the strength of Oulx serpentinite is given in Table 1 (see, further, Murrell and Ismail, 1976a), and in this case includes also the temperature at which the coefficient of sliding friction falls to half its room-temperature value (this is due to the buildup of a pore-water pressure in sealed specimens when decomposition of the serpentine occurs). The most sensitive indicator of the decomposition reaction seems to be a marked reduction in the strength of the rock. The strength falls to half its room-temperature value at a temperature of ~420°C (range 380–470°C), and there is a sharp break in the change of strength with temperature at this same temperature, in reasonable agreement with the temperature of ~450°C obtained by Scarfe and Wyllie (1967) for reaction (1). At a pressure of 0.662 GPa a somewhat higher temperature is indicated for the reaction (Table 1).

X-ray diffraction analysis of a quenched sample tested at 520°C, 0.138 GPa, showed little change in the diffraction pattern, though the 7.4 A.U. ($2\theta = 12°$) peak was weakened. At 620°C, 0.483 GPa, there was a further small change in the 7.4 A.U. peak, but at 670°C, 0.483 GPa, the serpentine pattern became very weak, and peaks corresponding to forsterite, talc, anthophyllite, enstatite, and quartz appeared (the strongest peaks being for forsterite at 3.88 A.U., 2.77 A.U., and 2.46 A.U.). At 720°C, 0.331 GPa, the main serpentine peak (at 7.4 A.U.) had disappeared completely, the forsterite pattern was strong, and talc peaks had also increased. At 770°C, 0.662 GPa, the serpentine pattern had disappeared completely, the forsterite pattern was still stronger, enstatite peaks were strong, and the talc peaks had weakened.* A thin section of a specimen tested at 720°C, 0.662 GPa, revealed the presence of radiating aggregates of fibrous crystals, which were presumably talc or anthophyllilte.

The data are compared with published phase diagrams in Fig. 5. There is good agreement with earlier and similar work of Raleigh and Paterson (1965), and the data are consistent with the phase boundaries obtained by Bowen and Tuttle (1949), Pistorius (1963), Yoder (1967), Kitahara et al. (1966), and Scarfe and Wyllie (1967) for reaction (1) above. There is also consistency with the work of Greenwood (1963) on the later stages of decomposition of the hydrous phases. The data are also in good agreement with the comprehensive phase diagram published by Hemley et al. (1977) for the system $MgO-SiO_2-H_2O$.

Serpentinized Peridotite

This came from a quarry (Cave Pietrisco, by courtesy of Ing. Vito Rotunno) at Caprie, about 1 km east of Condove in the Val di Susa to the south west of the Massif di Lanzo. (See the 1 : 50,000 map made by Nicolas, 1966.)

The rock is mainly composed of olivine and pyroxene phenocrysts (~98%), with olivine being dominant. From the peak heights of the X-ray powder diffraction pattern we estimate that there is no more than 3% pyroxene in the rock. About 60% of the olivine and pyroxene have been altered to serpentine and amphibole, respectively, whose crystals are small and fibrous. Some phenocrysts of olivine have rims of serpentine around them. There is about 1% of almost euhedral magnetite present, and another opaque, euhedral mineral, thought to be chromite, occasionally occupies fractures in the olivine phenocrysts. All the crystals are severely fractured and show a high degree of preferred orientation, which imparts a foliation to

---

* Tracings of the diffraction patterns may be obtained from the author.

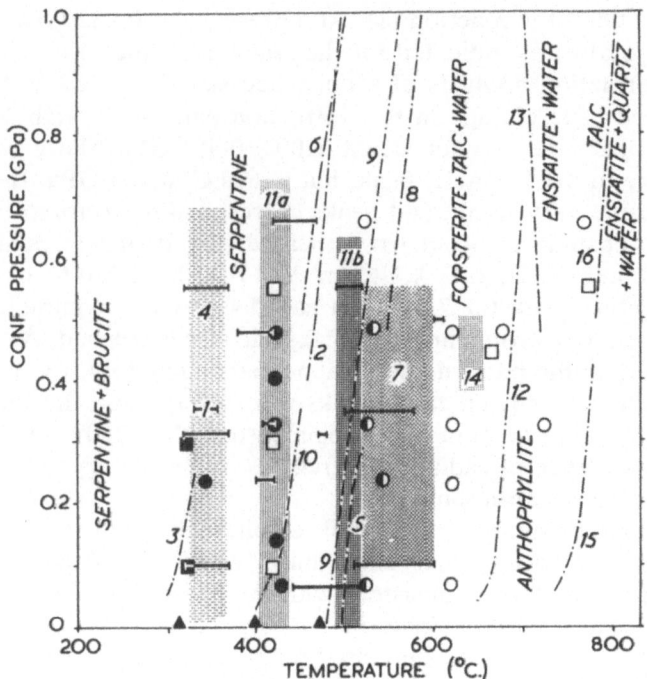

Fig. 5. Pressure–temperature stability fields for serpentine and its decomposition products. 1, 2, 3, 4 serpentine + brucite $\rightleftharpoons$ forsterite + water; 5, 6, 7, 8, 9, 10, 11 serpentine $\rightleftharpoons$ forsterite + talc + water; 12, 13, 14 forsterite + talc $\rightleftharpoons$ anthophyllite, enstatite; 15, 16 anthophyllite, talc $\rightleftharpoons$ enstatite + quartz + water. 1. Raleigh and Paterson (1965) (Fidalgo Island serpentinite + brucite); 2. Kitahara *et al.* (1966); 3. Scarfe and Wyllie (1967); 4. This work; 5. Bowen and Tuttle (1949); 6. Pistorius (1963); 7. Raleigh and Paterson (1965) (Tumut Pond and Cabramura antigorite–chrysotile serpentinites); 8. Kitahara *et al.* (1966); 9. Yoder (1967); 10. Scarfe and Wyllie (1967); 11a. This work, Oulx serpentinite (lizardite); 11b. This work, Caprie serpentinite (antigorite); 12. Greenwood (1963); 13. Kitahara *et al.* (1966); 14. This work; 15. Greenwood (1963); 16. Kitahara *et al.* (1966). Boundaries of stability fields found in this work (4, 11a, 11b, 14) and by Raleigh and Paterson (1965) (7) are shaded. Data from this work are indicated by the following symbols: Oulx serpentinite (lizardite): ● strong, ○ weak, ◐ intermediate, ▲ D.T.A.; Caprie serpentinite (antigorite + brucite), ■ strong, □ weak, ▲ D.T.A. Detailed D.T.A. data in Fig. 4.

the rock. Samples for the experiments reported here were cored perpendicular to the foliation.

X-ray diffraction showed that the serpentine was antigorite, which, taken together with the foliated character of the rock, implies metamorphism (Nicolas, 1969; Coleman, 1971).

The X-ray analysis also showed that a small amount of brucite was present. Assuming that the curves given by Hess and Otalora (1964) apply for antigorite as well as for lizardite, we compared the peak heights for brucite (4.77 A.U.) and antigorite (7.30 A.U.) and estimated that there is about 1% of brucite in the rock.

The DTA curve is shown in Fig. 4. The principal endothermic peak at ~770°C and the exothermic peak at ~824°C agree well with the values given by Faust and Fahey (1962) for antigorite. As observed by these authors, the shallow exothermic peak between 100°C and 500°C that is observed for lizardite and chrysotile does not occur with antigorite. The higher value of the main endothermic peak temperature for antigorite was also found by Weber and Greer (1965), who found values ranging between 725°C and 766°C.

An interesting feature is the shallow endothermic feature that extends between ~300°C and ~470°C and reaches a peak at ~408°C. This is probably associated with the dehydroxylation reaction involving brucite and serpentine (see below). Because there is only a small amount of brucite present, the main decomposition reaction for the serpentine takes place at a higher temperature, apparently starting at ~470°C and reaching its peak at ~770°C.

Brindley and Zussman (1957) studied the decomposition of antigorite heated in air by X-ray diffraction and found that it took place at a temperature of ~75°C higher than that for lizardite and chrysotile, with the h01 reflections weakening at 575–600°C and the serpentine pattern disappearing at 650–700°C.

When brucite is present an additional reaction to the seven discussed above can occur, as described by Bowen and Tuttle (1949), Pistorius (1963), Raleigh and Paterson (1965), Kitahara et al. (1966), Scarfe and Wyllie (1967), and Turner (1968):

$$\text{serpentine} + 2 \text{ brucite} \rightleftharpoons 4 \text{ forsterite} + 6H_2O.$$

Scarfe and Wyllie (1967) found that at pressures up to 0.3 GPa the reaction took place at 300–325°C, whereas earlier authors (for example Kitahara et al., 1966, whose experiments extended to pressures of 3 GPa) obtained temperatures of 400–500°C.

Data on the strength of the Caprie serpentinized peridotite are given in Table 1 (for details see Murrell and Ismail, 1976a). The strength and the coefficient of friction fall to half their room temperature values at a temperature of ~370°C, which is somewhat higher than the temperature found for the brucite reaction by Scarfe and Wyllie (1967) but agrees closely with the strength data of Raleigh and Paterson (1965) for heating times of 1 hour. On the other hand the break in the gradient of strength versus temperature occurred at ~320°C, in good agreement with the data of both Scarfe and Wyllie (1967) and Raleigh and Paterson (1965).

At higher temperatures there is embrittlement and a further drastic reduction in strength associated with the decomposition of serpentine in accordance with the reactions discussed above. The strength and friction fall to half their 420°C values at an average temperature of ~510°C.

X-ray analysis of samples tested at 0.310 GPa, 520°C, and at 0.103 GPa, 600°C, revealed little change in the diffraction pattern. At 0.448 GPa, 670°C, a new peak appeared at $2\theta = 6.15°$ (14.48 A.U.), and enstatite and quartz patterns also appeared, but otherwise there was little change (the antigorite peaks remained dominant). However, at 0.552 GPa, 780°C, the antigorite pattern had disappeared, the forsterite peaks were much stronger, and strong talc peaks had appeared. The $2\theta = 6.15°$ reflection had disappeared. (Tracings of the diffraction patterns may be obtained from the author.) The data are in accordance with the DTA evidence in indicating a higher temperature for the decomposition of antigorite.

The data for the brucite reaction are shown on the serpentine phase diagram in Fig. 5. Note that the curve (2) obtained by Kitahara *et al.* (1966) for this reaction is ~100 K higher than those obtained by other workers and falls in the region corresponding to reaction (1) above.

Trommsdorff and Evans (1972) have found field evidence of the sequence of reactions illustrated in Fig. 5 in their study of the contact metamorphism of the Malenco serpentinite (Italian Alps).

Chloritite

The chloritite came from Sauze d'Oulx, from the same outcrop as the serpentinite discussed above and from the south-facing side of the exposure, near the boundary with the schistes lustres.

The chemical analysis is very similar to that of the "blackwall" chlorite rock described by Chidester (1962, specimen W-38 from Waterbury mine), which is thought to mark the original contact of the Waterbury ultramafic body. Nicolas (1966) has described similar rock in an area to the northeast of the outcrop at Sauze d'Oulx referred to above. He also refers to segregations of chloritite contained in the serpentinite massif (Nicolas, 1969).

The rock is massive and shows little sign of fracturing. It consists of ~85% chlorite, together with ~4% euhedral magnetite, ~4% anhedral ilmenite (occasionally replaced, partially or totally, by sphene in fine aggregates). The chemical analysis suggests that the chlorite mineral is ripidolite (Hey, 1954; Foster, 1962), and this is confirmed by X-ray diffraction measurements.

The DTA curve is shown in Fig. 4. The principal endothermic peak at ~618°C and the narrow and sharp exothermic peak at ~841°C agree well with the values determined by Phillips (1963) for ripidolite (the chemical analysis of this ripidolite is also very similar to that of our chloritite rock). (The DTA curve is similar to that for clinochlore obtained by Grimshaw and Westerman; see Brindley and Ali, 1950. The latter authors showed that the endothermic reaction was associated with dehydration and the exothermic reaction represented recrystallization to form olivine.)

McOnie *et al.* (1975) show that at 0.207 GPa the decomposition products of chlorite depend on the $Fe/(Fe + Mg)$ ratio, but cordierite is always present. They write the following reactions:

1. chlorite $\rightleftharpoons$ H_2O + cordierite + olivine + spinel (for Mg-rich clinochlore).
2. chlorite $\rightleftharpoons$ H_2O + cordierite + olivine + hercynite
   chlorite $\rightleftharpoons$ H_2O + cordierite + olivine + magnetite
   chlorite $\rightleftharpoons$ H_2O + cordierite + orthoamphibole + magnetite (for ripido-lites increasingly rich in Fe).
3. chlorite $\rightleftharpoons$ H_2O + cordierite + quartz + magnetite (for Fe-rich daphnite).

However, above about 0.3–0.4 GPa cordierite has only a limited temperature field of stability, and McOnie *et al.* (1975) give the reactions:

4. chlorite $\rightleftharpoons$ H_2O + Fe-cordierite + Fe-gedrite + magnetite for Fe-rich daphnite above ~0.4 GPa, and T ~600°C) = H_2O + almandine + magnetite + quartz (for $T > 650$°C).
5. chlorite $\rightleftharpoons$ H_2O + forsterite + enstatite + spinel (for Mg-rich clinochlore above ~0.3 GPa).

Data on the strength of the Oulx chloritite are given in Table 1 (for further details see Murrell and Ismail, 1976a). There are two stages in the reduction in strength as temperature is increased. The cause of the large reduction between 20°C and 300°C is not yet understood. However, between ~300°C and 580°C the strength remains nearly constant. Above 580°C the strength falls very sharply, and the rock becomes very brittle.

X-ray analysis of a sample tested at 0.448 GPa, 670°C, showed little change in the diffraction pattern. However, in a sample tested at 0.138 GPa, 720°C, the ripidolite pattern had almost disappeared, and forsterite and spinel were found. At a pressure of 0.552 GPa and the same temperature (720°C), forsterite and spinel were also found, but the ripidolite pattern was still strong.

The dehydration curve for Fe-rich chlorite suggested by the present work is shown in Fig. 6. This curve is ~150 K lower than that determined by Fawcett and Yoder (1966) for Mg-rich chlorite (clinochlore) and ~50 K higher than that determined for Fe-rich chlorite (daphnite) by McOnie *et al.* (1975) (see also James *et al.*, 1976; Turnock, 1960). McOnie *et al.* (1975) have studied systematically the decomposition temperature of chlorites at a pressure of 0.207 GPa as a function of the $Fe/(Fe + Mg)$ ratio; from which we find that for a ratio of 0.8 (as for the rock studied in our experiments, which puts it on the boundary between daphnite and ripidolite in the Hey, 1954, classification; see also Deer *et al.*, 1962) the decomposition temperature is ~600°C, in good agreement with our proposed dehydration curve. We have also obtained evidence from deformation experiments on an altered granodiorite rock (Murrell and Ismail, 1976b) that the chlorite and quartz reaction occurs at a temperature of ~520°C (at a pressure of 0.448 GPa) for an Fe-rich chlorite $(Fe/(Fe + Mg)) = 0.83$, corresponding to a daphnite composition).

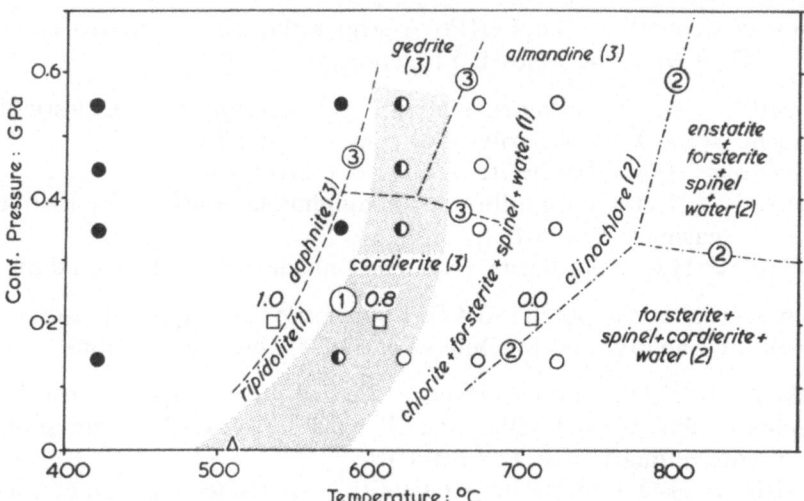

Fig. 6. Pressure–temperature stability fields for chlorite. 1. This work (ripidolite) (hatched area); 2. Fawcett and Yoder (1966) (clinochlore, Mg-rich); 3. Turnock (1960), McOnie *et al.* (1975) (daphnite, Fe-rich). ● strong, ○ weak, ◖ intermediate, △ D.T.A. ☐ decomposition as a function of Fe/(Fe + Mg) ratio (values 1.0, 0.8, and 0.0 shown) at 0.207 GPa pressure, from McOnie *et al.* (1975).

Examination of thin sections by optical microscopy revealed the occurrence of a major textural change at high temperatures and pressures and the presence of small crystals (~10 μm diameter) of new phases, probably olivine.

Discussion

Effect of Duration of Heating

The experiments discussed here were all of short duration, and phase equilibrium will not in general have been achieved except perhaps in the case of the gypsum–anhydrite transformation at temperatures well above the phase boundary. In addition, a common feature of the decomposition reactions of all four rocks studied here is the occurrence of metastable mineral assemblages, representing several stages of reaction. The formation of bassanite during the gypsum–anhydrite transformation is a particularly simple and clear example.

In the case of the chlorite rock the growth of new phases was sluggish, so that these phases were not identifiable by X-ray analysis, though the loss of the chlorite mineral (ripidolite) was observed. The maximum test temperature in this case was 720°C, which is above the endothermic DTA peak (centered at 618°C) associated with decomposition but below the exothermic peak (centered at 841°C) associated with recrystallization of new phases. In the undrained experiments, water lost from the chlorite structure is retained in the rock sample, and though much of this water occupies pores and cracks, thereby substantially reducing the rock strength, some of it may remain trapped in the cordierite structure if cordierite is formed. (Schreyer and Yoder, 1960 have shown that water loss from hydrous Mg-cordierite takes place over a very wide temperature range extending up to ~700°C, and that most of the water is molecular and unbound, possibly occupying ring channels in the cordierite structure, and only a small proportion is bound. Aines and Rossman (1984) show that unbound water exists in cordierite at 400–800°C). However, it must be assumed that eventually at some temperature >700°C all water would be lost from the cordierite structure and that no further phase change takes place until the solidus temperature is reached (Turner, 1968). The occurrence of melting in a granitic rock at a pressure of 552 MPa and a temperature of 670°C, associated with decomposition of chlorite in the rock (initiated at ~520°C), was noted by Murrell and Ismail (1976b).

In the case of the two serpentinite rocks studied by us there are several decomposition reactions, covering a temperature range from ~300°C to ~800°C. The lowest temperature reaction occurs when brucite is present. The main loss of water from the structure occurs at 400–600°C when the serpentine decomposes, but structural water remains in talc and in anthophyllite, and this is only lost at still higher temperatures.

Duration of heating will clearly affect the phase composition during an experiment in accordance with the kinetics of the decomposition reactions, including those of metastable phases, and this will influence the amount of water released into the rock. It may also affect the siting of molecular water, as in the case of any hydrous cordierite formed by decomposition of chlorites. Such structural, but unbound, water may not greatly influence the mechanical properties of the rock, but release of this water into cracks (e.g., at grain boundaries) would have a strong influence.

Effect of Deformation on Kinetics

Deformation probably changes the kinetics and increases the reaction rate through the influence of crystal defects. Wood and Walther (1983) have produced evidence to suggest that the rates of reactions of silicates with aqueous fluids follow zero-order kinetics controlled by the reacting surface area, in which a common rate constant applies for all silicates and for reactions involving dissolution, fluid production, or solid–solid transformations in the presence of a fluid of moderate to high pH. They show that prograde metamorphic dehydration reactions will then go to completion in a few tens or hundreds of years, and they also point out that an excess of surface defects produced by mechanical action also enhances the reaction rates. Cracking is likely to have been a major factor in the microscopic processes of the dehydration reaction in our experiments, by opening up fresh faces of the original hydrous mineral, free from the new phases (see Galwey *et al.*, 1981). Without cracking, the rate of reaction will become increasingly dependent on diffusion from or to the interface between the original hydrous phase and the new phase. The retention of water in the sealed samples will also tend to accelerate the growth of new phases and will enhance cracking due to deformation.

Effect of Differences between Pore Pressure and Confining Pressure

The conditions of our experiments, in which the rock samples were initially dry, were different from typical petrology experiments, in which samples are usually saturated with water (giving $P_{H_2O} = P_{total}$). The effect of such conditions has been discussed by Fyfe (1973). This question has been addressed theoretically and experimentally by Greenwood (1961) (see also Wyllie, 1962) and was considered by Raleigh and Paterson (1965) in the discussion of their deformation experiments on serpentinite. When the partial pressure of the water in the fluid phase (the pore pressure, in the parlance of this paper) differs from the pressure on the mineral phases (the total or confining pressure), then even if the system is a chemically closed one (that is, there is no loss or gain of matter from it—a condition described as "undrained" in rock mechanics; see Ismail and Murrell, 1976) the reaction is no longer univariant but divariant. It must be represented by an equilibrium surface in a space showing the three independent physical variables: confining pressure, pore pressure, and temperature. Greenwood (1961) shows that if the fluid phase (pore) pressure is less than the total (confining) pressure the dissociation temperature of the hydrate phase is lowered. Similarly in circumstances under which the fluid phase pressure was greater than the confining pressure, the dissociation temperature would be increased. Greenwood (1961) describes this latter condition as "not found" (his Figure 1(c)), but the only physical limitation on the condition is the mechanical strength of the solid phases.

  If the original rock samples in such experiments as ours were fully dense, that is, free of cracks and pores, then the pressure of the water phase released by the reaction under undrained conditions, either at the sample surface or at internal grain boundaries, would equal the total (confining) pressure (allowing for the flexibility of the jackets used).

  At low confining pressures, however, some open cracks and pores would be present, filled only with air at atmospheric pressure in experiments on dry rock, and the dissociation reaction would begin at the surfaces of these cracks and pores at temperatures lower than the water-saturated reaction temperature. The fluid-phase pressure would then stabilize at some pressure lower than the total (confining) pressure. Only when the temperature was raised to the water-saturated phase boundary value would the fluid-phase (pore) pressure become equal to the total (confining) pressure.

The pressure required to close cracks is $\sim 10\ K^*$, where K^* is the tensile strength (Digby and Murrell, 1976), so that for rocks cracks would tend to be closed at pressures above about 0.1–0.2 GPa. Only at total (confining) pressures lower than this (corresponding to depths down to ~ 8 km) would there exist fluid-phase (pore) pressures lower than the total (confining) pressure. Under these conditions, however, cracks and pores in the crustal rocks would tend to be interconnected to the surface, so that pore pressures would tend not only to be lower than the confining (total) pressure, but would also be maintained at a constant value at a given depth (assuming constant temperature). This latter is the "drained" condition (Ismail and Murrell, 1976). The reduction in strength under these conditions would be much less than at greater depths, where pore pressures would be equal to or greater than the total (confining) pressure.

As was pointed out by Greenwood (1961) in the cases where the fluid-phase (pore) pressure is less than the total (confining) pressure, the mineral grains are under localized nonhydrostatic (and nonuniform) stress and are not, therefore, in equilibrium. Pressure solution and recrystallization will tend to occur, so as to fill the pores, reduce the rock volume, and reduce the nonhydrostatic stress in the mineral grains. This will also occur more generally, even if the fluid-phase pressure equals the total (confining) pressure, so long as a macroscopic nonhydrostatic stress state exists in the rock. In the latter case subtle modes of deformation might be possible, involving the interaction of cracking processes resulting from the decomposition reaction and pressure solution effects that tend to relax the nonhydrostatic (shear) stresses. One interesting feature arising from the effect of a difference between fluid-phase (pore) pressure and total (confining) pressure is that under deep-burial conditions when a pore fluid, generated by a decomposition reaction, leaks away, tending to reduce the pore pressure, the decomposition temperature will tend to be reduced, so that at constant temperature the temperature over-step above the phase boundary will increase, and so will the reaction rate. Thus there is a tendency to buffer the fluid-phase (pore) pressures.

Influence of Dehydration Reactions on Strength and Deformation during Burial Metamorphism

The results of the deformation experiments strongly suggest that strength and deformation behavior can be very sensitive indicators of the initiation of decomposition reactions involving structural loss of a fluid phase. This is because of the generation of high pore-fluid pressures in cracks in rock when the rock is sealed (the undrained condition), resulting in marked loss of strength and embrittlement, together with dilatancy-hardening effects (Murrell and Ismail, 1976a). In principle, by carrying out experiments of longer duration the lower temperature limit of decomposition reactions could be studied even more closely, but our deformation experiments already show very good agreement with long-duration phase equilibrium studies.

This has important geological implications concerning styles of tectonic deformation (including seismicity), modes of metamorphic transformation, and chemical and mineral transport and accumulation in the crust, because of the sharp and early changes in strength and ductility that may occur when rocks are buried to a level at which prolonged heating causes phase changes involving release of structurally bound water.

Entrapment of Fluids Evolved during Metamorphism

The pore pressure created by prograde metamorphism depends on the kinetics of the transformation and of the diffusion of the pore fluid out of the rock. If the pore fluid is trapped in the source rock (which is undergoing metamorphism) because it is enclosed by rocks of low hydraulic conductivity, then the pore pressure will tend to rise until cracking and fracturing occurs as the effective principal compressive stresses are reduced—a hydraulic fracturing process. If the deviatoric stress in the rock formation is sufficiently high, shear fractures will form, but at low (or zero) deviatoric stress tensile fractures will form. Dilatancy owing to cracking and fracturing will tend to cause the pore pressure to drop, as the pore fluid comes to occupy a larger crack volume, so that effective compressive principal stresses will then increase and the rock will become stronger; this is dilatancy hardening (Ismail and Murrell, 1976). Any hardening process will tend to homogenize deformation and the distribution of pore fluid, while softening will localize the deformation and the pore fluid. Hardening also stabilizes the deformation and counteracts any tendency to seismic faulting. The deformation will tend to adjust to a condition of dynamic equilibrium depending on the deviatoric tectonic stress (or strain-rate), on the kinetics of pore-fluid generation, and on the kinetics of crack propagation and fracturing (which will depend on such factors as stress-corrosion and solution by the pore fluid) and of other modes of mass deformation such as plasticity (involving crystal deformation) or cataclastic flow (involving relative motions of grains at grain boundaries or intergranular spaces, such as cracks and pores, in which relative shear, dilation, or rotation may occur). The nucleation and growth of small grains of new phases may have a special role to play in cataclastic flow and may also contribute to the deformation by superplastic flow processes (Schmid *et al.*, 1977; Paterson, 1983).

The condition of trapped pore fluid is called the undrained condition and is the one under which the lowest strength may be observed and the deformation will be most widely distributed through the rock mass. Cracking and cataclastic flow are essential features of this deformation and might be observed at the macroscopic level as large-scale fractures or vein structures, but they could eventually result in a very finely divided new grain structure, probably with a fabric if the deviatoric stress is sufficiently high as more and more pore fluid is generated and deformation proceeds. If metamorphism proceeds for a sufficiently long period and the temperature increases, then in addition to the effect this will have on the kinetics of the various processes there will be an additional effect because of thermal expansion, especially of the pore fluid.

In due course, however, pore fluid will surely drain away. Hanshaw and Bredehoeft (1968) give some calculations of the diffusion of water from a gypsum layer, trapped between low permeability capping layers, and undergoing transformation to anhydrite, and show that with a hydraulic conductivity of 10^{-10} ms^{-1} (at the low end of the clay range; Bredehoeft and Hanshaw, 1968) in the cap rock the excess head of water, at a depth of ~1 km, would dissipate in ~50,000 a. However, it seems likely that the cap rock itself might also undergo hydraulic fracturing, perhaps induced by thermal expansion of the water, and this would increase the rate of dewatering of the rock undergoing prograde metamorphism (see Norris and Henley, 1976), because of the increased permeability. Fracturing of the cap rock will cause the pore pressure in the source rock to fall and that in the cap rock to rise, so that the source rock will tend to "harden" and the cap rock will tend to "soften." The softening of the cap rock will localize the deformation in fault zones, and this will probably cause the permeability and hydraulic conductivity for a given mass porosity (or water content) to increase (see Walther and Orville, 1982). For example, Turcotte and Schubert (1982) show that for a permeability model consisting of a cubic matrix of tubes of circular cross section (diameter δ) the permeability (k) is given by:

$$K = (\Phi/96).\delta^2$$

where Φ is the mass porosity. Thus K will increase rapidly with the pore tube diameter, so that a fault zone of large circular cross-section area will give a mass permeability much greater than a set of distributed tubular pores of smaller diameter, even if the total cross-sectional area of the tubes equals that of the fault (so that the mass porosity is the same in the two cases).

Another factor that affects permeability, and therefore dewatering, strength and deformation style, is the chemistry of the pore water. Burial to deeper levels where temperatures are higher will tend to increase chemical activity and reaction rates and to cause thermal expansion of the pore fluid, so that there will be a tendency for permeability to increase in the source rock, especially if the cap rock has an initially low permeability. However, once uplift, erosion, and cooling begins there will be a tendency for solute to be deposited from the cooling pore fluid, and this will clog the pores and reduce permeability. This process has been studied experimentally by Nur (1982). This points to another feature of the dewatering of a metamorphic source rock, and that is the likelihood of chemical change and the transport and deposition of minerals by migrating pore fluid, which will be enhanced by fault formation in the cap rocks.

Deformation Style during Metamorphic Dehydration

From the above arguments it seems that deformation in the metamorphic source rock will be of a distributed style because of dilatancy hardening and that a fabric will be produced, while in the cap rock, into which pore water flows from the source rock, localized deformation will take place in fault zones, fractures, and vein systems, because of strain softening caused by the increasing pore pressure. Moreover, since the temperature will be rising during the metamorphic process resulting from burial and there is a tempera-ture gradient across the source rock, a water-evolution peak will sweep through it, and there may be a tendency to create localized deformation features (fractures and vein systems) both in front of (at an earlier stage) and behind (at a later stage) the peak. Norris and Henley (1976) describe two such generations of cross-cutting mineral-filled vein systems in the Haast schists of Otago, New Zealand.

We may hypothesize that metamorphism on a regional scale is linked to lithosphere deformation, either in the form of stretching to form a sedimen-tary basin (McKenzie, 1978) or in the form of orogenic shortening in plate collision zones (Murrell, in press). In either case the crustal deformation (stretching and thinning, or shortening and thickening), and the burial of crustal rocks (by sedimentation or by nappe emplacement), may occur rela-tively rapidly, but the subsequent thermal adjustment occurs over a much longer period (typically ~50 Ma; see McKenzie, 1978; Murrell, in press). The burial history (and therefore the pressure–temperature history) of any given rock will also be affected by the sedimentation and erosion history of the crust in which it is located.

From the earlier discussion we conclude that prograde metamorphism involving dehydration will result in a sharp and early loss of strength, if the source rock is enclosed in a low permeability cap rock. Reduction of the pore pressure as dewatering proceeds will also be relatively rapid and will be accelerated by faulting of the cap rock. If the permeability of the immedi-ately surrounding rock formations is not very low a large excess of pore water may not build up in and adjacent to the source rock, but an excess head will build up at the nearest low permeability formation. The source rock may remain relatively strong (as in drained experiments described by Murrell and Ismail, 1976a, 1976b), but wherever there is a pore pressure the rock will be weakened. The deformation will be distributed in character. It seems likely that the process of prograde metamorphism involving dehydration and any consequent deformation and dewatering will therefore be relatively rapid compared with thermal relaxation times. Retrograde metamorphism during the uplift and erosional stages of orogeny will be slower and will probably be limited by lack of water resulting from loss of water during the prograde process.

Model for Texture Development during Burial Metamorphism

Broadly speaking, we hypothesize that during burial, following any earlier stages of consolidation, compaction, and diagenesis, any hydrated rock will first of all pass through a stage of fracture-and-vein formation as water evolves by dehydration from rocks at a deeper and hotter level. It will then itself undergo dehydration, when its internal pore pressure will pass through a peak, and deformation will be distributed and will produce a fabric if the deviatoric stress is sufficient. At the microscopic level the deformation will be dominated by cataclasis associated with the pressurized pore water and by cataclastic flow, but there may also be some element of crystal plasticity, especially associated perhaps with the small crystal nuclei of new phases. Finally, with deeper burial a further stage of fracture-and-vein formation will occur because of dehydration of the overlying, shallower rocks and at the same time grain growth and recrystallization of new phases will proceed. Insofar as water evolves in several stages and over a wide temperature interval, there could be a succession of stages of texture formation—the dominant one being associated with the peak in the evolution of water from the rock, followed by grain growth and recrystallization of new crystal phases.

Conclusions

1. A sharp change in the strength as a function of temperature is a sensitive indicator of the initiation of decomposition of hydrated minerals leading to the evolution of water in encapsulated (undrained) rocks.
2. From our experiments, such a sharp change in strength occurs in gypsum at $\sim 100°C$, in brucite-bearing serpentinite at $\sim 300°C$, in lizardite–serpentinite at $\sim 400°C$, in antigorite–serpentinite at $\sim 500°C$, and in a ripidolite–chlorite rock at $\sim 600°C$. The decomposition temperature is not strongly dependent on pressure for pressures greater than ~ 100 MPa. Experiments on rocks containing chlorites of different compositions (ripidolite, with $Fe:(Fe + Mg) = 0.8$, and daphnite, with $Fe:(Fe + Mg) = 0.83$) confirmed that the decomposition temperature decreases as the Fe content increases.
3. Unless melting occurs the style of deformation accompanying the dehy-

dration reaction in an encapsulated rock is a brittle one and involves not only a reduction in the fracture and frictional strength, but also a transition at high confining pressure from ductile to brittle behavior with a recurrence of stress-drops (strain-softening), accompanied, however, by dilatancy hardening.

4. If water is able to drain rapidly from the rock undergoing metamorphism the strength of the rock remains high, and the style of deformation will tend to differ from the style in undrained (encapsulated) rock.

5. Where dilatancy hardening occurs (in undrained, encapsulated rocks) the deformation is distributed and occurs by cataclastic flow, but with some contribution from other processes such as stress corrosion, pressure solution, superplasticity of finely divided new phases, and normal crystal plasticity.

6. Where low-temperature partial melting occurs (as in granodiorite) preceded by decomposition of a hydrous mineral, there is also a sharp change in strength in encapsulated rock. The strength of drained rock remains high under the same conditions, and melting will not take place.

7. During burial metamorphism of a hydrated rock distributed deformation will tend to take place at the peak of metamorphism and will be accompanied by grain growth and recrystallization of the new crystal phases, but it will be preceded by and followed by episodes of more heterogeneous deformation (fault and vein formation) owing to water evolved from underlying and overlying rock as the latter itself undergoes metamorphism.

8. A detailed understanding of the relationships between deformation and prograde metamorphism in which water is evolved has yet to be achieved, and quantitative relationships between dilatancy and deformation, pore pressure, and drainage conditions have not yet been obtained. Among the matters requiring further research we may mention the following:

   a. methods of determining the volume change during metamorphism and deformation at elevated pressures and temperatures, so that dilatancy and other contributions to the volume change can be estimated;

   b. methods of determining the pore pressure developed under undrained conditions (though it is difficult to envisage any form of direct measurement); and the changes of porosity and permeability that occur during metamorphism and deformation;

   c. more intensive study by optical and electron microscopy and X-ray methods of the structural and textural changes produced by experimental metamorphism, and of the contributions made by different modes of deformation, and in particular the contributions from cracking, pressure solution, superplasticity, and crystal plasticity;

   d. experiments under different loading and thermal conditions; for example, experiments at constant stress or strain (that is, under creep or stress-relaxation conditions) at fixed temperatures on the one hand or at a constant rate of temperature increase on the other. Such experiments could be used to learn more about the kinetics of the metamorphic process and the factors that affect it;

e. theoretical modelling of the development of water-filled cracks by metamorphism, of the subsequent textural development of the rock, and of the hydrodynamic conditions of the metamorphic process.

Acknowledgments

This work was carried out in part under grant GR/3/1880 from the Natural Environment Research Council. Thanks are due to our colleague N. J. Preston for assistance with the differential thermal analysis. We would also like to thank Professor J. Zussman (Manchester), Dr. M. S. Paterson (Canberra), Dr. R. Hall (U.C.L.), and the editors (Professor A. B. Thompson and Dr. D. C. Rubie) for critical reviews and comments. The Editor of *Tectonophysics* is thanked for permission to publish Figure 1, based on figures from Murrell and Ismail (1976a).

References

Aines, R. D., and Rossman, G. R. (1984) The high temperature behavior of water and carbon dioxide in cordierite and beryl. *Amer. Mineral.* **69,** 319–327.

Ball, M. C., and Taylor, H. F. W. (1963) The dehydration of chrystotile in air and under hydrothermal conditions. *Mineral. Mag.* **33,** 467–482.

Bowen, N. L., and Tuttle, O. F. (1949) The system $MgO-SiO_2-H_2O$. *Geol. Soc. Amer. Bull.* **60,** 436–460.

Bredehoeft, J. D., and Hanshaw, B. B. (1968) On the maintenance of anomalous fluid pressures. I. Thick sedimentary sequences. *Geol. Soc. Amer. Bull.* **79,** 1097–1106.

Brindley, G. W., and Ali, S. Z. (1950) X-ray study of the thermal transformations in some magnesian chlorite minerals. *Acta Cryst.* **3,** 25.

Brindley, G. W., and Hayami, R. (1965) Mechanism of formation of forsterite and enstatite from serpentine. *Mineral. Mag.* **35,** 189–195.

Brindley, G. W., and Zussman, J. (1957) A structural study of the thermal transformation of serpentine minerals to forsterite. *Amer. Mineral.* **42,** 461–474.

Chidester, A. H. (1962) Petrology and geochemistry of selected talc-bearing ultramafic rocks and adjacent country rocks in north-central Vermont. *U.S. Geol. Surv. Prof. Pap.* **345.**

Coleman, R. G. (1971) Petrologic and geophysical nature of serpentinites. *Geol. Soc. Amer. Bull.* **82,** 897–918.

Deer, W. A., Howie, R. A., and Zussman, J. (1962) *Rock Forming Minerals,* **3,** *Sheet Silicates.* Longmans, London.

Deer, W. R., Howie, A., and Zussman, J. (1966) *An Introduction to the Rock Form-ing Minerals*. Longmans, London.

Digby, P. J., and Murrell, S. A. F. (1976) The deformation of flat ellipsoidal cavities under large confining pressures. *Bull. Seism. Soc. Amer.* **66**, 425–432.

Faust, G. T., and Fahey, J. J. (1962) The serpentine-group minerals. *U.S. Geol. Surv. Prof. Pap.* **384-A.**

Fawcett, J. J., and Yoder, H. S. (1966) Phase relationship of chlorites in the system $MgO-Al_2O_3-SiO_2-H_2O$. *Amer. Mineral.* **51**, 353–380.

Foster, M. D. (1962) Interpretation of the composition and a classification of the chlorites. *U.S. Geol. Surv. Prof. Pap.* **414-A.**

Fyfe, W. S. (1973) Dehydration reactions. *Bull. Amer. Assoc. Petrol. Geol.* **57**, 190–197.

Fyfe, W. S., Price, N. J., and Thompson, A. B. (1978) *Fluids in the Earth's Crust*. Elsevier, Amsterdam. Preface.

Fyfe, W. S., Turner, F. J., and Verhoogen, J. (1958) Metamorphic reactions and metamorphic facies. *Geol. Soc. Amer. Mem.* **73.**

Galwey, A. K., Spinicci, R., and Guarini, G. G. T. (1981) Nucleation and growth processes occurring during the dehydration of certain alums: The generation, the development and the function of the reaction interface. *Proc. Roy. Soc. London* **A378**, 477–505.

Greenwood, H. J. (1961) The system $NaAlSi_2O_6-H_2O-argon$—total pressure and water pressure in metamorphism. *J. Geophys. Res.* **66**, 3923–3946.

Greenwood, H. J. (1963) The synthesis and stability of anthophyllite. *J. Petrol.* **4**, 317–351.

Griggs, D. T., and Handin, J. (1960) *Geol. Soc. Amer. Mem.* **79**, 347–364.

Hanshaw, B. B., and Bredehoeft, J. D. (1968) On the maintenance of anomalous fluid pressures. II. Source layer at depth. *Geol. Soc. Amer. Bull.* **79**, 1107–1122.

Hardie, L. A. (1967) The gypsum–anhydrite equilibrium at one atmosphere pressure. *Amer. Mineral.* **52**, 171–201.

Heard, H. C., and Rubey, W. W. (1966) Tectonic implications of gypsum dehydra-tion. *Geol. Soc. Amer. Bull.* **77**, 741–760.

Hemley, J. J., Montoya, J. W., Shaw, D. R., and Luce, R. W. (1977) Mineral equilibria in the $MgO-SiO_2-H_2O$ system. *Amer. J. Sci* **277**, 353–383.

Hess, H. H., and Otalora, G. (1964) Mineralogical and chemical composition of the Mayaguez serpentinite cores, in *A Study of Serpentinite*, edited by C. A. Burk, Natl. Acad. Sci.–Natl. Res. Council Publ. **1188**, 152–168.

Hey, M. A. (1954) A new review of the chlorites. *Mineral. Mag.* **30**, 277–292.

Holland, J. G., and Lambert, R. St.J. (1969) Structural regimes and metamorphic facies. *Tectonophysics* **7**, 197–217.

Hubbert, M. K., and Rubey, W. W. (1959) Role of fluid pressure in mechanics of overthrust faulting. I. *Geol. Soc. Amer. Bull.* **70**, 115–166.

Ismail, I. A. H., and Murrell, S. A. F. (1976) Dilatancy and the strength of rocks containing pore water under undrained conditions. *Geophys. J. Roy. Astr. Soc.* **44**, 107–134.

James, R. S., Turnock, A. C., and Fawcett, J. J. (1976) The stability and phase relations of iron chlorite below 8.5kb P_{H_2O}. *Contrib. Mineral. Petrol.* **56,** 1–25.

J.C.P.D.S. (1974) *Selected Powder Diffraction Data for Minerals.* 1st Ed. Publ. DBM-1-23. Joint Committee on Powder Diffraction Standards, Swarthmore, PA.

Kitahara, S., Takenouchi, S., and Kennedy, G. C. (1966) Phase relations in the system $MgO–SiO_2–H_2O$ at high temperatures and pressures. *Amer. J. Sci.* **264,** 223–233.

MacDonald, G. J. F. (1953) Anhydrite–gypsum equilibrium relations. *Amer. J. Sci.* **251,** 884–898.

McKenzie, D. P. (1978) Some remarks on the development of sedimentary basins. *Earth Planet. Sci. Lett.* **48,** 24–32.

McOnie, A. W., Fawcett, J. J., and James, R. S. (1975) The stability of intermediate chlorites of the clinochlore–daphnite series of 2kbar P_{H_2O}. *Amer. Mineral.* **60,** 1047–1062.

Murrell, S. A. F. (1958) In *Mechanical Properties of Non-metallic Brittle Materials,* edited by W. H. Walton, pp. 123–146. Butterworths, London.

Murrell, S. A. F. (1963) A criterion for brittle fracture of rock under triaxial stress and the effect of pore pressure on the criterion, in *Rock Mechanics,* edited by C. Fairhurst, pp. 563–577. Pergamon Press, Oxford.

Murrell, S. A. F. (1964a) The theory of the propagation of elliptical Griffith cracks under various conditions of plane strain or plane stress. Part I. *Br. J. Appl. Phys.* **15,** 1195–1210.

Murrell, S. A. F. (1964b) The theory of the propagation of elliptical Griffith cracks under various conditions of plane strain or plane stress. Parts II and III. *Br. J. Appl. Phys.* **15,** 1211–1223.

Murrell, S. A. F. (1965) The effect of triaxial stress systems on the strength of rocks at atmospheric temperatures. *Geophys. J. Roy. Astr. Soc.* **10,** 231–281.

Murrell, S. A. F. (1981) The rock mechanics of thrust and nappe formation. In *Thrust and Nappe Tectonics,* edited by K. R. McClay and N. J. Price, pp. 99–110. Blackwell, Oxford.

Murrell, S. A. F. (in press) The mechanics of tectogenesis in plate collision zones. Presented at the William Smith Symposium of the Geological Society of London, 1983. To be published in the Proceedings.

Murrell, S. A. F., and Digby, P. J. (1970) The theory of brittle fracture initiation under triaxial stress conditions. II. *Geophys. J. Roy. Astr. Soc.* **19,** 499–512.

Murrell, S. A. F., and Ismail, I. A. H. (1976a) The effect of decomposition of hydrous minerals on the mechanical properties of rocks at high pressures and temperatures. *Tectonophysics* **31,** 207–258.

Murrell, S. A. F., and Ismail, I. A. H. (1976b) The effect of temperature on the strength at high confining pressure of granodiorite containing free and chemically-bound water. *Contr. Mineral. Petrol.* **55,** 317–330.

Nicolas, A. (1966) Le complexe ophiolites-schistes-lustres entre Dora Maira et Grand Paradis. Tectonique et metamorphisme, II. Faculte des Sciences de Nantes.

Nicolas, A. (1969) Serpentinisation d'une lherzolite: Bilan chimique, implication tectonique. *Bull. Volcanol.* **32,** 499–508.

Norris, R. J., and Henley, R. W. (1976) De-watering of a metamorphic pile. *Geology* **4**, 333–336.

Nur, A. (1982) Processes in rocks with fluids at elevated pressure and temperature, in *High-Pressure Researchs in Geoscience,* edited by W. Schreyer, pp. 67–83. E. Schweizerbart'sche Verlagsbuchhandlung, Stuttgart.

Parrish, W., and Mack, M. (1963) Data for X-ray analysis. 2nd Ed., Vol. 1. Philips Technical Library, The Netherlands.

Paterson, M. S. (1973) Nonhydrostatic thermodynamics and its geologic applications. *Rev. Geophys. Space Phys.* **11**, 355–389.

Paterson, M. S. (1983) Creep in transforming polycrystalline materials. *Mechanics of Materials* **2**, 105–109.

Phillips, W. R. (1963) A differential thermal study of the chlorites. *Mineral. Mag.* **33**, 404–414.

Pistorius, C. W. F. T. (1963) Some phase relations in the systems $MgO–SiO_2–H_2O$, to high pressures and temperatures. *Neues Jb. Mineral., Monatsh.* **11**, 283–293.

Posnjak, E. (1938) The system $CaSO_4–H_2O$. *Amer. J. Sci.* **35A**, 247–272.

Raleigh, C. B. (1967) Tectonic implications of serpentinite weakening. *Geophys. J. Roy. Astr. Soc.* **14**, 45–51.

Raleigh, C. B., and Lee, W. H. K. (1969) In *Proceedings of the Andesite Conference,* edited by A. R. McBirney. Oregon Dep. Geol. Miner. Ind. Bull. **65**, 99–110.

Raleigh, C. B., and Paterson, M. S. (1965) Experimental deformation of serpentinite and its tectonic implications. *J. Geophys. Res.* **70**, 3965–3985.

Rubey, W. W., and Hubbert, M. K. (1959) Role of fluid pressure in mechanics of overthrust faulting, 2. *Geol. Soc. Amer. Bull.* **70**, 167–206.

Rutter, E. H. (1976) The kinetics of rock deformation by pressure solution. *Phil. Trans. Roy. Soc. London* **A283**, 203–219.

Rutter, E. H. (1983) Pressure solution in nature, theory and experiment. *J. Geol. Soc. London* **140**, 725–740.

Scarfe, C. M., and Wyllie, P. J. (1967) Serpentine dehydration curves and their bearing on serpentinite deformation in orogenesis. *Nature* (London) **215**, 945–946.

Schmid, S. M., Boland, J. N., and Paterson, M. S. (1977) Superplastic flow in finegrained limestone. *Tectonophysics* **43**, 257–291.

Schreyer, W., and Yoder, H. S. (1960) Hydrous Mg-cordierite. *Carnegie Inst. Washington Yearbook* **59**, 91–94.

Trommsdorff, V., and Evans, B. W. (1972) Progressive metamorphism of antigorite schist in the Bergell tonalite aureole (Italy). *Amer. J. Sci.* **272**, 423–437.

Turcotte, D. L., and Schubert, G. (1982) *Geodynamics: Applications of Continuum Physics to Geological Problems.* Wiley, New York.

Turner, F. J. (1968) *Metamorphic Petrology: Mineralogical and Field Aspects.* McGraw-Hill, New York.

Turnock, A. C. (1960) The stability of iron chlorites. *Carnegie Inst. Washington Yearbook* **59**, 98–103.

Walther, J. V., and Orville, P. M. (1982) Volatile production and transport in regional metamorphism. *Contrib. Mineral. Petrol.* **79**, 252–257.

Weber, J. N., and Greer, R. T. (1965) Dehydration of serpentine: Heat of reaction and reaction kinetics at PH_2O = 1 atm. *Amer. Mineral.* **50,** 451–464.

Wood, B. J., and Walther, J. V. (1983) Rates of hydrothermal reactions. *Science* **222,** 413–415.

Wyllie, P. J. (1962) The effect of "impure" pore fluids on metamorphic dissociation reactions. *Mineral. Mag.* **33,** 9–25.

Yoder, H. S. (1955) Role of water in metamorphism. *Geol. Soc. Amer. Spec. Pap.* **62,** 505–524.

Yoder, H. S. (1967) Spilites and serpentinites. *Rep. Dir. Geophys. Lab. Carnegie Inst.* **65,** 269–279.

Zaccagna, D., Mattirolo, E., and Franchi, S. (1914) Carta Geologica d'Italia. Oulx sheet. FO.54, 1 : 100,000. Novara.

Zen, E-an (1965) Solubility measurements in the system $CaSO_4$–$NaCl$–H_2O at 35°, 50° and 70°C and one atmosphere pressure. *J. Petrol.* **6,** 124–164.

Chapter 9
The Permeation of Water into Hydrating Shear Zones

E. H. Rutter and K. H. Brodie

Introduction

Very little is known about the *in situ* permeability of medium- to high-grade metamorphic rocks. Several studies have been made of fluid permeation at room temperature in crystalline rocks, over a wide range of hydrostatic pressures, and including the modifying effects of cataclastic deformation (e.g., Brace *et al.*, 1968; Brace, 1980; Zoback and Byerlee, 1975). More recently, experimental studies of high-temperature permeability and its modification through hydrothermal solution, transport, and cementation have been made on granite (Morrow *et al.*, 1981) and permeability determined for low-permeability materials such as fault gouge (Byerlee, 1983). The range of *in situ* permeability values available (to depths of 2–3 km) appear to be consistent with values obtained in laboratory studies (Brace, 1980). Such studies have been stimulated through high-level radioactive waste disposal programs, the need to understand fluid flow around seismically hazardous fault zones, and in connection with various techniques for the storage and recovery of energy resources in rocks (Brace, 1980).

Walther and Orville (1982) have considered the kinetics of permeation of water through large volumes of crustal rocks undergoing devolatilization, and Etheridge *et al.* (1983) consider that permeability may be sufficiently high in such situations ($\sim 10^{-17}$ m^2) to admit large-scale convective circulation in the mid-crust. In this contribution we consider the implications for rock permeability of the often limited degrees of retrogressive metamorphism (hydration) commonly associated with the development of shear zones in otherwise high-grade metamorphic rocks.

Shear zones in rocks deformed under high-pressure–temperature conditions are often selective sites of metamorphic transformations involving hydration and metasomatism (e.g., Beach, 1973). In contrast, the surrounding, less or nondeformed host rocks are often unaltered. This implies an enhancement of water permeation rate along a shear zone relative to permeation sideways into the undeformed rock. Permeability enhancement may arise simply because of the grain size and shape contrast between host rock and shear zone rock, or it may depend upon active creation of permeable void volume as a result of the deformation mechanisms. Cataclasis, whether dominant or present as a subsidiary deformation mechanism, necessarily involves dilatancy, which in turn renders the resistance to deformation sensitive to mean pressure. For cataclasis in deep-seated deformation of rocks, it is often inferred that accompanying high pore fluid pressures are necessary to keep the resistance to flow down to geologically realistic levels (generally considered to be less than 2 or 3 kb). The time relationships between deformation and hydration are crucial to the question of the role of metamorphic reactions in the deformation of rocks in hydrated shear zones, and hydration can only occur if P_{H_2O}/σ_3 reaches a sufficiently high level and if a sufficient quantity of water is supplied.

Permeability Estimation

It is difficult to measure permeability values lower than 10^{-22} m² (approximately 10^{-10} darcys), and the *in situ* permeability of many metamorphic rocks may be several orders of magnitude lower than this. To provide some feeling for the magnitude of permeability values, that of Westerly granite at 20°C and 400 MPa confining pressure is about 5×10^{-22} m² (Brace *et al.,* 1968), that of Tennessee sandstone (6% porosity) at 20°C and 150 MPa confining pressure is 3×10^{-18} m², and at 500°C, 2×10^{-19} m² (Higgs, 1981). A very approximate estimate of permeability under high-grade metamorphic conditions can be made based on the observation that around the hydrated shear zones cutting a pyroxene granulite facies metabasic dyke at Scolpaig, N. Uist, Scotland, hydration has not penetrated much more than 1 m laterally away from shear zone networks, which are of the order of 10 to 100 m in length, and individual shear zones are of the order of 0.1 m in width.

For the one-dimensional decay of a pore fluid pressure transient δP on a background pore pressure P_0, applied at time $t = 0$ at point $x = 0$, the characteristic distance, x, over which the pressure transient decays to about $0.5\delta P$ after time t, is given by:

$$x^2 = Kt/\phi\beta\eta \qquad (1)$$

where K is rock permeability and η is fluid viscosity. ϕ represents the water storage capacity of the rock at constant pressure and equals the porosity when the fluid and the rock do not interact chemically. β depends on the rock compressibility, porosity and fluid compressibility, but at sufficiently high temperatures when the fluid is water, fluid compressibility is much greater than that of rock, in which case β may be approximated by fluid compressibility. Low fluid storage capacity favors the spreading of a fluid pressure transient throughout a rock mass, because a smaller quantity of fluid must be moved to produce a given pressure change. The importance of possible permeability enhancement through dilation during cataclasis is off-set to some extent by this effect.

Porosity under high-grade metamorphic conditions is likely to be very small, less than 0.1%. To hydrate completely a basic rock in the amphibolite facies requires about 1 wt% water. At 600°C and 100 MPa pore pressure the specific volume of water is about 3.0×10^{-3} m^3 kg^{-1} (Kennedy and Holser, 1966). Neglecting the volume changes in the solid phases, the rock therefore soaks up about 10% of its own volume in water during hydration. Thus before hydration has gone to completion the water storage capacity through hydration far exceeds that resulting from porosity. A propagating interface region between hydrated and anhydrous rock, which will result from the permeation of water into anhydrous rock, will therefore behave like a sink for water, resulting in a step in an otherwise smoothly varying pore-pressure gradient. The faster the hydration kinetics relative to the permeation rate, the steeper will be the pressure transient.

The problem of the propagation of the hydration interface into anhydrous rock is therefore a complicated one, involving variation in the physical prop-erties of the system with time and distance. The general effect of the tran-siently increased water storage capacity at the hydration/dehydration inter-face is to reduce the velocity of the characteristic distance, x. For our rough estimate we will therefore assume water storage capacity, ϕ, to be equal to the hydration storage, understanding that the effect will be to overestimate permeability. Taking the compressibility of water (from $P-V-T$ data) at 600°C and 100 MPa to be 2×10^{-2} MPa^{-1}, and the viscosity to be 5×10^{-11} MPa.s (Dudziak and Franck, 1966), x has been plotted against t, for various K values in Fig. 1. If we assume, for example, that hydration has only spread a distance of the order of 10 m in 10 Ma, or 1 m in 10^5 years, the apparent permeability is of order 10^{-26} m^2, a permeability value about 10^{-6} of that of Westerly granite at room temperature and moderate confining pressure, and about 10^{-3} of the lowest measurable permeability values. In the absence of hydration, so that water storage would be due only to porosity, similar permeation distances and times would arise from permeabilities of only $\sim 10^{-28}$ m^2.

It is interesting to compare such rate of permeation figures with rates of grain-boundary diffusion. Brace et al. (1968) give a relation between perme-ability, K, and diffusion coefficient, D_v, as:

$$K = D_v \eta \beta \qquad (2)$$

D_v is, however, a volume diffusion coefficient. Replacing D_v by D_{GB}, the grain-boundary diffusion coefficient, the effective permeability will be low-ered by a factor of approximately w/d, where w is the effective grain-bound-ary width and d is the mean grain size. This accounts for the reduction in the area of the "window" through which diffusion can occur. The magnitude of

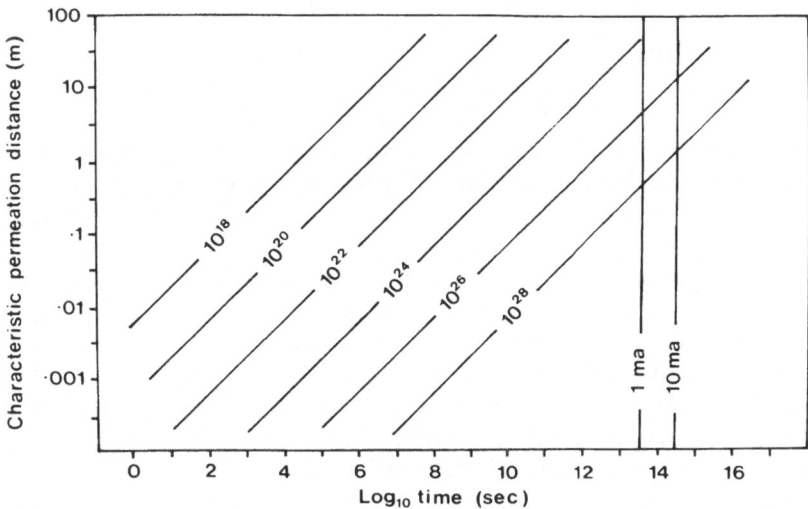

Fig. 1. Characteristic permeation distance (x) versus time (t) contoured for various values of permeability (K) at 600°C and fluid pressure = 100 MPa. $x = (Kt/\phi\beta\eta)^{1/2}$, where ϕ is fluid storage (10%), β is fluid compressibility (2×10^{-2} MPa), and η is fluid viscosity (5×10^{-11} MPa.s). K is in m^2 and t is in seconds.

D_{GB} for transport through strongly adsorbed aqueous films within grain interfaces has been discussed by Rutter (1976, 1983), who argues that D_{GB} will be degraded by a factor of $\sim 10^4$ relative to its value in bulk water at the appropriate temperature, because of the structuring of the adsorbed aqueous film. Assuming $D_{GB} \sim 10^{-12}$ m^2 s^{-1}, taking $w \sim 10^{-9}$ m and $d = 10^{-3}$ m (1 mm), and using the values taken above for η and β, $K \sim 10^{-30}$ m^2. This is slow relative to the permeability values given above but would become competitive at grain sizes of the order of 1 μm over a length scale less than 10 m. In the same way that volatile transport by grain boundary diffusion will be more rapid in finer grained material because a greater density of diffusion pathways is present, we might expect that, all other factors remaining constant, grain-size reduction would enhance permeability. In the example of the Scottish shear zones above, volatile permeation *along* shear zones has clearly been between 10 and 100 times more effective than *across* them. A grain-size reduction of 10 to 100 times by whatever mechanism might be expected to be able to account for the observed permeation anisotropy. Such a degree of grain-size reduction is in fact observed in the shear zones, thus the observation of preferred localization of hydration along these finer grained shear zones tells us nothing of whether the hydration was syntectonic or posttectonic.

From the approximate calculations above we conclude:

1. Under conditions of medium- to high-grade static metamorphism, the permeability of medium-grained rocks can be very low, around 10^{-26} m^2.

This is about 10^{-3} of the lowest measurable permeability values and allows fluids to permeate only a few meters in time periods of the order of 1 Ma in medium-grained rocks. Volatile phase transport via grain boundary diffusion may be about 10,000 times less effective than permeation through pores. This conclusion is not meant to imply, of course, that there are not geological situations in which considerably greater values of permeability obtain in metamorphic rocks (e.g., Etheridge *et al.*, 1983). It seems likely that permeability in mid/deep crustal environments varies widely, probably over 10 orders of magnitude (likely to be determined mainly by effective confining pressure) so that *generalization* about the *in situ* permeability of metamorphic rocks is probably unwise.

2. All other factors remaining constant, permeation should be generally more effective at finer grain sizes (though the reverse sometimes seems to be true, e.g., Peters, 1963, and would imply possibly tighter grain-boundary structure in finer grained material).

Discussion

Timing of Hydration in Retrogressive Shear Zones

Subject to the assumption that grain-size reduction does not change the properties of pores and grain boundaries, commonly observed degrees of grain-size reduction (e.g., 10 : 1) correspond with the degree of permeation enhancement along hydrous shear zones. Thus preferred hydration along shear zones can often be explained equally satisfactorily by postdeformation fluid ingress except where, for example, an incongruent pressure solution (Beach, 1982) type of microstructure provides unequivocal evidence for *syntectonic* hydration. The localization and development of the shear zones themselves may have been entirely by mechanical processes (e.g., by cataclasis, or by plastic deformation with dynamic recrystallization) without the deformation being causally linked with the chemical metamorphism, and would provide an example of deformation-enhanced metamorphism rather than the reverse. In many retrogressive shear zones, especially where there has been posttectonic grain coarsening, it is difficult or impossible to deduce deformation/hydration time relations, and hence the role of the metamorphism in the deformation, if any.

Control of Propagation Rate of Hydrous Shear Zones by Fluid Permeation

If in certain cases it can be assumed or inferred that deformation and hydration *did* proceed together, or at least that creep deformation depended on the attainment of a particular value of P_{H_2O} near the tip of a propagating shear zone, there may be stress conditions under which creep rate or rate of shear zone growth is governed principally by fluid permeation rate. A localized shear zone probably propagates like a spreading shear-mode crack in an elastic medium, but with a viscous fluid in the plane of the shear zone and driven by the state of intensified stress in the material ahead of the crack tip. The rate of growth is governed by the kinetics of the flow processes in the approximately homogeneously stressed material within the zone behind the tip, and the breakdown processes in the tip region itself.

Consider a two-dimensional shear zone of length l and thickness y, which propagates into the host material at velocity dl/dt, driven by an applied shear stress τ_a (Fig. 2). The resistance to permanent deformation of the material within the shear zone is $\tau_f(\dot{\gamma})$. $\dot{\gamma}$ is the shear strain rate of the material within the shear zone. The distribution of shear displacement δx along the zone is given by:

$$\delta x = 2G^{-1}(\tau_a - \tau_f)[(l/2)^2 - x^2]^{1/2} \tag{3}$$

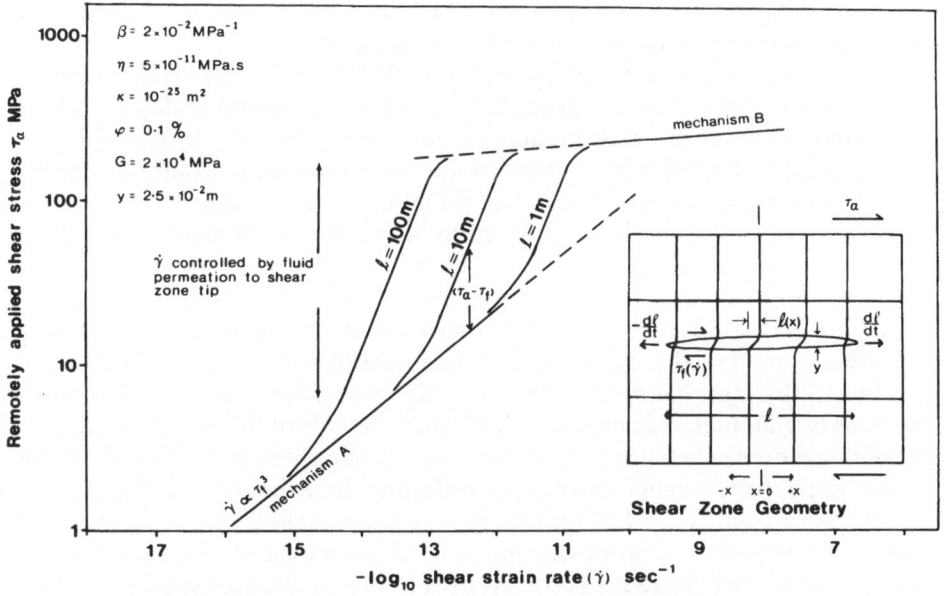

Fig. 2. Graph illustrating the concept of control of local deformation rate $(\dot{\gamma})$ in a growing shear zone by rate of fluid permeation along the shear zone. Deformation mechanism A is assumed to be pore pressure sensitive but does not result in significant permeability change. Mechanism B is either pore pressure insensitive or results in a large permeability increase. The rheological behavior shown for mechanisms A and B is purely hypothetical. Curves are shown in the region labelled "$\dot{\gamma}$ controlled by permeation to shear zone tip," calculated for various shear zone lengths and for the physical properties indicated. Within the shear zone the remotely applied shear stress is relaxed to τ_f as shown. Inset: diagram showing shear zone geometry and definitions of terms.

(Rudnicki, 1980), where G is the shear modulus. The difference between the shear stress applied remotely, τ_a, and the shear flow resistance across the shear zone is accounted for by the state of intensified stress in the elastic material surrounding the tips of the zone. The stress intensity factor, k, is given by:

$$k = (8/\pi)(\tau_a - \tau_f)l^{1/2} \qquad (4)$$

from which the stress components may be obtained by the standard methods of fracture mechanics (e.g., Rudnicki, 1980).

During the period of shear zone propagation the maximum displacement, and hence shear strain for a given shear zone width, is fixed by the elastic properties of the host rock and the length of the shear zone. For example, the shear strain, γ, will range between 0.1 and 1.0 when the shear zone is 10 m long and .025 m wide for stresses 10 MPa $\leq (\tau_a - \tau_f) \leq$ 100 MPa when $G = 2 \times 10^4$ MPa. Finite shear strains in shear zones may often exceed such values, and it seems likely that higher strains develop after the propagation stage when the above simple geometry has been destroyed, for example, by the intersection of differently oriented shear zones.

The problem of a coupling between pore fluid permeation and the spreading of a shear zone in a porous material has been explored in detail by Rice and Simons (1976) for deformation by cataclastic processes. We will here take a simplified approach in order to illustrate how local strain rate can be controlled by the kinetics of fluid permeation. We will assume that shear zone propagation depends upon the attainment of a particular value of P_{H_2O} at the tip of the shear zone, and that water only reaches the tip by permeation along the zone from the center. Thus dl/dt is governed by the permeation rate. We are not concerned primarily with *how* water gains access to the shear zone, but the occurrence of localized hydration along shear zones testifies to the fact that it does. We are only concerned here with the consequences of that fact. We may also note that the nature of the "breakdown" process, whereby the rock is rendered weak at the shear zone tip and behind is not important, except insofar as different deformation processes may modify permeability in different ways. At high applied stresses cataclasis may be the breakdown process, and the local deviatoric stresses for this will be very sensitive to pore pressure. At lower applied stresses weakening may be achieved by the facilitation of plasticity or by diffusive transfer processes involving hydration; in either case some particular pore pressure is assumed to be required.

For simplicity we assume that the shear zone grows at the same rate as the characteristic permeation distance, and that the physical properties K and ϕ are *not* modified through shearing, so that

$$dl/dt = K/2l\phi\beta\eta \qquad (5)$$

Combining Eq. (3) and Eq. (5) we obtain for the shear strain rate $\dot{\gamma}$ in the shear zone,

$$\dot{\gamma} = (\tau_a - \tau_f) K/4Gy\phi\beta\eta[(l/2)^2 - x^2] \tag{6}$$

The maximum strain rate will occur at the center of the zone, at $x = 0$, decaying to zero at the tips. Equation (6) is represented graphically in Fig. 2, which shows how a range of stress, temperature, strain rate, and rock property conditions can lead to the rate of deformation being controlled largely by the fluid permeation rate. The parallel phenomenon of a range of stress intensities for which crack velocity in a brittle material is controlled by the kinetics of water vapor permeation along the crack to the tip is well known in the literature of fracture mechanics (e.g., Atkinson, 1982).

Summary

The kinetic problems of hydration of basic rocks in shear zones have been considered, and constraints provided by a natural shear zone occurrence used to make an order of magnitude estimate of permeability to fluid flow. This has been used as a basis for discussion of timing relationships between deformation and metamorphism. Only in certain circumstances can the timing relations between deformation and hydration in retrogressive shear zones be determined; thus it is not always possible to infer that deformability is facilitated by metamorphism. For those cases in which a direct relationship exists, it can be argued that conditions may arise whereby rate of deformation is controlled by rate of ingress of water.

Acknowledgments

We are grateful to Fraser Wigley, David Rubie, and Alan Thompson for discussion and critical comments. This work was supported through U.K. Natural Environment Research Council grants GR3/3548 and GR3/3848.

References

Atkinson, B. K. (1982) Subcritical crack propagation in rocks; theory, experimental results and applications. *J. Struct. Geol.* **4**, 41–56.

Beach, A. (1973) The mineralogy of high temperature shear zones at Scourie, N.W. Scotland. *J. Petrol.* **14**, 231–248.

Beach, A. (1982) Deformation mechanisms in some cover thrust sheets from the external French Alps. *J. Struct. Geol.* **4**, 137–150.

Brace, W. F. (1980) Permeability of crystalline and argillaceous rocks. *Int. J. Rock Mech. Min. Sci.* **17**, 241–251.

Brace, W. F., Walsh, J. B., and Frangos, W. T. (1968) Permeability of granite under high pressure. *J. Geophys. Res.* **73**, 2225–2236.

Byerlee, J. D. (1983) Permeability of fault zones. *U.S. Geol. Surv. Open File Rep.* **83-525**, 319–320.

Dudziak, K. H., and Franck, E. U. (1966) Messungen der Viskositat des Wassers bis 560°C und 3500 bar. Ber Bunsenges. *Phys. Chem.* **70**, 1120–1128.

Etheridge, M. A., Wall, V. J., and Vernon, R. H. (1983) The role of the fluid phase during regional metamorphism and deformation. *J. Met. Geol.* **1**, 205–226.

Higgs, N. G. (1981) Mechanical properties of ultrafine grained quartz, chlorite and bentonite in environments appropriate to upper crustal earthquakes. Ph.D. Thesis, Texas A & M University.

Kennedy, G. C., and Holser, W. T. (1966) Pressure–volume–temperature and phase relations of water and carbon dioxide, in *Handbook of Physical Constants,* edited by S. P. Clark, Jr. Geol. Soc. Am. Mem. **97**, 371–383.

Morrow, C., Lockner, D., Moore, D., and Byerlee, J. D. (1981) Permeability of granite in a temperature gradient. *J. Geophys. Res.* **86**, 3002–3008.

Peters, T. (1963) Mineralogie und Petrologie des Totalp Serpentins, bei Davos. *Schweiz. Min. Pet. Mitt.* **43**, 529–685.

Rice, J. R., and Simons, D. A. (1976) The stabilization of spreading shear faults by coupled deformation-diffusion effects in fluid infiltrated porous materials. *J. Geophys. Res.* **81**, 5322–5334.

Rudnicki, J. W. (1980) Fracture mechanics applied to the Earth's crust. *Ann. Rev. Earth Plan. Sci.* **8**, 489–525.

Rutter, E. H. (1976) The kinetics of rock deformation by pressure solution. *Phil. Trans. Roy. Soc. London,* **A283**, 203–219.

Rutter, E. H. (1983) Pressure solution in nature, theory and experiment. *J. Geol. Soc. London* **140**, 725–740.

Walther, J. V., and Orville, P. M. (1982) Volatile production and transport in regional metamorphism. *Contrib. Mineral. Petrol.* **79**, 133–145.

Zoback, M. D., and Byerlee, J. D. (1975) The effect of microcrack dilatancy on the permeability of Westerly granite. *J. Geophys. Res.* **80**, 752–755.

Chapter 10
The Possible Effects of Deformation on Chemical Processes in Metamorphic Fault Zones

R. P. Wintsch

Introduction

The influence of deformation on structural aspects of petrology is formally recognized in the classification of metamorphic rocks based on the degree of foliation development (hornfels and granofels versus slate, phyllite, schist, and gneiss) and in the modification of textures during syntectonic mineral growth (e.g., Misch, 1969; Rosenfeld, 1970). The influence of deformation on chemical aspects of petrology is less fully explored but is evident in processes like pressure solution (e.g., Walton et al., 1964; Kerrich et al., 1977) and cleavage development (e.g., Knipe, 1981) and may drive some reactions in high-grade metamorphic rocks (Wintsch, 1975, 1981). Deformation affects chemical processes through its influence (usually positive) on chemical potential of solids (Reitan, 1977). In metamorphic rocks containing an aqueous fluid, these chemical potentials are linked to the activities of aqueous ions in the grain boundary region. Gradients in activities of aqueous species (and chemical potential) can cause diffusive mass transfer (DMT) along the grain-boundary region. This causes the redistribution of chemical components and hence minerals in the rock.

The potential role of deformation in modifying chemical potential is explored here by examining theoretically some of the sinks of mechanical energy (elastic strain energy, plastic strain energy associated with lattice defects, surface energy, and heat) that become the sources of energy for chemical work. The mechanical energy is transformed to chemical energy by reaction of deformed crystals with the aqueous grain-boundary fluid. This fluid is considered as a critical link between deformational and chemical processes because it is necessary for the operation of some metamorphic reaction mechanisms (Carmichael, 1969; Wyart, 1975; Matthews, 1980) and some deformation processes (pressure solution).

Metamorphic Reaction Mechanism

A large number of metamorphic reactions have been proposed, but the textures preserved in rocks frequently are not consistent with the stoichiometry or even the reactants predicted from theoretical phase equilibria. This has led Carmichael (1969), Fisher (1970), and Kwak (1974), among others, to describe metamorphic reactions as a series of replacement and ion exchange reactions involving many phases. These replacement reactions occur by dissolution of the mineral reactant into the grain-boundary fluid and precipitation of the product mineral from this fluid. On the scale of a thin section or hand specimen, the net reactions may be balanced for all nonvolatile components. At the scale of the individual reaction site, however, it is required that ions are transported by a diffusive mechanism to and from other replacement sites. This makes the replacement reactions sensitive to fluctuations in the local activity of many components. As shown below, such fluctuations may easily be induced by deformational processes.

Role of Stress and Strain

Deformation may influence which metamorphic reaction will occur and where it will take place. Through its effect on the local activities of some aqueous species, deformation helps determine which minerals will dissolve and/or will be replaced (Beach, 1979; Rutter, 1983). Deformation may also influence the site of reactions by (1) affecting the distribution of sites of favored dissolution, nucleation, and precipitation; (2) by establishing gradients in chemical potential of aqueous species; and (3) by providing pathways for enhanced diffusive mass transfer (DMT).

Sites of dissolution can occur preferentially in zones of stress or strain concentration. High differential stress causes an increased mean stress and thus an increased solubility that leads to pressure solution (e.g., de Boer, 1977). High plastic strain (high dislocation density) increases lattice energy, which increases solubility (Bosworth, 1981). Grain size reduction increases surface energy, which also enhances solubility (e.g., Enüstün and Turkevich, 1960). A portion of the work of deformation may be dissipated as heat, which also increases silicate solubility in aqueous fluid. All of these processes are favored by localized high stresses, the distribution of which in a deforming rock may define areas of high chemical potential relative to undeformed areas (e.g., Robin, 1978; Rutter, 1983).

The most obvious strain-related sites of precipitation are regions of extension such as "pressure shadows" and extension veins. Smaller and less conspicuous sites can be created in a deforming aggregate by heterogeneous deformation (e.g., Knipe, 1981; Knipe and Wintsch, this volume). Other precipitation sites may be created by the preferential dissolution of crystals or parts of crystals with high, nonuniform defect densities (Engelder, 1982).

Strain in the deforming rock may generate pathways of relatively easy diffusive mass transfer (e.g., Yund *et al.*, 1981). These paths are established by an increased population of (1) dislocations, (2) subgrain walls, (3) grain boundaries, (4) surfaces between solids and fluids, and (5) pore space. The first three can occur through the generation and movement of dislocations during syntectonic recrystallization. Such recrystallization and concomitant grain-size reduction are particularly important because they involve the production of new grain boundaries. Movement of these grain boundaries during recrystallization allows reactions to take place at the migrating interface, which reduces the dependence of such reactions on lattice diffusion (Knipe, 1981). The impact of these pathways on DMT will depend on the type of path, its frequency, orientation, and duration. The last variable is especially difficult to assess because grain growth during static periods may destroy these paths and hence the evidence for their existence.

Thus the stress and strain patterns developed in a rock during deformation can affect all the processes by which metamorphic reactions occur. As shown in the following discussion, the chemical potential differences established by deformation processes may exceed 100 cal/mol. Thus in actively deforming shear zones nonhydrostatic stress, dislocations at high density or grain-size reduction may be more important than heat in driving mineral reactions.

Sources of Energy

The sources of energy most commonly considered effective in driving metamorphic reactions are heat and hydrostatic pressure. Other sources of energy that may be important at constant temperature are chemical work and different forms of mechanical work. Chemical work can be important especially in immature sedimentary rocks where metastable minerals and incompatible mineral associations may be juxtaposed at deposition. In prograde metamorphism, overstepping of metamorphic reactions leads to chemical metastability, and retrograde metamorphic conditions universally lead to metastability through "understepping."

Mechanical work done on a rock by deforming it may be stored as some combination of elastic strain energy, energy associated with defects, surface energy, and heat. In addition, there is an energy source resulting from differences in hydrostatic pressure ($\bar{V}(P_2 - P_1)$) where $\bar{V}$ and P are molar volume and pressure, respectively, which leads in part to pressure solution. These energy sinks, including the $\bar{V}dP$ term, become potential souces of energy driving chemical reactions. The relative importance of these energy sources is impossible to calculate at this time but will depend on environmental (confining pressure, temperature, fluid pressure, fluid composition) and rheological (mineralogy, preferred orientation, preexisting structure) factors. In order to assess the magnitude of these chemical and mechanical sources, each is considered separately in the following sections.

To facilitate quantitative comparison of the various energy sources, an activity diagram showing some mineral stabilities in equilibrium with an aqueous fluid is presented in Fig. 1. The diagram is calculated using the computer program SUPCRT and the thermodynamic data given by Helgeson *et al.* (1978). As well as the classical aqueous activity variables of the diagram (a_{K^+}/a_{H^+} and $a_{SiO_{2(aq)}}$ upper and right-hand full scales), the calculated chemical potentials of K_2O and SiO_2 are also given (left-hand and lower scales). The mechanical energy sources are indicated on the axes external to the diagram and are referred to an equilibrium (strain-free) assemblage muscovite–microcline–quartz–fluid at 300°C and 200 MPa hydrostatic pressure (point A) as defined by the reaction:

$$\text{muscovite} + 6SiO_{2(aq)} + 2K^+ = 3 \text{ microcline} + 2H^+ \tag{1}$$

Chemical Energy

Metamorphic reactions can be driven by the metastability of a mineral or a mineral assemblage. For the limited chemical system of Fig. 1, several metastable polymorphs of SiO_2 and $KAlSi_3O_8$ give a quantitative measure of such chemical energies involved. At 300°C and 200 MPa, the stable potassium feldspar polymorph is microcline, and its stability field is shown as a solid line (phase boundary) in Fig. 1. A stable equilibrium among quartz, muscovite, and microcline would buffer the composition of the fluid at point A. Sanidine has a free energy ~700 cal/mol higher than microcline at these conditions, and its smaller stability field with respect to microcline–muscovite equilibria (reaction 1), is given by the dashed line in Fig. 1. The metastability of the SiO_2 polymorphs with respect to quartz results from their relatively higher free energies. The calculated free energies of quartz, chalcedony, cristobalite, and amorphous SiO_2 at 300°C and 200 MPa are -207.492, -207.136, -206.753, and -206.133 Kcal/mol, respectively. The grain-boundary fluid in equilibrium with quartz would have an aqueous activity of SiO_2 of 0.021 (or $10^{-1.6795}$) mol/kg. The family of fluid compositions in equilibrium with quartz is given by the dashed line labeled QUARTZ SATURATION (Fig. 1). The equilibria of H_2O and chalcedony, cristobalite, and amorphous silica (all metastable with respect to H_2O–quartz equilibrium) are indicated by the horizontal dashed curves (Fig. 1). Thus the increased metastability of the SiO_2 polymorphs with respect to quartz leads to their increased solubility. In the case of K-feldspar, the fluid in equilibrium with sanidine and muscovite rather than microcline and muscovite (reaction 1) will define a higher K^+/H^+ aqueous activity ratio at the same activity of aqueous SiO_2, or a higher aqueous activity of SiO_2 at the same K^+/H^+ aqueous activity ratio. Unlike the SiO_2 polymorphs whose free energies and

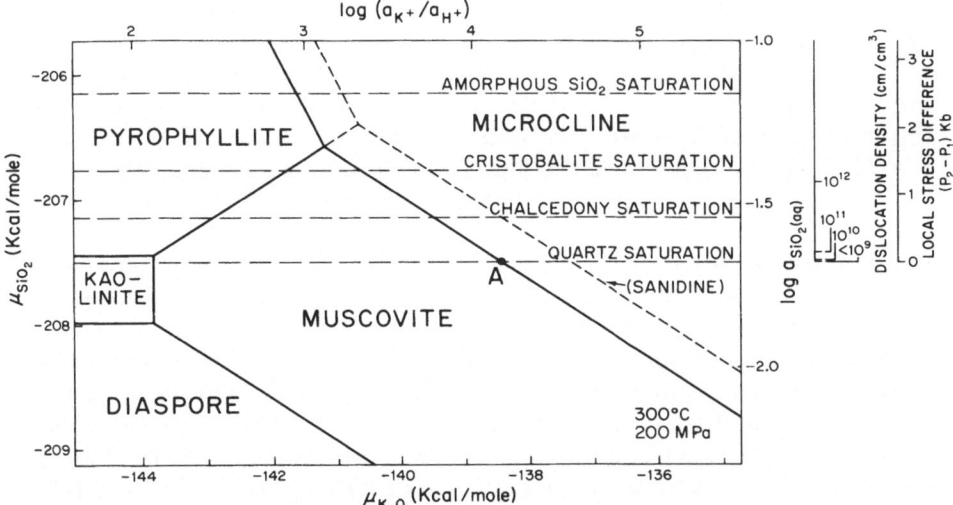

Fig. 1. A chemical potential diagram showing the relationships among chemical potential (μ) and aqueous activity (a) in the system K_2O–Al_2O_3–SiO_2–H_2O–HCl calculated using the computer program SUPCRT and the data compiled by Helgeson *et al.* (1978). The solubilities of quartz, chalcedony, cristobalite, and amorphous SiO_2 are indicated by the long dashed lines. The smaller stability field of sanidine relative to microcline is indicated by the short dashed line. The contributions to the chemical potential of SiO_2 in the fluid by high dislocation density in quartz and of local stress difference are indicated on the exterior right of the diagram. They are referred to an equilibrium assemblage of fluid and coarse-grained muscovite, microcline, and quartz (point A). See text for other details.

solubilities increase in steps, the continuous structural series between microcline and sanidine can lead to a continuous change in free energy (Helgeson *et al.*, 1978, p. 136) and hence also a continuous change in solubility. In this case an infinite number of intermediate metastable states are possible, and the region between the muscovite–microcline and the muscovite–sanidine phase boundaries could be contoured for structural state of the feldspar.

Elastic Strain Energy and Local Stress Differences

A well-known and thoroughly studied deformation mechanism affecting chemical processes is "pressure solution," the dissolution of crystals under nonhydrostatic stress into an aqueous fluid, and transport of these components by a diffusive process (e.g., Rutter, 1983). In low-grade metamorphic rocks, pressure solution is an important deformation mechanism that leads to a reduction of porosity and a change in grain shape and contributes to the development of tectonic cleavages and stylolites. The dissolved material may form new structures by precipitating in pressure shadows or in veins (see Kerrich, 1977, for review).

The driving force for pressure solution comes from the elastic strain energy stored in the crystal and the difference in the normal stress across different faces of the grain. For a one-component crystal, the chemical potential of that component in the fluid (μ) will be (e.g., Robin, 1978):

$$\mu = \mu_0 + (\bar{F} - \bar{F}_0) + P(\bar{V}^s - \bar{V}_0^s) + (P - P_w)\bar{V}, \tag{2}$$

where $\bar{F}$ and $\bar{V}^s$ are the molar Helmoltz energy and volume in the stressed state, respectively, $\bar{F}_0$ and $\bar{V}_0^s$ are the same in the reference state of T of interest and uniform hydrostatic pressure (P_w), P is the pressure at a point of interest on the solid surface, and $\bar{V}$ is the molar volume. The internal strain energy, represented by the second and third terms on the right, is usually small relative to the fourth term except where the surface is exposed only to the hydrostatic pressure (i.e., where $P = P_w$) (see Robin, 1978, p. 1386) and even in this case the solubility increase will be less than 0.1% (deBoer, 1977). Thus the fourth term in Eq. 2 comes close to describing the total contribution of local stress differences (<100 MPa) on the chemical potential of the crystal in the fluid. The calculated increase in μ_{SiO_2} as a function of nonhydrostatic stress at $P = 200$ MPa is given on the right-hand scale of Fig. 1. In the general case the increase in P could lead to an increase in μ of hundreds of calories (deBoer, 1977), whereas in the case of spherical grains in a rock of high porosity an increase of more than 1 Kcal is theoretically possible (Robin, 1978).

Defect Energy

The energy stored in dislocations and in twin and subgrain boundaries can be large enough to drive chemical reactions (Brodie, 1980; Wintsch and Dunning, 1983). The energy stored in natural materials is usually small but is nonetheless geologically significant because it is responsible for the dynamic recrystallization of strained crystals in deformed rocks (Nicolas and Poirier, 1976, p. 63ff). In aqueous environments the metastability caused by this strain energy increases the rates of solution and the solubility of the deformed crystals (Petrovich, 1981a, 1981b; Bosworth, 1981), and such "strain solution" with accompanying diffusive mass transfer may be an important deformation mechanism in some tectonic settings (Engelder, 1982; Wintsch and Dunning, 1983; Meike, 1983). The increase in solubility resulting from dislocations in quartz has been calculated by Wintsch and Dunning (1983), and the results are shown in the right-hand scale external to Fig. 1.

In moderately deformed materials, the density of free dislocations rarely exceeds 10^9 cm/cm^3, and thus the solubility increase is trivial (see Fig. 1). However, the density of dislocations in "tangles" and in subgrain and grain boundaries in mantles is significant because dissolution occurs at surfaces. Thus higher aqueous activities may be caused by the dissolution of material with a high density of dislocations concentrated near surfaces. This in turn will establish local chemical potential gradients and could modify local replacement reactions. According to the calculations of Wintsch and Dunning (1983), a dislocation density of $10^{11}/cm^2$ will raise the free energy of quartz by ~ 65 cal/mol, which will lead to a solubility increase of 6% at 300°C and 200 MPa. This increase in solubility is equivalent to that caused by an increase in temperature of 7°C or an increase in pressure of 10 MPa.

Surface Energy

Surface energy is commonly disregarded in petrology because it makes such a small contribution to the total energy of most geological systems. However, it may be important where processes are dominated by phenomena occurring at phase boundaries, such as nucleation and dissolution, and may influence microstructural evolution, mineral or partial melt distribution and association, and sector zoning (e.g., Vernon, 1975; Stumm and Morgan, 1970, p. 445ff; deVore, 1963; Holister, 1970; Jurewicz and Watson, 1984). The amount of energy stored on a surface/cm^2 is not well known because of the large sensitivity of experimental measurements to chemical environment (e.g., Parks, 1984). The energy associated with a surface is a maximum when all bonds remain broken and no adsorption occurs, as in a perfect vacuum. The hydroxylation of the charged Si,O, and other ions on the surfaces of silicates lowers the energy of the surface by producing, in part, SiOH groups, which also changes the composition of the surface. Surface energy may be further lowered by the adsorption of additional H_2O molecules the oxygen of which is bonded to the H in SiOH groups (e.g., Anderson and Wickersheim, 1964). For example, the surface energy of quartz is reduced from $\sim 4.8 \times 10^{-5}$ cal/cm^2 in a vacuum to $\sim 8 \times 10^{-6}$ cal/cm^2 in liquid water (estimates of Parks, 1984). Thus the surface energy of quartz may be lowered by 4×10^{-5} cal/cm^2, and this energy is dissipated primarily as heat and as a change in fluid composition. This compositional change occurs in a closed system through the adsorption of H_2O onto the surface, which leaves the solution more concentrated in solutes.

In the case of muscovite, Parks (1984) used the data in Bryant et al. (1963) to calculate the surface energy in a vacuum to be 1.22×10^{-4} cal/cm^2. This is close to the value of 1.20×10^{-4} cal/cm^2 calculated by Orowan (1933) from the data of Obreimoff (1930). In contrast to these values, the surface energy of muscovite in liquid H_2O is only about 2.63×10^{-6} cal/cm^2 (Bailey and Kay, 1967; Bailey and Daniels, 1972). These data suggest that the surface energy of muscovite could be reduced by as much as 98% by the interaction of H_2O with the broken bonds on the muscovite surface. As in the example of quartz, this energy is dissipated in part as heat and in part as chemical work, but the chemical work is complicated by incongruent dissolution of the muscovite (see below). Because of these complications, that part of the surface energy dissipated as chemical work cannot be calculated with confidence and is not included in Fig. 1.

The high efficiency of transfer of the surface energy of muscovite and quartz to the fluid by adsorption of water shows that the surface energy contribution to the free energy of these crystals *in the presence of water* is small, and that the increased solubility of small crystals should be correspondingly small. For example, the contribution of surface energy to a cubic grain of quartz 1 μm on a side is only 11 cal/mol SiO_2. Such small contributions are too small to show on Fig. 1. Some of the surface energy transferred to the fluid, however, may be important in destroying chemical equilibrium and driving reactions (see below).

Heat

The portion of the mechanical work done on a rock during deformation and not stored as elastic strain energy, plastic strain energy, or surface energy is dissipated as heat. What portion of the total energy appears as heat is an important question that will clearly depend, in part, on which deformation mechanisms are operating. The consequence of this heat addition will naturally be a rise in the temperature of the deforming rock. Thermomechanical softening and strain concentration are potential geological consequences. What temperature increase takes place is directly related to the stress and the strain rate and inversely related to the heat capacity and thermal conductivity of the rock and the width of the fault zone, and will be limited by endothermic processes in the rock such as mineral reactions and microstructural changes (Brun and Cobbold, 1980; Lachenbruch, 1980). In narrow (a few cm) fault zones where the stress, strain rate, and displacement are large, temperatures may rise by as much as 1000°C, where frictional melting and pseudotachylite may occur (e.g., Wallace, 1976; Passchier, 1982).

In the ductile regime, the possible temperature increases are not well established. Theoretical calculations suggest that temperature rises of up to 150°C may be possible (Lachenbruch, 1980), except where strain rate is constant (Poirier et al., 1979). Most calculations include the assumption that all mechanical energy is converted to heat (e.g., Brun and Cobbold, 1980). This ignores the heat consumed in endothermic reactions, in microstructural modification, and in heat transfer by circulating fluids. Thus in rocks undergoing prograde metamorphic dehydration reactions, such calculations may overestimate temperatures obtained by shear heating. As summarized by Scholz (1980), the results of several field studies suggest that shear heating can occur in nature. Some of the evidence for this comes from an assessment of mineral geothermometers, but evidence for a close approach to chemical equilibrium that such calculations assume is often not strong (Barton and

England, 1979). Several recent studies of highly deformed rocks have shown that chemical equilibrium is not attained in many cases (e.g., Brodie 1980; Kerrich *et al.*, 1980; Anderson, 1983; Wintsch and Knipe, 1983), which jeopardizes some of the interpretations of natural shear heating. The quantitative estimation of shear heating in polymineralic rocks is very complicated and deserves much more attention, especially in mechanically and chemically dynamic environments.

Of the several energy sources discussed above, heat is the most difficult to assess quantitatively because of the many factors that affect the rate of heat generation and the rate of heat dissipation. As is the case in prograde metamorphism, the heat added to a rock during deformation may have the effect (1) of rendering some minerals metastable, which drives mineral reactions and (2) of contributing to the activation energy for the coarsening of grains and annealing of defects. The latter process may seriously limit the lifetime of microstructures, and frustrates attempts to identify coeval mineral assemblages and microstructures. The chemical work done by surface reactions is also very difficult to quantify and at this point cannot be related to the surface energies calculated above.

The sources of energy available for driving chemical reactions (other than heat and surface energy) are compared in the energy diagram of Fig. 1. Relative to an equilibrium among strain-free, coarse-grained muscovite, microcline, quartz, and aqueous fluid (reaction 1), at point A, several different sources of energy are indicated. Chemical energy stored in the metastable polymorphs is indicated by the dashed curves. The increased solubility of quartz as a function of dislocation density is given on the right side of the diagram. A dislocation density of $5 \times 10^{11}/cm^2$ would yield a solubility equivalent to that of chalcedony. The increased solubility of SiO_2 resulting from local stress differences is given on the extreme right-hand scale, and again, is referred to point A. A solubility equivalent to that of chalcedony at 300°C and 200 MPa would require a local stress difference of 65 MPa. Thus in all cases some of the energy dissipated by deformation is transferred to the fluid and is manifested as an increase in the chemical potential of the constituent components.

To assess the relative geochemical significance of these deformational processes is difficult because the partitioning of energy among the various sources cannot be done with confidence. Situations where these effects could be important are easier to discern. The densities of free dislocations in most natural materials are not high enough to increase mineral solubilities by more than 0.1% but the densities in tangles or subgrain walls may be large enough to increase solubilities by >10%. Maximum deviatoric stresses in the crust probably fall between 5 and 40 MPa (Etheridge, 1983). In the lower range of stresses, the driving force for pressure solution of, for example quartz, will be only 27 cal/mol quartz, but in the extreme case, as may occur especially in local sites of high stress, the driving force for dissolution may exceed 200 cal/mol. The contribution of increased surface energy to dynamic metamorphic systems will be trivial for grain sizes >50 μm but may be substantial where grain size falls to below 1 μm. Thus none of the above sources of energy can be expected to be as significant as temperature in static, contact metamorphic environments, because both the magnitude and the concentration of the mechanical energy sinks are too small to affect metamorphic reactions.

In an active, long-lived shear zone, however, both the magnitudes of the stresses and strains as well as their concentration in time and space may be large enough to modify the chemistry and mineralogy of the deformed rock significantly. In shear zones that accommodate many repeated strain events (Sibson, 1980; Passchier, 1982) this will be particularly true. Successive strain events would in part deform new grains and lead to repeated dissolution and precipitation. Evidence for such cyclic precipitation has been found in some mylonites (Wintsch and Knipe, 1983).

Applications

In the preceding section it was shown that in dynamic metamorphic environments mechanical energy dissipated during deformation can provide the source of energy driving chemical reactions. The volume of rock affected by these deformational processes is probably small, however, except in shear zones where repeated strain events may cause reaction in a relatively large portion of the fault zone. In the following examples two situations are examined in detail to show how mechanical energy sinks may be manifested as chemical energy sources.

Defect Energy

In the more deformed mantles of some grains, feldspars may have dislocation densities of 10^{10} cm^{-2} (Sodre Borges and White, 1980) and may locally exceed 10^{11} cm^{-2} (White, 1975, Fig. 5b: Knipe and White, 1979; Sacerdoti et al., 1980). In an example of a deformed feldspar crystal from Lewisian derived mylonites along the Moine thrust, Knipe and Wintsch (this volume) observed dislocation densities locally reaching 2×10^{11} cm^{-2}. The strain energies associated with dislocations in feldspar are probably 20% larger than in quartz (because of larger Burgers vectors), and the increase in lattice energy resulting from a dislocation density of 2×10^{11} cm^{-2} in quartz (130 cal/mol) provides a minimum estimate for the excess free energy of this strained feldspar. Figure 2 shows the effect of such excess free energy on the potential equilibrium in a mylonite assemblage. Equilibrium among unstrained quartz, muscovite, microcline, and fluid is represented by point A (see also Fig. 1). The metastable equilibrium among strained microcline, unstrained

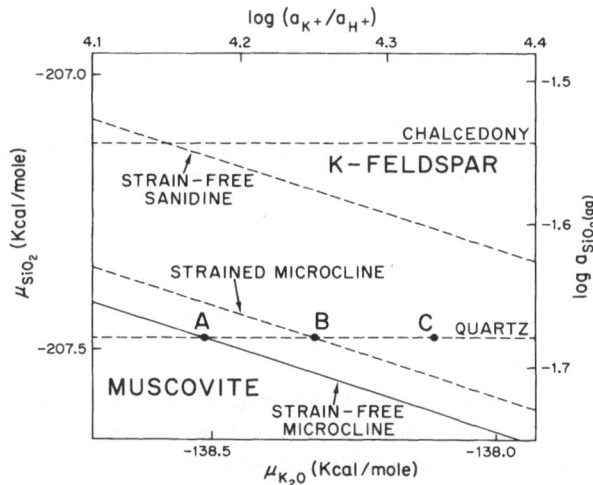

Fig. 2. A chemical potential diagram, enlarged from Fig. 1, comparing the effects of chemical metastability (chalcedony, strain-free sanidine) and high defect density-induced metastability (strained microcline) on a stable strain-free muscovite–microcline–quartz–fluid assemblage (point A). See text for details.

quartz and muscovite and fluid is given by point B. At equilibrium the fluid composition would be buffered at composition A by the undeformed assemblage, but composition A is entirely within the muscovite stability field relative to the deformed (metastable) assemblage. The greater solubility of the strained microcline relative to the undeformed microcline leads to the shift in the deformed microcline–muscovite stability field boundary (Fig. 2) and causes the replacement of the microcline by muscovite via the reaction:

$$3 \text{ microcline} + 2\text{H}^+ = \text{muscovite} + 6 \text{ SiO}_{2(aq)} + 2\text{K}^+. \qquad (1)$$

Thus the defect energy in the deformed microcline is capable of driving the above replacement reaction even in a static environment.

If the local stress differences causing the core-mantle structure were 5 MPa, this would lead to another 130 cal/mol of metastability. Acting together the sources of energy would sum to 260 cal/mol and displace the muscovite–microcline boundary to point C, Fig. 2. Thus in this example, plastic strain and possibly also local stress differences are capable of causing the dissolution of feldspar.

In such a situation several geochemical consequences are possible. The dissolution of the deformed microcline would produce a fluid supersaturated with respect to undeformed microcline, and strain-free microcline would precipitate at the same site. If precipitation at this site were retarded, for example, by a locally high stress, then the associated chemical potential gradients would drive the DMT of the components of microcline away from this site and toward a more favorable precipitation site. The chemical composition of the precipitating feldspar may be different from that dissolving because of ion exchange with alkalis in the fluid. In the analogous example of plagioclase, changes in mineral composition are common (e.g., White, 1975; Sodre Borges and White, 1980; Kerrich et al., 1980; Brown et al., 1981; Wintsch and Knipe, 1983; Anderson, 1983; Watts and Williams, 1983). It is also possible that the components of microcline could redistribute themselves as muscovite and quartz (reaction 1), an example of reaction softening (e.g., White and Knipe, 1978).

Surface Energy

The above calculations show that in the presence of water the contribution of surface energy to the free energy of the crystal is small in all but extreme cases. Knipe and Wintsch (this volume) describe some microbreccia zones in muscovite-rich folia from some mylonites from the Moine thrust. The grain size in one such zone was as small as 1.0×0.1 μm, where the surrounding grains are 10×30 μm. From the discussion above, 1.2×10^{-4} cal/cm^2 is dissipated by the creation of muscovite surface, but only ~2% of this energy contributes to the free energy of the crystal in the presence of water. In this microbreccia the grain-size reduction would dissipate 1585 cal/mol muscovite, but the contribution of the surface energy to the bulk muscovite would only be 32 cal/mol.

How the remaining energy is partitioned between chemical work and heat is not known, but the cation exchange of K^+ from the broken surface with H^+ in the water (Garrels and Howard, 1959):

$$K\text{-(surface)} + H_2O = H\text{-(surface)} + KOH \qquad (3)$$

can be identified as a major source of chemical work done. This exchange will clearly increase the aqueous K^+/H^+ activity ratio in the fluid, and hence also the chemical potential of K_2O as defined by the reaction (Helgeson, 1968):

$$K_2O + 2H^+ = H_2O + 2K^+ \qquad (4)$$

Calculations of the surface energy reduction caused by the above surface exchange reaction cannot be done uniquely because of the several variables involved, especially the volume and starting composition of the exchanging fluid. For the purpose of exploring the consequences of this chemical work on aqueous fluid/fault rock systems, a calculation is made following the procedure of Wintsch (1975), assuming a starting $a_{K^+}/a_{H^+} = 10^{4.2}$ (point A, Fig. 1), $a_{K^+} = 10^{-2}$ mol/kg, $a_{H^+} = 10^{-6.2}$ mol/kg, starting muscovite flake $30 \times 30 \times 10$ μm, final muscovite $1.0 \times 1.0 \times 0.1$ μm. Only exchange of H^+ for K^+ on (001) surfaces is allowed. At low temperature (Garrels and Howard, 1959) and probably also at high temperatures (Wintsch et al., 1980) the exchange of H^+ for K^+ is essentially complete and is therefore assumed.

Activity coefficients are ignored and total dissociation of aqueous complexes is assumed. If this exchange occurs in 1 g of fluid, the K^+/H^+ activity ratio increases by only 0.05%, equivalent to only 0.57 cal/mol increase in the chemical potential of K_2O at the conditions of Fig. 1. If the exchange occurs in 10^{-3} g H_2O, then the activity ratio and the chemical potential increase by 200% and 840 cal/mol, respectively.

These surface energy contributions are compared and assessed in Fig. 3. Equilibrium among fluid and coarse-grained muscovite, microcline, and quartz is represented by point A. The increased surface area of muscovite lowers its stability and causes a reduction in size of the muscovite stability field (dashed line at point B, Fig. 3). The chemical work done by ion exchange with 10^{-3} g of a fluid of composition A is indicated by point C. The position of point C is highly dependent on the boundary conditions of the calculation. In this calculation 10^{-3} g fluid exchanges with 2.6×10^{-8} gm muscovite. Making the fluid/solid ratio less extreme would increase the calculated effect so that point C can be considered a conservative estimate of the chemical work of ion exchange. The geochemical consequences of this are an increase in the chemical potential of K_2O with either (1) the stabilization of feldspar with respect to muscovite (reaction 1), or (2) the transport of aqueous species down their chemical potential gradient. The former is an example of reaction hardening, the reverse process of reaction softening described by White and Knipe (1978) and Dixon and Williams (1983).

As in this example the fluid/solid ratio potentially limits the chemical effects of the other mechanical energy sinks as well; the question could be raised whether sufficient volume of deformed material is available to satu-

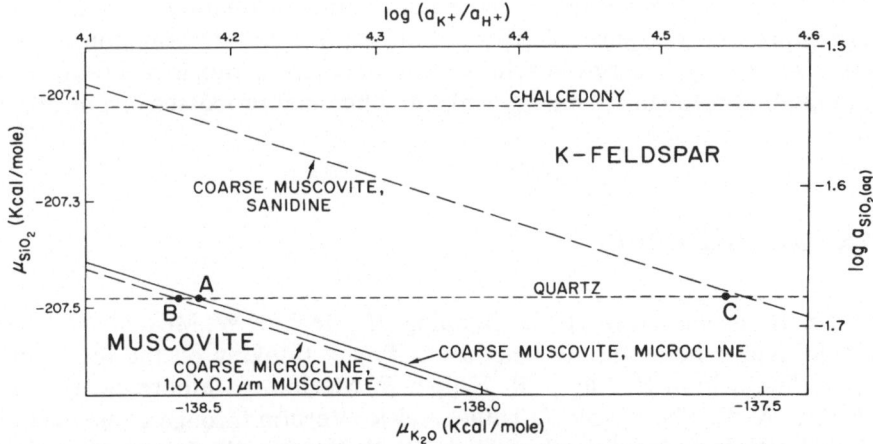

Fig. 3. A chemical potential diagram, enlarged from Fig. 1, showing the effect of surface energy of fine-grained muscovite on the stability field of muscovite (long dashed line and point B). A calculated example of the effect of surface exchange on the composition of the fluid is indicated by point C (see text for details).

rate the available fluid in a deforming rock. Consider, for instance, a 1 μm^3 portion of a quartz crystal to have a dislocation density of 10^{11} cm dislocation/cm^3. The increase in solubility of this strained quartz over strain-free quartz at 300°C and 200 MPa is calculated to be 5.9% or 0.0012 mol $SiO_{2(aq)}$ kg^{-1}. Because 1 μm^3 of quartz is only 4.4×10^{-14} mol or 1.6×10^{-12} g quartz, the deformed portion of this crystal can saturate only 4×10^{-12} kg H_2O. This yields a fluid/deformed quartz mass ratio of 1500 : 1 or a volume ratio at 300°C and 200 MPa of 5600 : 1. Thus a 1 μm^3 volume of quartz with a dislocation density of 10^{11} cm^{-2} could theoretically saturate all the fluid in a 5.6×10^6 μm^3 volume of rock with a porosity of 0.1%.

The geochemical implications of this are that the chemical potential of SiO_2 may deviate from the value established by a strain-free quartz over much larger volumes of rock than the small region of the deformed grain.

Conclusions

Deformational processes can provide the source of energy driving chemical or mineral reactions. It may change both the nature of the reactions and the distribution of minerals in the deforming rock. In weakly deformed prograde metamorphic rocks, the relative significance of mechanical sources of energy are likely to be small, but in highly and repeatedly deformed fault zones, these effects may be substantial. An important conclusion of these results is that chemical equilibrium is unlikely to be achieved in activity deforming rocks and may not even be closely approached in rocks where stresses or strains are large. Much more work on the chemical composition of bulk rocks and single grains from fault zones in relation to rock types, temperature and pressure of deformation, strain, and deformation mechanism is necessary before the relative and absolute geologic significance of the individual mechanical energy sinks on chemical processes can be ascertained.

Acknowledgments

Conversations with B. Bayly, J. Dunning, E. Merino, W. Means, and especially R. Knipe have greatly helped clarify my thinking on the topics discussed. The help of B. Bayly, R. Knipe, E. Merino, C. Moore, A. Owens, D. Rubie, A. B. Thompson, J. Tullis, and R. Vernon through comments on earlier drafts is gratefully acknowledged. T. Brown, W. Moran, J. Tolen, and B. Hill helped with manuscript preparation. This work was partially supported by a grant from Research Corporation and by NSF grant EAR-8313807.

References

Anderson, Jr., J. H., and Wickersheim, K. A. (1964) New infrared characterization of water and hydroxyl groups on silica surfaces. *Surf. Sci.* **2**, 252–260.

Anderson, J. R. (1983) Petrology of a portion of the Eastern Peninsular Ranges mylonite zone, Southern California. *Contrib. Mineral. Petrol.* **84**, 253–271.

Bailey, A. I., and Daniels, H. (1972) Einfluss des Mediums auf die Grenzflachenenergie in System Glimmer/Flussigkeit/Daumpf. *Kolloid Z. u. Z. Polymere* **250**, 148–151.

Bailey, A. I., and Kay, S. M. (1967) A direct measurement of the influence of vapour, of liquid, and of oriented monolayers on the interfacial energy of mica. *Proc. Roy. Soc. London* **A301**, 47–56.

Barton, C. M., and England, P. C. (1979). Shear heating at the Olympos (Greece) thrust and the deformation properties of carbonates at geological strain rates. *Geol. Soc. Amer. Bull.* **90**, 483–492.

Beach, A. (1979) Pressure solution as a metamorphic process in deformed terrigenous sedimentary rocks. *Lithos* **12**, 51–58.

Boer, R. B. de (1977) On the thermodynamics of pressure solution—interaction between chemical and mechanical forces. *Geochim. Cosmochim. Acta* **41**, 249–256.

Bosworth, W. (1981) Strain induced preferential dissolution of halite. *Tectonophysics* **78**, 509–525.

Brodie, K. H. (1980) Variations in mineral chemistry across a shear zone in phlogopite peridotite. *J. Struct. Geol.* **2**, 265–272.

Brown, W. L., Macandiere, J., Ohnenstetter, D., and Ohnenstetter, M. (1981) Ductile shear zones in a meta-anorthosite from Harris, Scotland: Textural and compositional changes in plagioclase. *J. Struct. Geol.* **2**, 281–287.

Brun, J. P., and Cobbold, P. R. (1980) Strain heating and thermal softening in continental shear zones: A review. *J. Struct. Geol.* **2**, 149–158.

Bryant, P. J., Taylor, L. M., and Gutshall (1963) Cleavage studies of lamellar solids in various gas environments, in Transaction of the Tenth National Vacuum Symposium of the American Vacuum Society, edited by G. H. Bancroft, 21–26.

Carmichael, D. M. (1969) On the mechanism of prograde metamorphic reactions in quartz bearing pelitic rocks. *Contrib. Mineral. Petrol.* **20**, 244–267.

DeVore, G. W. (1963) Compositions of silicate surfaces and surface phenomena. *Contrib. Geol.* **2**, 21–37.

Dixon, J., and Williams, G. (1983) Reaction softening in mylonites from the Arnaboll thrust, Sutherland. *Scott. J. Geol.* **19**, 157–168.

Engelder, T. (1982) A natural example of the simultaneous operation of free-face dissolution and pressure solution. *Geochim. Cosmochim. Acta* **46**, 69–74.

Enüstün, B. V., and Turkevich, J. (1960) Solubility of fine particles of strontium sulfate. *J. Am. Chem. Soc.* **82**, 4502–4509.

Etheridge, M. A. (1983) Differential stress magnitudes during regional deformation and metamorphism: Upper bound imposed by tensile fracturing. *Geology* **11**, 231–234.

Fisher, G. W. (1970) The application of ionic equilibria to metamorphic differentiation: An example. *Contrib. Mineral. Petrol.* **29,** 91–103.

Garrels, R. M., and Howard, P. (1959) Reactions of feldspar and mica at low temperature and pressure, in *Clays and Clay Minerals,* edited by A. Swineford, pp. 68–88. Pergamon Press, New York.

Helgeson, H. C. (1968) Evaluation of irreversible reactions in geochemical processes involving minerals and aqueous solutions. I. Thermodynamic relations. *Geochim. Cosmochim. Acta* **32,** 853–877.

Helgeson, H. C., Delany, J. M., Nesbitt, H. W., and Bird, D. K. (1978) Summary and critique of the thermodynamic properties of rock-forming minerals. *Amer. J. Sci.* **278A,** 1–229.

Holister, L. S. (1970) Origin, mechanism, and consequences of compositional sector-zoning in staurolite. *Amer. Mineral.* **55,** 742–766.

Jurewicz, S. R., and Watson, E. B. (1984) Distribution of partial melt in a felsic system: The importance of surface energy. *Contr. Mineral. Petrol.* **85,** 25–29.

Kerrich, R. (1977) An historical review and synthesis of research on pressure solution. *Zbl. Geol. Paläont.* **1977,** 512–550.

Kerrich, R., Fyfe, W. S., Gorman, B. E., and Allison, I. (1977) Local modification of rock chemistry by deformation. *Contrib. Mineral. Petrol.* **65,** 183–190.

Kerrich, R., Allison, I., Barrett, R. L., Moss, S., and Starkey, J. (1980) Microstructural and chemical transformations accompanying deformation of granite in a shear zone at Mieville, Switzerland, with implications for stress corrosion cracking and superplastic flow. *Contrib. Mineral. Petrol.* **73,** 221–242.

Knipe, R. J. (1981) The interaction between deformation and metamorphism in slates. *Tectonophysics* **78,** 249–272.

Knipe, R. J., and White, S. H. (1979) Deformation in low grade shear zones in the Old Red Sandstone, S. W. Wales. *J. Struct. Geol.* **1,** 53–66.

Kwak, T. A. P. (1974) Natural staurolite breakdown reactions at moderate to high pressures. *Contrib. Mineral. Petrol.* **44,** 57–80.

Lachenbruch, A. M. (1980) Frictional heating, fluid pressure, and the resistance to fault motion. *J. Geophys. Res.* **85,** 6097–6112.

Matthews, A. (1980) Influences of kinetics and mechanism in metamorphism: A study of albite crystallization. *Geochim. Cosmochim. Acta* **44,** 387–402.

Misch, P. (1969) Paracrystalline microboudinage of zoned grains and other criteria for synkinematic growth of metamorphic minerals. *Amer. J. Sci.* **267,** 43–63.

Meike, A. (1983) Microstructure of stylolites in limestone (abstract). *Geol. Soc. Amer. Abstr. Progrs.* **15,** 642.

Nicolas, A., and Poirier, J. P. (1976) *Crystalline Plasticity and Solid State Flow in Metamorphic Rocks.* Wiley, New York.

Obreimoff, J. W. (1930) The splitting strength of mica. *Proc. Roy. Soc. London* **127A,** 290.

Orowan, E. (1933) Die Zugfestigkeit van Glimmer und das Problem der techischen Festigkeit. *Zeit. Physik.* **82,** 235–266.

Parks, G. A. (1984) Surface and interfacial energies of quartz. *J. Geophys. Res.* **89,** 3997–4008.

Passchier, C. W. (1982) Pseudotachylyte and the development of ultramylonite bands in the Saint-Barthélemy Massif, French Pyrenees. *J. Struct. Geol.* **4,** 69–79.

Petrovich, R. (1981a) Kinetics of dissolution of mechanically comminuted rock-forming oxides and silicates. I. Deformation and dissolution of quartz under laboratory conditions. *Geochim. Cosmochim. Acta* **45,** 1665–1674.

Petrovich, R. (1981b) Kinetics of dissolution of mechanically comminuted rock-forming oxides and silicates. II. Deformation and dissolution of oxides and silicates in the laboratory and at the Earth's surface. *Geochim. Cosmochim. Acta* **45,** 1675–1686.

Poirier, J. P., Bouchey, J. L., and Jones, J. J. (1979) A dynamic model for a seismic ductile shear zone. *Earth Planet. Sci. Lett.* **45,** 441–453.

Reitan, P. H. (1977) Energetics of metamorphic reactions. *Lithos* **10,** 121–128.

Robin, P-Y. F. (1978) Pressure solution at grain-to-grain contacts. *Geochim. Cosmochim. Acta* **42,** 1383–1389.

Rosenfeld, J. L. (1970) Rotated garnets in metamorphic rocks. *Geol. Soc. Amer. Spec. Paper* **129,** 105p.

Rutter, E. M. (1983) Pressure solution in nature, theory and experiment. *J. Geol. Soc. London* **140,** 725–740.

Sacerdoti, M., Labernardière, H., and Gandais, M. (1980) Transmission electron microscope (TEM) study of geologically deformed potassic feldspars. *Bull. Minéral.* **103,** 148–155.

Scholz, C. H. (1980) Shear heating and the state of stress on faults. *J. Geophys. Res.* **85,** 6174–6184.

Sibson, R. M. (1980) Transient discontinuities in ductile shear zones. *J. Struct. Geol.* **2,** 165–171.

Sodre Borges, F., and White, S. H. (1980) Microstructural and chemical studies of sheared anorthosites, Roneval, South Harris. *J. Struct. Geol.* **2,** 273–280.

Stumm, W., and Morgan, J. J. (1970) *Agnatic Chemistry.* Wiley Interscience, New York.

Vernon, R. H. (1975) *Metamorphic Processes, Reactions and Microstructure Development.* Halshed Press, New York.

Wallace, R. C. (1976) Partial fusion along the Alpine Fault Zone, New Zealand. *Geol. Soc. Amer. Bull.* **87,** 1225–1228.

Walton, M., Mills, A., Hansen, E. (1964) Compositionally zoned granitic pebbles in three metamorphosed conglomerates. *Amer. J. Sci.* **262,** 1–25.

Watts, M. J., and Williams, G. D. (1983) Strain geometry, microstructure and mineral chemistry in metagabbro shear zones: A study of softening mechanisms during progressive mylonitization. *J. Struct. Geol.* **5,** 507–517.

White, S. (1975) Tectonic deformation and recrystallization of oligoclase. *Contrib. Mineral. Petrol.* **50,** 287–304.

White, S. H., and Knipe, R. J. (1978) Transformation- and reaction-enhanced ductility in rocks. *J. Geol. Soc. London* **135,** 513–516.

Wintsch, R. P. (1975) Feldspathization as a result of deformation. *Geol. Soc. Amer. Bull.* **85,** 35–38.

Wintsch, R. P. (1981) Syntectonic oxidation. *Amer. J. Sci.* **281,** 1223–1239.

Wintsch, R. P., and Dunning, J. D. (1983) The role of defect density on "strain solution" (abstract) *Trans. Am. Geophys. Union (EOS)* **64,** 319.

Wintsch, R. P., and Knipe, R. (1983) Growth of a zoned plagioclase in a mylonite. *Geology* **11,** 360–363.

Wintsch, R. P., Merino, E., and Blakely, R. F. (1980) Rapid quench hydrothermal experiments in dilute chloride solutions applied to the muscovite–quartz–sanidine equilibrium. *Amer. Mineral.* **65,** 1002–1011.

Wyart, J. (1975) The mechanism of the action of water in hydrothermal reactions. *Fortschr. Miner.* **52,** 169–176.

Yund, R. A., Smith, B. M., and Tullis, J. (1981) Dislocation-assisted diffusion of oxygen in albite. *Phys. Chem. Minerals* **7,** 185–189.

Chapter 11
Deformation with Simultaneous Chemical Change: The Thermodynamic Basis

B. Bayly

Introduction

The purpose of this contribution is to put forward a proposal about the theory of chemical kinetics and deformation: When a material is under non-hydrostatic stress, the chemical potential of a component at a point in the material has a range of values according to which plane in the material one considers. The idea that the normal compressive stress on a plane has this kind of variability has long been accepted; here it is proposed that the chemical potential is linked to the normal stress in such a way that it too is many valued. For any single plane p it is proposed that:

$$\mu_a^p = U_a + \sigma^p V_a - TS_a \tag{1}$$

where σ^p is the normal compressive stress on plane p, μ_a^p is the chemical potential of component a in joules/mol for that plane, and the other symbols have their usual meanings. For all planes simultaneously:

$$\mu_a(\underset{\sim}{n}) = \underset{\sim}{n} \underset{\approx}{M} \underset{\sim}{n} \quad \text{and} \quad \underset{\approx}{M}_a = U_a \underset{\approx}{I} + \underset{\approx}{\sigma} V_a - TS_a \underset{\approx}{I}$$

where $\underset{\sim}{n}$, $\underset{\approx}{I}$, and $\underset{\approx}{\sigma}$ are a unit vector, a unit tensor, and a stress tensor, respectively. These equations generalize Kamb's proposal about interfaces (1959) to noninterface planes.

The contribution has three parts: First, the scope of current theory is surveyed and it is argued that there is something lacking; second, the new proposal is illustrated by an example; and third, in the appendix, the objections that immediately spring to mind are considered. It is concluded that the new proposal does not conflict with any part of current theory and is in fact a natural extension, despite seeming unfamiliar. For readers who have difficulty believing this, attention is directed to the concept of an *associated equilibrium state* (see Appendix); though rarely discussed, this concept is the key to unifying nonhydrostatic stress conditions with the existing body of thermodynamic theory.

Applied to metamorphic rocks, the proposal frees us to write equations for any stress state, whether hydrostatic or not; for materials of all intermediate viscosities between inviscid liquids and infinitely viscous solids; and for the combined effects of simultaneous diffusion, creep, and change of phase. In particular, the materials at a crystal boundary become tractable; we can consider a crystal as having a skin of more disordered, more deformable, but still crystalline, material and the adjacent fluid as having high or low viscosity according to conditions in the immediate vicinity, and continue to relate diffusion rates, for example, to thermodynamic quantities while the materials deform. The algebra may be arduous but the proposal just introduced seems to provide the key to getting started.

The Scope of Current Theory

An example that illustrates the state of affairs is a perthite equilibrated at, say, 500°C and 500 MPa. Let the pressure be changed to 600 MPa; then we have some idea how to estimate new values for μ^{Ab} in the potassic phase and μ^{Ab} in the sodic phase; if a diffusive exchange of sodium and potassium is detected experimentally, we can then suggest what potential difference is the driving agent for the diffusion. But now, instead of that program, let the compressive stress be changed to 600 MPa in only one direction, being maintained at 500 MPa in two other orthogonal directions. Again, diffusion will be initiated but a theoretical basis for comparison is more difficult to come by. Current theories all proceed by *assuming* that the original equilibrium state remains undisturbed in some respect and, on that basis, calculating the system's rate of response in other respects. There has so far been no theory admitting the possibility that the change of stress disturbs equilibrium in all respects, and that creep, diffusion, and change in the bleb's volume-fraction all occur. The new proposal admits this possibility and, it is hoped, will soon provide nonzero estimates of the rates (work in progress). Examples of other work embodying the idea of equilibrium include Li *et al.* (1966), Larché and Cahn (1978), Green (1980), and Lehner and Heidug (1983).

The Present Proposal: Preliminary Results

As an illustration of the type of result, we focus on a very small region at the margin of a bleb as described above, in comparison with which the bleb can

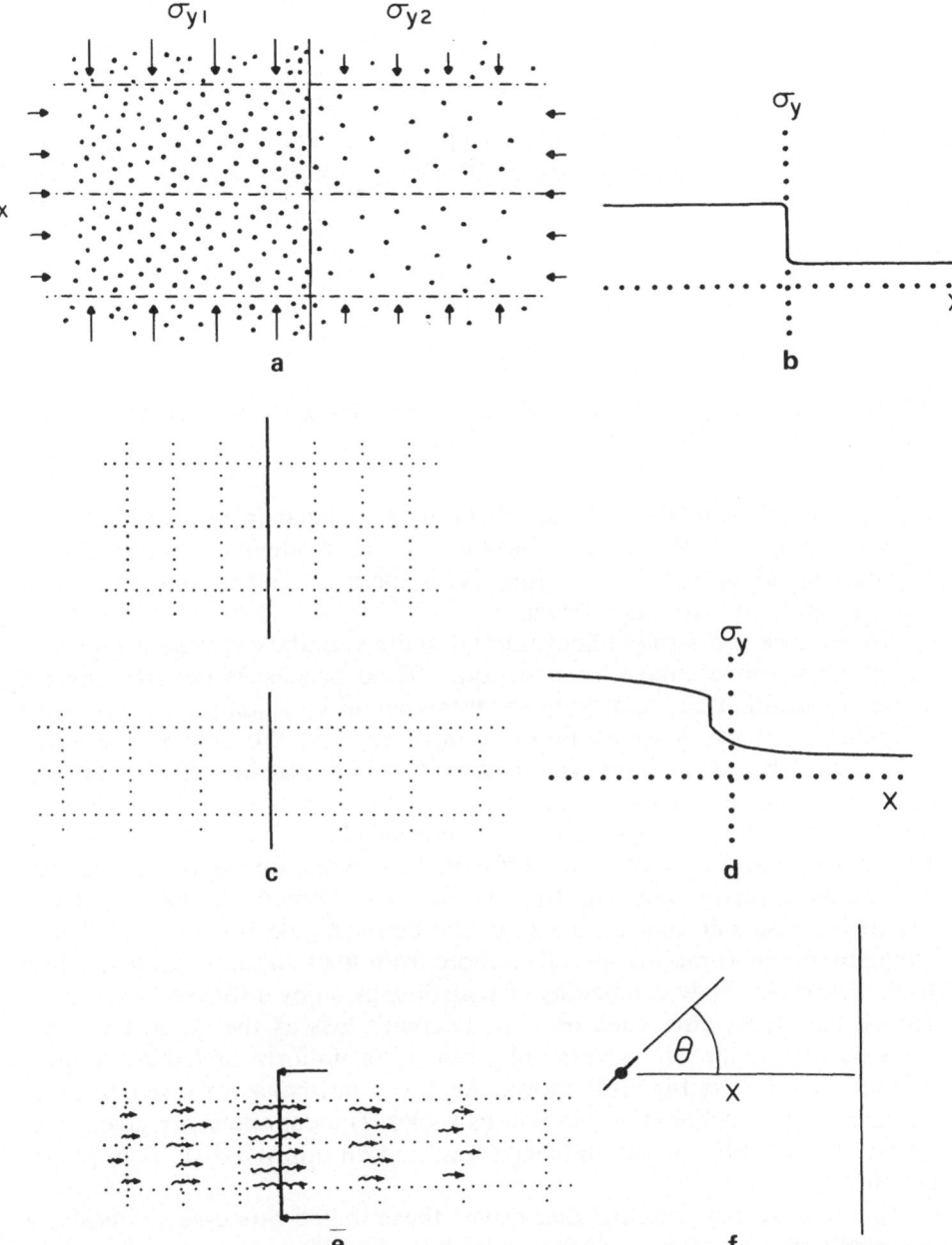

Fig. 1. Diffusion at the boundary between two pure one-component polymorphs. (a) The phase boundary and stresses envisaged. (b) The stress profile in absence of diffusion. (c) The initial and later states of an imaginary grid in the material. (d) The stress profile if material diffuses across the boundary. (e) The initial grid from (c) as modified by diffusion of material as well as strain. (f) The angle θ.

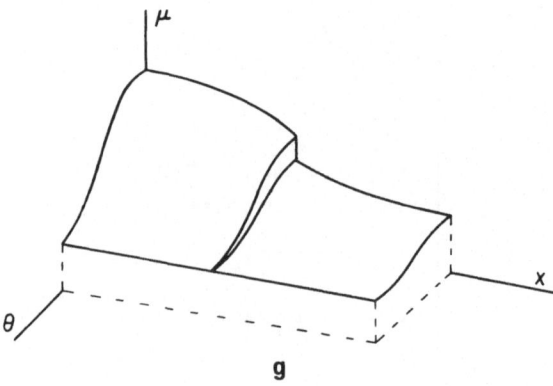

Fig. 1. (continued) (g) The potential surface showing gradients that drive creep, diffusion, and change of phase.

be considered infinitely wide and its boundary planar instead of curved, as shown in Fig. 1(a). We assume that there are no gradients or displacements of material along the Z direction. With these simplifications and more, results can be derived as follows.

We assume that straight lines normal to the boundary remain straight and parallel but move closer to each other. If no processes occurred except creeping strain at constant volume, there would be a jump in stress at the boundary, but stress would be uniform throughout the volumes on either side (Fig. 1(b)), and strain would be uniform across the whole field (Fig. 1(c)). If, however, material migrates across the boundary, the stress profiles become curved (Fig. 1(d)) and cells of material change volume. On one side, the cells diminish in volume by diffusive loss, while on the other side they expand by diffusive gain (Fig. 1(e)). Diffusion and creep are linked. Cells on the high-stress side that are close to the boundary do not need to change shape by creep as rapidly as cells remote from the boundary because, close to the boundary, they can readily change dimension by diffusive loss as well. Hence the stress difference $\sigma_y - \sigma_x$ becomes less as the boundary is approached from the high-stress side; but σ_x is uniform and thus σ_y must become smaller, as Fig. 1(d) shows. At this point, behavior resembles that described by Fletcher (1982) in porous rocks. The equations are given elsewhere (Bayly, 1983) in preliminary form, and an improved set is in preparation.

To visualize the potential that drives these linked processes, consider a series of points at different distances from the boundary and, at each point, a series of planes with different orientations (Fig. 1(f)). The entire set of planes can be displayed using an axis for distance x and an axis for orientation θ. The present proposal assigns a chemical potential to each plane at every point, giving a potential surface on the x, θ plane. If we apply the foregoing ideas to a pure one-component material that exists in two phases, instead of to something as complicated as feldspar, the potential surface can be drawn (Fig. 1(g)). The gradient along θ drives material from planes of one orientation into planes with another orientation at the same point, giving what we normally call *strain*, while the gradient along x drives the diffusion along x. The step in the potential surface is an *affinity* in the sense of de Donder (1928) that gives a finite rate of change of phase; in our simple example, this equals the rate at which material diffuses to the boundary from one side and diffuses away from it on the other. The stress profile in Fig. 1(d) and the potential profile along x in Fig. 1(g) adjust themselves so that the gradients and the step-height are compatible in this respect. A last feature of the potential surface is the straight horizontal line at the bottom; the height of this line is the potential noted by Kamb (1959) and also by Gibbs (1878, p. 196). The arguments that prove that this line must be straight and horizontal are sound (Paterson, 1973, pp. 363–367), but to focus only on this portion of the set-up is to miss out the interesting parts.

Further Applications

The situation illustrated in Fig. 1 is the simplest that contains the behaviors of interest. But, if it is established by considering that situation that Eq. (1) has no inconsistencies, the range of further applications becomes rather wide. If there are temperature gradients or composition gradients at an interface as well as stress gradients, the effect on the potential surface can be readily assessed using standard equations; also the effect of elastic strain on U, or the effect of variation in the concentration of defects (Wintsch, this volume) can be added: The power of the concept of chemical potential to combine the competing effects of several variables is already well known. If the difficulty of using chemical potential in nonhydrostatic states is in fact overcome by the present proposal, a large field of opportunities opens up: The stressed perthite bleb leads on to the stressed porphyroblast; growth rate at a face under high normal stress and growth rate at a face under low normal stress can be compared, the possibility of a composition contrast explored, and so on.

A particularly interesting feature of the present proposal is its bearing on pore-fluid chemistry. The properties of a pore fluid in a stressed crystal aggregate have been a source of difficulty for many years. To assume the pore fluid inviscid makes the stress distribution in the aggregate unrealistic; at the other extreme, to assume the pore fluid to be stationary despite gradients of stress is an oversimplification of another kind. But the present proposal permits discussion of all intermediate viscosities; at a highly compressed interface between two crystals, the only pore fluid or "less ordered intergranular phase" must have very high viscosity, whereas at a less highly compressed interface in the same aggregate, the less ordered phase can have a lower viscosity, higher water content, etc. At the most highly compressed localities, one seems to need in fact a mechanism for constant regeneration of a high-viscosity, less-ordered intergranular phase; and once one has a theory for chemical effects in such highly viscous materials, possible regeneration mechanisms can be postulated. The first idea is to revisit Fig. 1, imagining now that the phase of lower viscosity is the intergranular phase; the second idea is that the step can be replaced by a steep but continuous change in properties. Then Fig. 1 begins to describe those few nanometers where the defect-rich surface skin of the crystalline phase merges with the less-ordered, but still highly ordered, perhaps glassy, intergranular material—that in turn merges, as one travels sideways along the intergranular surface film, into a less-ordered, more watery, more mobile fluid. It is suggested that these intermediate viscosities must exist and that their mechanics and chemistry become fully accessible to study if chemical potential and nonhydrostatic stress are considered simultaneously in the manner proposed. However, the theory is at present in only an embryonic state, and the first test against observations still in the future.

Acknowledgments

I have been greatly helped by many colleagues who have discussed the topic in general terms. For comments on the present text, I am indebted to Robert Wintsch, Martin Casey, and Ernest Perkins.

Appendix

The purposes of the Appendix are to illustrate the idea of an *associated equilibrium state,* to consider how the present proposals relate to Gibbs' treatment of a solid under nonhydrostatic stress, and to consider possible difficulties in the present approach.

Associated Equilibrium State

The concept is illustrated by an example. Consider a long straight bar of material along which some minor constituent J is able to diffuse, and let a concentration gradient exist so that J is currently diffusing (Fig. 2(a)). The most versatile way of predicting diffusion rates is through the chemical potentials in Fig. 2(c). These are perfectly well defined potentials for a series of equilibrium states, and these states (Fig. 2(b)) are *closely associated* with

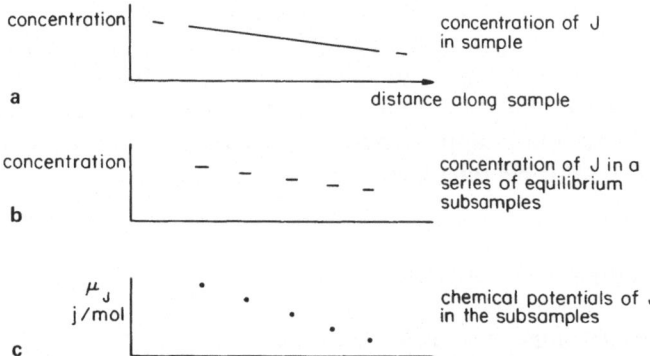

Fig. 2. Thermodynamics of a continuous disequilibrium system. A fundamental postulate of nonequilibrium thermodynamics states that the diffusion rate of component J in the real disequilibrium system, Fig. 2(a), can be effectively predicted from the potentials in Fig. 2(c).

the real disequilibrium profile of interest. It is hard to overemphasize the importance of Fig. 2(b) in relating Figs. 2(c) and 2(a) (and surprising that the topic is given so little attention in textbooks; but see de Groot 1951, p. 11). Turning to a single point in a material under nonhydrostatic stress, we use an exactly analogous process: For five planes suffering normal stresses σ_i ($i = 1, 2, 3 \ldots 5$) we imagine five material samples under separate hydrostatic pressures p_i, each equal to a σ_i. From these equilibrium states, we establish five potentials μ_i and then suggest that the equilibrium potentials μ_i are a good basis for predicting behavior in the real, disequilibrium, nonhydrostatic condition.

(The concept of *local equilibrium* is sometimes useful but in the present context it is better avoided. The disequilibrium in Fig. 2(a) is linked to the gradient in concentration, and hence a small portion shows disequilibrium to just as great an extent as a large portion. It is the jump to Fig. 2(b) that introduces equilibrium, not the focussing down on a small local region.)

Relation to the Work of Gibbs

Gibbs discussed a solid cube whose three pairs of faces touch three pairs of fluid-filled cells at pressure p', p'', and p''' (1878, p. 194ff.). Gibbs showed that for equilibrium with respect to the tendency of the solid to dissolve, it is necessary for the chemical potential of the material of the solid in the three fluids to be μ', μ'', and μ''' where $\mu^i = \mu_0 + p^i V$ ($i = '$, $''$, or $'''$). Gibbs said nothing about the chemical potential of the material of the solid *in the solid*. But the current proposal is that under Gibbs' assumptions the chemical potential in the solid is the same at every point, and at every point the chemical potential is close to the tensor $\underset{\sim}{M}$ that has μ', μ'', and μ''' as its principal components. Thus the current proposals do not conflict with Gibbs' treatment but cover what he left as an unfilled gap. We should note the difference between "close to" and "equal to"; but in Gibbs' treatment the fluids are inviscid and the solid is infinite in viscosity, and it is suggested that as this condition is approached, the approximation implied by "close to" will be found to become exact.

Difficulties in the Proposal

Points that might appear as difficulties are:

1. "Chemical potential is a *quantity of energy* and is thus intrinsically incapable of varying with orientation." In the definition $\mu_a = [\partial G/\partial n_a]_{p,T,n_i}$ ($i \neq a$) the chemical potential is a quantity of energy per mole of component added to a thermodynamic system. For two planes at a single point, the thermodynamic system the material enters by crossing one plane is not the same as the system it could enter by crossing the other plane. The difficulty is not encountered if we emphasize that the virtual transfer of material must involve *crossing the bounding surface* of a thermodynamic system; it is to this crossing that the directional quality attaches, to the denominator δn_a rather than to the numerator δG.

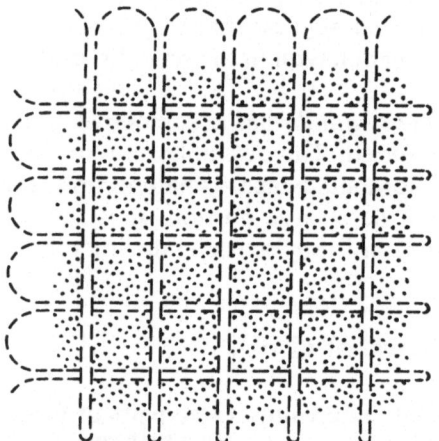

Fig. 3. Isovolumetric strain as an adding-and-subtracting process. We imagine a continuum to be laced with planar channels so that, treating the continuum as a thermodynamic system, material can be introduced into the system across horizontal interfaces, expanding the continuum vertically, while at an equal rate, material is withdrawn across vertical interfaces, allowing the continuum to contract horizontally. The present proposal associates an energy flux with this two-way material flux that is exactly the energy flux we would assign if the change of dimensions occurred by viscous flow instead.

2. "If chemical potential has a range of values at a single point, one can envisage a cycle of adding and subtracting material that closes and yet liberates energy." Continuing from point 1, to liberate energy one would need to add material to a system across one plane and remove an equal amount from the same system across a different plane. These two steps would change the system's dimensions, and the energy liberated would not come mysteriously from nowhere: It would come from the external system that supplies the nonhydrostatic stress state, as does the heat that is liberated by viscous deformation. In fact, the add-and-subtract operation would be wholly equivalent to a viscous deformation if the adding and subtracting were distributed over a uniform infinite set of parallel planes, and conversely, it is useful to imagine a viscous deformation as this kind of uniformly distributed adding and subtracting, as indicated by Fig. 3.

References

Bayly, M. B. (1983) Chemical potential in viscous phases under nonhydrostatic stress. *Phil. Mag.* **A47**, L39–44.

de Donder T. (1928) *L'affinité*. Gauthier-Villars, Paris.

de Groot, S. R. (1951) *Thermodynamics of Irreversible Processes*. North-Holland Publishing Company, Amsterdam.

Fletcher, R. C. (1982) Coupling of diffusional mass transport and deformation in a tight rock. *Tectonophysics* **83**, 275–291.

Gibbs, J. W. (1878) On the equilibrium of heterogeneous substances. *Trans. Connecticut Academy* **3**, 343–524; reprinted in *Scientific Papers of J. W. Gibbs*. 1961. Dover Publications, Inc., New York.

Green, II, H. W. (1980) On the thermodynamics of non-hydrostatically stressed solids. *Phil. Mag.* **41**, 637–647.

Kamb, W. B. (1959) Theory of preferred crystal orientation developed by crystallization under stress. *J. Geol.* **67**, 153–170.

Larché, F. C., and Cahn, J. W. (1978) Thermochemical equilibrium of multiphase solids under stress. *Acta Metal.* **26**, 1579–1589.

Lehner, F. K., and Heidug, W. (1983) On the thermodynamics of coherent phase transitions in solids. Brown University Report, Providence, R.I.

Li, J. C. M., Oriani, R. A., and Darken, L. S. (1966) The thermodynamics of stressed solids. *Zeitsch. für Physikalische Chemie Neue Folge* **49**, 271–290.

Paterson, M. S. (1973) Non-hydrostatic thermodynamics and its geologic applications. *Rev. Geophys. and Space Phys.* **11**, 355–389.

Index